ENVIRONMENTAL HEALTH PROCEDURES

Clay's Library of Health and the Environment

An increasing breadth and depth of knowledge is required to tackle the health threats of the environment in the 21ˢᵗ century, and to accommodate the increasing sophistication and globalization of policies and practices.

Clay's Library of Health and the Environment provides a focus for the publication of leading-edge knowledge in this field, tackling broad and detailed issues. The flagship publication *Clay's Handbook of Environmental Health*, now in its 18ᵗʰ edition, continues to serve environmental health officers and other professionals in over thirty countries.

Series Editor:
Bill Bassett: Honorary Fellow, School of Postgraduate Medicine and Health Sciences, University of Exeter, and formerly Chief Executive and Director of Environmental Health and Housing, Exeter City Council, UK

Editorial Board:
Xavier Bonnefoy: Regional Adviser, European Centre for Environment and Health, World Health Organization, Bonn, Germany
Don Boon: Director of Environmental Health and Trading Standards, London Borough of Croydon, UK
David Chambers: Head of the Schools of Social Sciences and Law, University of Greenwich, UK
Michael Cooke: Environmental Health and Sustainable Development consultant, UK, formerly Chief Executive of the CIEH

ENVIRONMENTAL HEALTH PROCEDURES

Sixth edition

W.H. BASSETT
DMA, FCIEH

*Honorary Fellow, School of Postgraduate Medicine
and Health Sciences, University of Exeter.
Formerly Chief Executive and Director of Housing and
Environmental Health, Exeter City Council, UK*

London and New York

First published 1983 by Chapman & Hall
Second edition 1987
Third edition 1992
Fourth edition 1995

Fifth Edition 1998 by E & FN Spon

Sixth edition first published 2002
by Spon Press
11 New Fetter Lane, London EC4P 4EE

Simultaneously published in the USA and Canada
by Spon Press
29 West 35th Street, New York, NY 10001

Spon Press is an imprint of the Taylor & Francis Group

© 1983, 1987, 1992, 1995, 1998, 2002 W.H. Bassett

Typeset in 10/12 Times
by Graphicraft Limited, Hong Kong
Printed and bound in Great Britain by
St Edmundsbury Press, Bury St Edmunds, Suffolk

British Library Cataloguing in Publication Data
A catalogue record for this book is available from the British Library

Library of Congress Cataloging in Publication Data
A catalog record for this book has been requested

ISBN 0-415-25719-0

Contents

Flow charts

Subject index of procedures

N.B. FC refers to the flowchart number given to each diagram

Abandoned refuse
  Removal – RD(A)A 1978
  FC113

Abandoned vehicles
  Removal – RD(A)A 1978
  FC112a and b

Accidents
  Notification of HASAWA 1974
  FC73

Accumulations
  (see also litter and waste)
  Abandoned refuse (in open air)
  – RD(A)A 1978 FC113
  Abandoned trolleys – EPA 1990
  FC40
  Abandoned vehicles – RD(A)A
  1978 FC112a and b
  Accumulation of rubbish
  (seriously detrimental) – PHA
  1961 FC9
  As statutory nuisances – EPA
  1990 FC33a
  Cleansing of filthy or verminous
  premises – PHA 1936/61
  FC10
  From demolition of buildings –
  BA 1984 FC20
  Removal of controlled waste
  from land – EPA 1990
  FC32
  Removal of noxious matter –
  PHA 1936 FC8

Animal boarding establishment
  Licensing – ABEA 1963 FC94

Air pollution
  Authorization etc. of
  prescribed processes – EPA
  1990 FC27
  Burning of crop residues – EPA
  1990 FC44
  Control of grit and dust from
  furnaces – CAA 1993 FC61
  Height of chimneys
  – serving furnaces – CAA 1993
  FC63
  – not serving furnaces – CAA
  1993 FC64
  Obtaining information about
  atmospheric pollution – EPA
  1990 FC66
  Prohibition of dark smoke from
  chimneys – CAA 1993 FC59
  Prohibition of dark smoke from
  industrial and trade premises –
  CAA 1993 FC60

Air quality
  LA reviews – EA 1995 FC57
  Air quality management areas –
  EA 1995 FC58

Anti-social behaviour
  Anti-social behaviour orders –
  CDA 1998 FC118

Approvals
  Meat products establishments –
  FSA 1990 FC80
  Minced meat premises – FSA
  1990 FC82
  Fishery product establishments –
  FSA 1990 FC86

Preface

In addition to the usual updating throughout the book, this edition contains a number of more extensive alterations.

Introduction

A new introductory chapter outlines the background against which environmental health procedures are enforced and operated. This chapter contains information on local authority enforcement policies and the national framework that shapes these including the Enforcement Concordat, Best Value and the enforcement prescriptions of bodies that oversee local enforcement including the Health and Safety Commission and the Food Standards Agency. The enforcement options available to local authorities are indicated and explained and there are sections dealing with covert surveillance, delegation, powers of entry and notice requirements. The role of the Commissioner for Local Administration is also included.

Integrated Pollution Prevention and Control (IPPC)

A further new chapter deals with the various procedures contained in the system of IPPC introduced by the Pollution Prevention and Control Act 1999 and the Pollution Prevention and Control Regulations 2000. This chapter is supported by 10 new flow charts detailing the main procedures contained in this legislation.

Because this system will run in tandem with the procedures in part 1 of the Environmental Protection Act 1990 for several years, these are retained and updated together with a section explaining how it is intended to absorb them into the new IPPC controls.

Air quality management

The procedures involved in the local authority reviews of air quality and the establishment of air quality management areas have been updated to include the recent technical and administrative advice from central government.

Food Safety

The procedure for licensing under the Food Safety (General Food Hygiene) (Butchers' Shops) Amendment Regulations 2000 has been added and the section dealing with the various controls of fishery products and shellfish has been rewritten, together with its flow charts, to include the provisions of the Food Safety (Fishery Products and Live Shellfish) Regulations 1998 and 1999.

Anti-social behaviour orders

A section has been introduced to deal with these orders which are made under the Crime and Disorder Act 1998 and which are increasingly being used by local authorities to deal with a variety of situations that often include matters of direct concern to environmental health departments.

House and area improvement

The changes to be made by the Regulatory Reform (Housing Assistance) (England and Wales) Order 2002 to the procedure dealing with renewal areas is included. The same Order introduces major change to the way in which LAs may assist in the repair and improvement of housing but, since house renovation grants will continue to be available for sometime, they are still included in this edition.

I am very grateful for the continuing assistance of the environmental health department of the South Hams District Council in sourcing legislative change and to Christopher C.H. Fry, MPhil, BA(Chem), MCIEH, DMS, LRSC, FRSH, AIEMA, Environmental Health Consultant and Associate Environmental Auditor, for his assistance in the updating of the chapters dealing with IPPC, air quality management and contaminated land.

W.H. Bassett
Exeter
2002

Extract from the preface
to the first edition (1983)

The idea for this work came after many exasperating years of having to wade through Acts of Parliament or legal encyclopediae in order to find the legal answer to a fairly simple practical problem which faced me as an environmental health officer. The author has also been conscious of a need for a reference text for both practising and pupil environmental health officers, community physicians and others which gives basic information about environmental health procedures in a simple form. It has also been suggested to the author that other people outside the environmental health profession including lawyers, law centre administrators, surveyors, estate agents, elected local authority members and members of the general public who become involved in environmental health problems would welcome a procedural analysis of this type.

This book attempts to meet all those needs and provide a means of identifying procedures relating to the many facets of environmental health practice.

Each of the main procedures has been reduced to a diagrammatic form and included with it is text on the basic issues concerned with that procedure. By this means the steps involved in the different procedures can be quickly followed and individual stages identified. Related procedures have been cross-referenced. At the beginning of each chapter, the general procedural provisions and definitions relating to the procedures contained in that chapter are indicated. Where a procedure differs either procedurally or in definition this is indicated either in the diagram or in the text.

W.H. Bassett
Exeter
1983

Abbreviations

The following abbreviations have been used:

ABE	animal boarding establishment
ABEA	Animal Boarding Establishments Act
AO	authorized officer
AQMA	Act Quality Management Area
BA	Building Act
B of DA	Breeding of Dogs Act
BS	British Standard
CA	clearance area
CAA	Clean Air Act
CDA	Crime and Disorder Act
CDSC	Communicable Disease Surveillance Centre (Public Health Laboratory Service)
CEHO	chief environmental health officer
CI	Chief Inspector
CIEH	Chartered Institute of Environmental Health
CJPOA	Criminal Justice and Public Order Act
CO	closing order
CoG	code of guidance
CoP	code of practice
CPA	Control of Pollution Act
CPO	compulsory purchase order
CSA	Caravan Sites Act
CSCDA	Caravan Sites and Control of Development Act
CS and DPA	Chronically Sick and Disabled Persons Act
DAN	Deferred Action Notice
DC	district council
D and COA	Deregulation and Contracting Out Act
DEFRA	Department of the Environment, Food and Rural Affairs
D(F of L)A	Dogs (Fouling of Land) Act
DHA	district health authority
DO	demolition order
DoE	Department of the Environment
DoH	Department of Health
DPA	Disabled Persons Act
DTLR	Department of Transport, Local Government and the Regions

DWAA	Dangerous Wild Animals Act
EA	enforcing authority
EA	Environment Agency
EA	Environment Act
EHO	environmental health officer
EN	enforcement notice
EPA	Environmental Protection Act
EPO	emergency prohibition order
FA	fire authority (Housing Act, Chapter 15)
FA	food authority (FSA, Chapter 12)
FC	flow chart
FSA	Food Safety Act or Food Standards Agency as appropriate
GGN	General Guidance Note
GIA	general improvement area
GLA	Greater London Authority
GRS	Group repair scheme
HA	Housing Act
HAA	Housing Action Area
HGCRA	Housing Grants, Construction and Regeneration Act
HASAWA	Health and Safety at Work etc. Act
HELA	HSE/LA Liaison Group
HIMO	house in multiple occupation
HSC	Health and Safety Commission
HSE	Health and Safety Executive
JP	Justice of the Peace
KY	knackers' yard
LA	licensing authority (Licensing, Chapter 13)
LA	local authority
LA	litter authority (Litter, Chapter 6)
LACOTS	Local Authorities Coordinating Body on Food and Trading Standards
LAN	litter abatement notice
LAO	litter abatement order
LAWDC	local authority waste disposal company
LCA	Land Compensation Act
LGA	Local Government Act
LGHA	Local Government and Housing Act
LG(MP)A	Local Government (Miscellaneous Provisions) Act
LNRH	late night refreshment house
MAFF	Ministry of Agriculture, Fisheries and Food
M & D(G) Regs	Milk and Dairies (General) Regulations
MI	meat inspector
MO	management order
M and QA	Mines and Quarries Act

M(SD) Regs. 1989	Milk (Special Designation) Regulations 1989
M(S and S) Regs	Meat (Sterilization and Staining) Regulations
MTA	Minded to take Action Notice
NA	Noise Act
NAA	National Assistance Act
NAZ	noise abatement zone
NMHS	National Meat Hygiene Service
NRA	National Rivers Authority
NSNA	Noise and Statutory Nuisance Act
OBO	obstructive building order
OSRPA	Offices, Shops and Railway Premises Act
OVS	official veterinary surgeon
PAA	Pet Animals Act
PDPA	Prevention of Damage by Pests Act
PF	prescribed form
PHA	Public Health Act
PH(CoD)A	Public Health (Control of Disease) Act
PH(ID) Regs	Public Health (Infectious Diseases) Regulations
PLA	principal litter authority
PM(H) Regs	Poultry Meat (Hygiene) Regulations
PM(H, I and E) Regs	Poultry Meat (Hygiene, Inspection and Export) Regulations
PN	prohibition notice
PO	prohibition order
PP and CA	Pollution Prevention and Control Act
RA	renewal area
RD(A)A	Refuse Disposal (Amenity) Act
REA	Riding Establishments Act
RF and OFMA	Rag Flock and Other Filling Materials Act
RIDDOR	Reporting of Injuries, Diseases and Dangerous Occurrences Regulations
RSA	Radioactive Substances Act
RTA	Road Traffic Act
SCA	smoke control area
SEPA	Scottish Environmental Protection Agency
SH	slaughterhouse
SHA	Slaughterhouses Act
SLN	street litter notice
SMDA	Scrap Metal Dealers Act
SoS	Secretary of State
STA	Sunday Trading Act
SU	sewerage undertaker
TA	Theatres Act
TCPA	Town and Country Planning Act
U/T	undertaking
VME	Vehicles, machinery and equipment

WCA	waste collection authority
WDA	waste disposal authority
WIA	Water Industry Act
WML	waste management licence
WRA	waste regulation authority
WU	water undertaker

Chapter 1

INTRODUCTION

ENVIRONMENTAL HEALTH LAW

The procedures interpreted in this book are based on primary and secondary legislation as it applies in England. In some cases they will also apply in the rest of the UK but this varies because of the separate legislative powers of the National Assemblies in Northern Ireland, Scotland and Wales. In Northern Ireland and Scotland the primary legislation as well as the secondary legislation may be different but the Welsh Assembly has powers only in respect of the making of secondary legislation e.g. regulations.

In each procedure the applicability of the procedure to the rest of the UK is indicated in the section headed 'Extent'.

ENFORCING AUTHORITIES

There is no single, legal definition of an environmental health authority and responsibility for the enforcement of each procedure is identified either in the particular piece of primary legislation or in the regulations made under it. Normally these will be found in each chapter, or procedure where they differ within a chapter, by reference to the definition of 'local authority' or 'enforcement authority'.

Generally the lowest tier of local authority for the area, excluding parish councils, enforces the environmental health law dealt with in this book. These are:

(a) In England outside of London
   (i)   Metropolitan and unitary councils and
   (ii) District councils in areas where there is a county council.
(b) In London:
   (i)   London borough councils
   (ii) The Common Council of the City of London
   (iii) The Sub-Treasurer of the Inner Temple and the Under-Treasurer of the Middle Temple.
(c) In Northern Ireland, the city councils, borough councils and district councils.

(d) In Scotland, the unitary authorities and
(e) In Wales, county councils and county borough councils in the single tier system.

THE FRAMEWORK FOR ENFORCEMENT POLICIES

In undertaking the enforcement of environmental health legislation and forming their individual enforcement policies LAs need to take account of a series of initiatives taken at national level which seek to guide their enforcement attitudes. The most significant are outlined below and it will be seen that in total they form a very substantial influence over LA enforcement activity.

Better Regulation

A Better Regulation Task Force (BRTF) was set up in September 1997 to advise the Government on action needed to improve the effectiveness and credibility of government regulation. A Working Group on Consumer Affairs deals with environmental health laws. The Task Force has agreed the following as the Principles of Good Regulation:

- Transparency – in that the need for regulation is clearly defined, regulations are simple and clear and those being regulated understand their obligations.
- Accountability – of regulators to government, citizens and Parliament with consultation before decisions are taken and a well published and accessible appeals procedure in place.
- Targeting – so that regulation is aimed at the problem rather than a scatter-gun or universal approach.
- Consistency – including even enforcement by relevant authorities.
- Proportionality – in that enforcement action should be related to the seriousness of the offence.

The Enforcement Concordat

This agreement between the Local Authority Associations and the Government was reached in March 1998 and is supported by the BRTF. The Concordat is not mandatory at present and each LA was expected to adopt its provisions individually. However, because of a low adoption rate, consideration is being given to making it mandatory. It deals with:

- Standards of level of service and performance.
- Openness in the provision of information and the use of plain language.
- Helpfulness in working with those affected by environmental health laws.
- Well publicized procedures for dealing with complaints about enforcement.
- Proportionality to ensure that enforcement action is related to risk levels.

- Consistency by enforcing in a fair, equitable and consistent manner.
- Procedures to guide the enforcement action of individual officers.

Best Value

This concept involves continually improving how all functions, including those involving enforcement, are undertaken by an authority. The scheme is given statutory force in Part I of the Local Government Act 1999 and involves processes that ensure functional reviews include challenge, comparison, consultation and competition. Benchmarking, which is the continuous process of measuring services against leaders in the field allowing the identification of best practices, is central to all aspects of this review process.

The Best Value Inspectorate, an arm of the Audit Commission, monitors the scheme including visits to LAs.

Performance Management

An Introductory Guide to Performance Management in Local Authority Trading Standards and Environmental Health Enforcement Services is a best practice guide published by the Better Regulation Unit in 1999. It aims to assist LAs with best value targeted performance management and includes matters relating to enforcement. The guide identifies the following themes as being central to good performance:

- Protection of the wider agenda i.e. councils leading and energizing local communities
- Transparency and consistency
- Quality and value
- Delivery and review

The Food Standards Agency

The Agency took over from MAFF most of its previous responsibilities under food safety legislation (see the Food Standards Act 1999 (Transitional and Consequential Provisions and Savings) (England) Regulations 2000) and also sets and monitors standards of performance for food authorities in enforcing food safety laws. They may make reports on that performance, including guidance on improving standards, and direct that the FA should publish, within a specified time, details of actions they will take to comply. They may also require each FA to provide information and to make its records available for inspection.

In 2000 the Agency published a Framework Agreement on Local Authority Food Law Enforcement which applies to the whole of the UK. This has four main elements:

- The standard – which sets out the requirements for the planning, management and delivery of LA food law enforcement services.

- Service planning guidance – on matters that include:
  1. Service aims and objectives
  2. Background
  3. Service delivery
  4. Resources
  5. Quality assessment
  6. Review
- A Monitoring scheme – which sets out the arrangements for the Agency to obtain enforcement information from the LAs and is built on the earlier arrangements under the Official Control of Foodstuffs Directive (89/397/EEC).
- An Audit scheme – which provides for a rolling programme of audits by the Agency.

In addition, statutory Codes of Practice issued under sect. 40 of the FSA 1990 contain mandatory provisions as to how and when certain actions are to be taken. These are identified in the appropriate procedures.

The Health and Safety Commission (HSC)

The HSC have advised LAs that the following elements are essential for a LA to discharge its duty as an enforcing authority under HASAWA:

(a) a clear published statement of enforcement policy and practice;
(b) a system for prioritized planned inspection activity;
(c) a service plan;
(d) capacity to investigate accidents and respond to complaints;
(e) arrangements for benchmarking performance with peer LAs;
(f) a trained and competent inspectorate;
(g) liaison and co-operation with the Lead Authority Partnership Scheme.

Section 18(4) of the HASWA 1974 requires LAs to act in accordance with guidance issued under that section by the HSC. A revised set of such guidance was issued to LAs in September 2001 which gives such statutory guidance on the matters listed above. The HSCs Statement on Enforcement Policy was published in January 2002, and which LAs are required to reflect in their policies, the main elements of which are:

(a) Proportionality – relating risks to enforcement action;
(b) Consistency of approach;
(c) Transparency – helping duty holders to understand what is expected of them;
(d) Targeting – of inspections to those areas containing the most serious risks and
(e) Prosecution – taking account of HSC guidance in reaching decisions.

Under the aegis of HELA a Protocol for Inter-authority Auditing of Local Authorities' Management of Health and Safety Enforcement Guidelines for LAs was issued in June 2000. This sets out how LAs should arrange for auditing by peer LAs or by an external auditor. The suggested audit framework is designed to help LAs assure themselves that they are making an effective contribution to the HSC's strategic themes and continuing aims and that they are complying with the HSC's mandatory guidance under sect. 18.

LOCAL AUTHORITY ENFORCEMENT POLICIES

Working within the directions and advice contained in the framework described above, each enforcing authority should adopt its own enforcement policy to guide departments and individual officers as to the way in which enforcement is to be carried out. This should include how decisions are to be reached in individual cases as to which enforcement option to adopt.

Such policies will also include provisions for dealing with disputes and complaints arising from enforcement activity.

ENFORCEMENT OPTIONS

The flow charts in this book indicate only the statutory remedies/provisions prescribed in the legislation that contains the particular procedure. There will also be a number of other remedies/provisions available depending upon the particular circumstances and the possibilities are given below. In addition there are a number of statutory remedies that are available generally and these are also included here.

No action

In some procedures once a LA is satisfied that a particular set of circumstances exists it is required by law to take the next step of statutory action, usually the service of a statutory notice, for example in the case of the existence of a statutory nuisance under sect. 80(1) of the EPA 1990. These situations are usually identified in the legislation by the use of the word 'shall' in respect of action required of the local authority. In other cases the action to be taken, or a decision to take no action, is at the discretion of the LA.

Oral warning

It may be felt that, in the circumstances of a particular contravention, it can be adequately remedied by speaking to the person responsible and asking for action to be taken. Such actions should be recorded.

Informal letters/notices

These may be used to confirm the existence of contraventions and to ask for them to be remedied. Informal notices are in effect a letter given the form of a notice but nevertheless are not part of a statutory procedure and no offences are committed by not complying with them.

Statutory notices

These are notices served under the provisions of a particular procedure. They must be in accordance with the requirements of that procedure, and with the general provisions covering such notices, and are legally enforceable. Non-compliance will lead to the possibility of prosecution and/or the taking of further steps specified in the legislation. They are the most usual form of statutory remedy used in environmental health law.

Formal cautions

These are applicable to the enforcement of all criminal law and may sometimes be appropriate in the enforcement of environmental health legislation. Their purpose is to deal quickly and simply with less serious offences, diverting them from unnecessary court action and reducing the chances of re-offending. The detailed provisions are contained in Home Office circular 18/1994 and in the National Standards for Cautioning (Revised) which are attached to that circular. The essential elements are:

1. Cautioning decisions are at the discretion of the enforcing authority.
2. Cautions should be used only where they are likely to be effective and their use is appropriate for the offence.
3. They should not be used for indictable-only offences.
4. Consideration should be given to:
   • The nature and extent of the harm caused by the offence
   • Whether the offence was racially motivated
   • Any involvement of a breach of trust
   • The existence of a systematic and organised background
   • The views of the persons offended against.
5. Cautions should be recorded.
6. Multiple cautioning of the same offender should not generally be used.
7. Before a caution is administered
   • There must be evidence of the offender's guilt
   • The offender must admit the offence
   • The offender must understand the significance of the caution and give informed consent to being cautioned
   • Consideration must be given to the public interest.
8. Officers giving the caution, which is of course done in person, should hold a position of seniority.

Whilst continued offending after the receipt of a formal warning is not an offence in itself, the fact will both guide subsequent decisions regarding that individual person, organisation or company and will also be taken into account by a court in any later proceedings.

Prosecutions

Environmental health law is a branch of criminal law and prosecution of offenders in the courts is the usual last stage for most of the procedures dealt with here. The decision to prosecute or not is a matter of discretion for the LA, guided by its'own enforcement policy.

Cases taken for offences against Housing Acts procedures are normally a matter initially for the county court whereas most other environmental health procedures are dealt with at the initial stages by the magistrates' court. Higher courts will become involved in the event of appeals against decisions made by these courts although there are situations, e.g. with the abatement of statutory nuisances, where a LA may proceed directly to the High Court in order to seek a remedy.

Closure/withdrawal of licence etc

It is sometimes the case that, in addition to prosecution to exact a penalty, the LA may give consideration to the cancellation of the licence/approval it may have issued authorizing the activity where the offence has taken place e.g. approval of meat products establishments and pet shop licences.

COVERT SURVEILLANCE

In some situations it may be necessary for LA enforcement officers to act in a covert manner in the investigation of possible breaches of environmental health legislation e.g. the illegal sale of food that is unfit for human consumption. The Regulation of Investigatory Powers Act 2000 provides a legal basis within which such investigation can take place including the interception of communications, the acquisition and disclosure of communications data, the carrying out of surveillance and the use of covert human intelligence sources.

The Investigatory Powers (Prescription of Office Ranks and Positions) Order 2000 requires that such surveillance carried out by LAs must be by officers of at least Assistant Chief Officer status or by an officer designated as being responsible for the management of an investigation. Explanatory Notes accompany the Act and regulations and it is intended to support these with a code of practice, a draft of which was published for consultation in August 2001.

DELEGATION OF AUTHORITY

The Local Government Act 1972 allows a LA to delegate its powers and responsibilities to committees, sub-committees and officers. In respect of environmental health law enforcement a clear scheme of delegation to officers is vital if investigation and remedial action is to be swift and effective. Such a scheme will require the formal approval of the individual LA. It will include matters like the exercise of powers of entry and inspection, the service of statutory notices and the institution of legal proceedings. There is also a need for the scheme to include the designation of officers to whom the Act or regulations concerned give authority to act e.g. the designation of proper officers, authorized officers and officers who will be Inspectors under the HASAWA 1974.

POWERS OF ENTRY

The powers of entry and investigation that relate to each procedure are given either in the section on General Procedural Provisions at the beginning of each chapter or within the text relating to the procedure itself.

SERVICE OF NOTICES BY LAs

The LGA 1972 makes provision for the service of notices etc. by LAs and, unless otherwise excluded by a particular provision of another Act, is available in addition to methods of service available under other legislation.

'(1) Subject to subsection (8) below, subsections (2) to (5) below shall have effect in relation to any notice, order or other document required or authorized by or under any enactment to be given to or served on any person by or on behalf of a local authority or by an officer of a local authority.

(2) Any such document may be given to or served on the person in question either by delivering it to him, or by leaving it at his proper address, or by sending it by post to him at that address.

(3) Any such document may:
(a) in the case of a body corporate, be given to or served on the secretary or clerk of that body;
(b) in the case of a partnership, be given to or served on a partner or a person having the control or management of the partnership business.

(4) For the purposes of this section and of section 26 of the Interpretation Act 1889 (service of documents by post) in its application to this section, the proper address of any person to or on whom a document is to be given or served shall be his last known address, except that:
(a) in the case of a body corporate or their secretary or clerk, it shall be the address of the registered or principal office of that body;
(b) in the case of a partnership or a person having the control or management of the partnership business, it shall be that of the principal officer of the partnership;
and for the purposes of this subsection the principal office of a company registered outside the United Kingdom or of a partnership carrying on

business outside the United Kingdom shall be their principal office within the United Kingdom.

(5) If the person to be given or served with any document mentioned in subsection (1) above has specified an address within the United Kingdom other than his proper address within the meaning of subsection (4) above as the one at which he or someone on his behalf will accept documents of the same description as that document, that address shall also be treated for the purposes of this section and section 26 of the Interpretation Act 1889 as his proper address.

(6) ... (Repealed)

(7) If the name or address of any owner, lessee or occupier of land to or on whom any document mentioned in subsection (1) above is to be given or served cannot after reasonable enquiry be ascertained, the document may be given or served either by leaving it in the hands of a person who is or appears to be resident or employed on the land or by leaving it conspicuously affixed to some building or object on the land.

(8) This section shall apply to a document required or authorized by or under any enactment to be given to or served on any person by or on behalf of the chairman of a parish meeting as it applies to a document so required or authorized to be given to or served on any person by or on behalf of a local authority.

(9) The foregoing provisions of this section do not apply to a document which is to be given or served in any proceedings in court.

(10) Except as aforesaid and subject to any provision of any enactment or instrument excluding the foregoing provisions of this section, the methods of giving or serving documents which are available under those provisions are in addition to the methods which are available under any other enactment or any instrument made under any enactment.

(11) In this section "local authority" includes a joint authority and a police authority established under section 3 of the Police Act 1996.'

(Sect. 233)

AUTHENTICATION OF DOCUMENTS

The following provisions of the LGA 1972 will apply unless a particular piece of legislation has provisions which cover these issues.

'(1) Any notice, order or other document which a local authority is authorized or required by or under any enactment (including any enactment in this Act) to give, make or issue may be signed on behalf of the authority by the proper officer of the authority.

(2) Any document purporting to bear the signature of the proper officer of the authority shall be deemed, until the contrary is proved, to have been duly given, made or issued by the authority of the local authority.

In this subsection the word 'signature' includes a facsimile of a signature by whatever process reproduced.

(3) Where any enactment or instrument made under an enactment makes, in relation to any document or class of documents, provision with respect to the matters dealt with by one of the two foregoing subsections, that subsection shall not apply in relation to that document or class of documents.

(4) In this section "local authority" includes a joint authority and a police authority established under section 3 of the Police Act 1996.'

(Sect. 234)

INFORMATION REGARDING OWNERSHIPS ETC.

Where a LA, in performing functions under any legislation, requires information regarding interests in any land it may serve a notice, specifying the land and the legal provision containing the particular function under which it is acting, requiring the person concerned to declare, within not less than 14 days, the nature of his interest and the names and addresses of all persons whom he believes have interests.

The notice may be served on any of the following:

(a) the occupier;
(b) the freeholder, mortgagee or lessee;
(c) the person directly or indirectly receiving the rent;
(d) the person authorized to manage the land or arrange for its letting.

Failure to comply with the notice or the making of false statements carries a maximum penalty of level 5 on the standard scale (LG(MP)A 1976 sect. 16).

These powers are given to county councils, district councils, London borough councils, the Common Council and the Council of the Isles of Scilly and to police authorities, joint authorities, parish and community councils (LG(MP)A 1976 sect. 44(1)).

THE COMMISSIONER FOR LOCAL ADMINISTRATION
(THE OMBUDSMAN)

Each procedural flow chart indicates the stages at which the courts may become involved in dealing with appeals, prosecutions etc. In addition to this, any LA action, or lack of it, is subject to investigation by the Local Government Ombudsman. Whilst the Ombudsman cannot challenge the decision itself e.g. not to serve a statutory notice, he may do so in respect of the way in which the decision was reached. Whilst he does not have power to change the LA decision, the Ombudsman can issue a public report criticizing the authority and is also able to indicate a level of compensation which he feels ought to be made.

FURTHER READING

This book is an interpretation of particular legislative procedures that are used in environmental health work. For a study of the wider legislative context in which these procedures are to be exercised the reader is referred to:

1. *Legal Competence in Environmental Health*, Terence Moran, 1997, E & FN Spon, London.
2. *Clay's Handbook of Environmental Health 18th edition*, W.H. Bassett, Editor 1999, E & FN Spon, London.

Chapter 2

PUBLIC HEALTH ACTS

GENERAL PROCEDURAL PROVISIONS

The following general provisions are those which are normally applicable to actions under the PHAs 1936 and 1961 and are also generally applicable to procedures being taken under the CAA 1993 and the PDPA 1949. Unless otherwise indicated they will apply to the procedures in this chapter.

Local authority

Local authorities for the purpose of the procedures in this chapter are:
- (a) district councils including metropolitan and unitary councils;
- (b) London borough councils;
- (c) the Common Council of the City of London;
- (d) the Sub-Treasurer of the Inner Temple and the Under-Treasurer of the Middle Temple;
- (e) in Wales, councils of county and county borough councils (PHA 1936 sect. 1(2)).

Port health authorities may have the powers under the Public Health Acts assigned to them (PH(CoD)A 1984 sect. 3) and in these cases LAs within the port health district cease to discharge those functions.

Extent

The procedures in this chapter apply to England and Wales but do not apply to Scotland or to Northern Ireland (see the Public Health (Ireland) Acts 1878–1907) (PHA 1936 sect. 347(2)).

Notices

- (a) **Form**. All notices must be in writing (PHA 1936 sect. 283) but, with minor exceptions noted in the text, there are no prescribed forms.
- (b) **Authentication**. Notices must be signed (includes a facsimile signature) by the proper officer or any officer so authorized in writing (PHA 1936 sect. 284).

(c) **Service**. This may be effected by one of the following:
   (i) delivery to the person;
   (ii) in the case of service on an officer of the council, leaving it or sending it in a prepaid letter addressed to him at his office;
   (iii) in the case of any other person, leaving it or sending it in a prepaid letter addressed to him at his usual or last known residence;
   (iv) for an incorporated company or body, by delivering it to its secretary or clerk at its registered or principal office, or sending it in a prepaid letter addressed to him at that office;
   (v) in the case of an owner by virtue of the fact he receives the rackrent as agent for another person, by leaving it or sending it in a prepaid letter addressed to him at his place of business;
   (vi) in the case of the owner or occupier of a premises, where it is not practicable after reasonable enquiries to ascertain his name and address or where the premises is unoccupied, by addressing it to the 'owner' or 'occupier' of the premises (naming them) to which the notice relates and delivering it to some person on that premises, or, if there is no one to whom the notice can be delivered, by affixing it or a copy of it, to a conspicuous part of the premises (PHA 1936 sect. 285).

Notices served in any of these ways will be held in later proceedings to have been properly served, but it is possible to serve in some other way and prove service in subsequent proceedings. In particular, the methods of service set out in sect. 233 LGA 1972 (page 8) are also available to notices under the PHAs. Proof of receipt and/or service of notices should be obtained.

Notices requiring the execution of works

These provisions (known as the Part 12 provisions) apply to any procedures where the section concerned specifically indicates this to be the case and this is shown in the procedures where appropriate.

(a) **Content**. The notice must indicate the nature of the works required and state the time within which they are to be executed. The time allowed should not be less than the period for appeal, i.e. 21 days.
(b) **Appeals**. Appeal against these notices may be made to a magistrates' court within 21 days of service on any of the following grounds:
   (i) the notice is not justified;
   (ii) there has been some material informality, defect or error in, or in connection with, the notice;
   (iii) The LA has refused unreasonably to approve alternative works;
   (iv) the works required are unreasonable in character or extent or are unnecessary;
   (v) the time allowed for compliance is not reasonable;

    (vi) it would have been equitable for the notice to have been served
        on the occupier instead of the owner or vice versa;
  (vii) that some other person ought to contribute towards the cost of
        compliance.

(c) **Enforcement**. Following failure to comply with a notice to execute
works within the time specified, subject to appeal, the LA may, at its
discretion, carry out the work itself and recover expenses reasonably
incurred from the person on whom the notice was served. In addition,
the person concerned is liable to a fine not exceeding level 4 on the
standard scale and a maximum daily penalty of £2 for a continuing
offence (PHA 1936 sects. 290 and 300).

Recovery of costs

Costs incurred by LAs in carrying out works in default of an owner
are recoverable both by proceeding in the County Court and by the debt
becoming a charge on the property. They may be recovered from an owner
by instalments and with interest (PHA 1936 sect. 291).

Power of entry

An authorized officer of a council, producing if required a duly authenticated document showing his authority, has a right to enter any premises at
all reasonable hours for the following purposes:

  (a) ascertaining if there has been a contravention of this Act;
  (b) seeing if circumstances exist which would authorize or require the
      council to take any action or execute work;
  (c) to take any action or execute any work authorized or required;
  (d) generally for the council's performance of its functions.

For premises other than a factory, workshop or workplace at least 24
hours notice must be given before entry may be demanded.
  Where:

  (a) admission has been refused; or
  (b) refusal is apprehended; or
  (c) the premises are unoccupied; or
  (d) the occupier is temporarily absent; or
  (e) the case is one of urgency; or
  (f) application for admission would defeat the object of entry;

the LA, following notice of intention to any occupier, may make application to a JP by way of sworn information in writing, and the JP may
authorize entry by an authorized officer, if necessary by force. The authorized officer may take with him any other persons as may be necessary but
must leave any unoccupied premises as secure as he found them (PHA 1936
sect. 287).

The penalty for obstructing any person acting in the execution of these Acts, regulations and orders is a maximum level 1 on the standard scale and £20 daily for continuing offences (PHA 1936 sect. 288).

Where an owner is prevented from executing any work which he has been required to undertake, he may apply to a magistrates' court for an order to the occupier to permit execution of the works (PHA 1936 sect. 289).

Appeals

Unless a longer period is specified, any appeals to magistrates' courts are to be made by way of complaint for an order within 21 days from the date on which the council's requirement, notice, etc. was served. Notices and other documents must state the rights of appeal and the time within which appeal may be made (PHA 1936 sect. 300).

Persons aggrieved by decisions of the magistrates' court may appeal to the Crown Court except where there is a provision for arbitration (PHA 1936 sect. 301).

Power of a LA to execute work on behalf of owners or occupiers at their request

A LA, by agreement with an owner or occupier of any premises, may undertake on his behalf works which it has required him to carry out under the Act or undertake sewer or drain construction, alteration or repair which the owner or occupier is entitled to execute. The costs are chargeable to the owner or occupier (PHA 1936 sect. 275).

DEFINITIONS

The following definitions are applicable to all procedures in this chapter unless otherwise indicated:

Drain means a drain used for the drainage of one building or any buildings or yards appurtenant to buildings within the same curtilage.

House means a dwelling-house, whether a private dwelling-house or not.

Owner means the person for the time being receiving the rackrent of the premises . . . whether on his own account or as agent or trustee for any other person, or who would so receive the same if those premises were let at a rackrent.

Prejudicial to health means injurious, or likely to cause injury, to health.

Premises includes messuages, buildings, lands, easements and hereditaments of any tenure.

Private sewer means a sewer which is not a public sewer.

Public sewer means a sewer for the time being vested in a sewerage undertaker in its capacity as such, whether vested in that undertaker by virtue of a scheme under Schedule 2 to the WA 1989 or Schedule 2 to the WIA 1991 or under sect. 179 of that Act of 1991 or otherwise.

Sewer does not include a drain (see definition above) . . . but . . . includes all sewers and drains used for the drainage of buildings and yards appurtenant to buildings.

Workplace does not include a factory . . . but save as aforesaid includes any place in which persons are employed otherwise than in domestic service (PHA 1936 sect. 343).

REPAIRS TO SEWERAGE, DRAINAGE AND SANITARY CONVENIENCES

REPAIR OF DRAINS, PRIVATE SEWERS, ETC. BY LAs

Reference

Public Health Act 1961 sect. 17 (as substituted by sect. 27 Local Government (Miscellaneous Provisions) Act 1982).

Scope

The procedure applies to any drain, private sewer, water closet, waste pipe or soil pipe which is:

(a) not sufficiently maintained and kept in good repair; and
(b) can be sufficiently repaired at a cost not exceeding £250 (sect. 17(1)).

Works by LAs on land belonging to statutory undertakers and held or used by them for those purposes is not authorized unless it affects houses or buildings used as offices or showrooms other than those forming part of a railway station (sect. 17(10) and (11)).

The procedure under this section to deal with blocked or stopped-up (as opposed to defective) drains, private sewers, etc. is covered separately in FC3.

Persons concerned

Notices are to be served on the person or persons concerned which means:

(a) in relation to a water closet, waste pipe or soil pipe, the owner or occupier of the premises; and
(b) in relation to a drain or private sewer, any person owning any premises drained by means of it and also, in the case of a sewer, the owner of the sewer (sect. 17(2)).

Notices

Notices must be in writing (PHA 1936 sect. 283) but are not subject to the provisions of sect. 290.

FC1 Repair of drains, private sewers, etc. by LAs

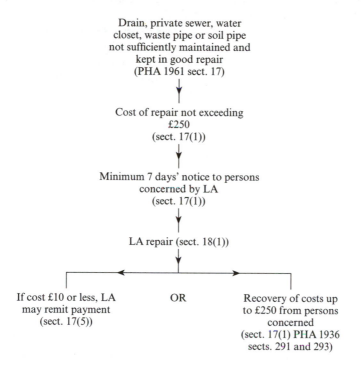

Drain, private sewer, water
closet, waste pipe or soil pipe
not sufficiently maintained and
kept in good repair
(PHA 1961 sect. 17)

↓

Cost of repair not exceeding
£250
(sect. 17(1))

↓

Minimum 7 days' notice to persons
concerned by LA
(sect. 17(1))

↓

LA repair (sect. 18(1))

↓

| If cost £10 or less, LA may remit payment (sect. 17(5)) | OR | Recovery of costs up to £250 from persons concerned (sect. 17(1) PHA 1936 sects. 291 and 293) |

Notes
1. For alternative procedure, see BA 1984 sect. 59, FC12.
2. For blocked drains and private sewers, see FC3.
3. There is no appeal against a notice served under sect. 17, neither is there a penalty for non-compliance.
4. LAs may clear obstructions in drains or private sewers by agreement before the expiration of the notice (PHA 1961 sect. 22).

Recovery of costs

Costs not exceeding £250 incurred by the LA are recoverable from the persons concerned (as defined above) in such proportions, if there is more than one person concerned, as the LA may determine. Costs of £10 or less may be remitted (sect. 17(4) and (5)).

 In any proceedings to recover the LA's expenses of up to £250, the court must consider whether or not the LA was justified in concluding that the drain, private sewer, etc. was not sufficiently maintained and kept in good repair and decide whether any apportionment between different persons is fair (sect. 17 (6)).

Definitions

See page 15.

BLOCKED PRIVATE SEWERS

Reference

Local Government (Miscellaneous Provisions) Act 1976 sect. 35.

Scope

This procedure allows a notice requiring the unblocking of a sewer to be served on owners or occupiers of all premises draining into it (sect. 35(1)).

Notices

Notice may be served on each of the persons who is an owner or occupier of premises served by the sewer or on any of these as the LA sees fit. The time specified within which the works are required must not be less than 48 hours but can be longer at the discretion of the LA (sect. 35(1)). Notices must be in writing (PHA 1936 sect. 283).

Notices apportioning costs

Following works in default, the LA must specify by notice to those persons in receipt of the notice to clear the blockage, the amount to be recovered from each and indicate which other persons are to be charged and the amount. In apportioning cost, the LA may have regard to any matters relating to the cause of the blockage and to any agreements relating to the cleansing of that private sewer (sect. 35(3)). Particular regard should be paid to any determinations under sect. 22 BA 1984 dealing with the drainage of buildings in combination.

FC2 Blocked private sewers

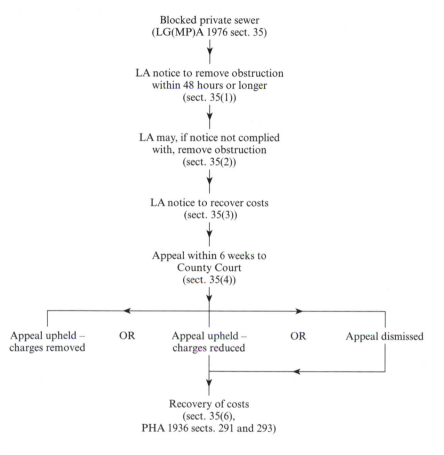

Notes

1. There is no appeal against the notice to remove the blockage under sect. 35(1) and neither is there a penalty for non-compliance.
2. For defective private sewers, see FC1 and 3.

Appeals against notices recovering costs

In considering an appeal that whole or part should be paid by some other person, the county court may either dismiss the appeal or order the whole or part to be paid by other owners or occupiers of premises served by the sewer, provided that such owners or occupiers have been given 8 days' notice of the appeal (sect. 35(5)).

Definitions (see also page 15)

Notice means a notice in writing (LG(MP)A 1976 sect. 44).
Owner in relation to any land, place or premises means a person who, either on his own account or as agent or trustee for another person, is receiving the rackrent of the land, place or premises or would be entitled to receive it if the land, place or premises were let at a rackrent (LG(MP)A 1976 sect. 44).

BLOCKED DRAINS, PRIVATE SEWERS, ETC.

Reference

Public Health Act 1961 sect. 17 (as substituted by sect. 27 Local Government (Miscellaneous Provisions) Act 1982).

Scope

This procedure may be applied to any drain, private sewer, water closet, waste pipe or soil pipe on any premises which is stopped-up. There is no exclusion here for premises owned by statutory undertakers as there is in relation to other parts of sect. 17 (see FCl).

Because sect. 17 appears to limit the service of the notice to the owner or occupier of the premises at which the drain, etc. is stopped-up, blocked private sewers are best dealt with under sect. 35 (LG(MP)A) since owners of other premises draining into the sewer may be involved (FC2).

Notices

Notices requiring remedy of the defect within 48 hours (this period is specified by the section and should not be shorter or longer) must be in writing, but are not subject to sect. 290 PHA 1936 provisions and are to be served on either the owner or occupier of the premises on which the drain, private sewer, etc. is stopped-up (sect. 17(3)).

Recovery of costs

If the cost to the LA of undertaking the works in default does not exceed £10, recovery may be waived otherwise costs are recoverable from persons

FC3 Blocked drains, private sewers, etc.

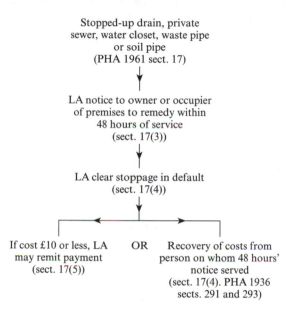

Stopped-up drain, private
sewer, water closet, waste pipe
or soil pipe
(PHA 1961 sect. 17)

LA notice to owner or occupier
of premises to remedy within
48 hours of service
(sect. 17(3))

LA clear stoppage in default
(sect. 17(4))

| If cost £10 or less, LA may remit payment (sect. 17(5)) | OR | Recovery of costs from person on whom 48 hours' notice served (sect. 17(4). PHA 1936 sects. 291 and 293) |

Notes

1. At the request of the owner or occupier a LA may cleanse or repair drains, water closets, sinks or gullies and recover costs. This could be done before the expiration of the 48 hours' notice by agreement (PHA 1961 sec. 22).
2. There is no appeal against a notice served under sect. 17, neither is there a penalty for non-compliance.
3. For defective drainage see FC1 and 2.
4. For alternative procedure for blocked private sewers see LG(MP)A 1976 sect. 35, at FC2.

on whom the notices were served (sect. 17(5)). In any proceedings to recover expenses, the magistrates' court may inquire:

(a) whether any requirement in the notice was reasonable; and
(b) whether the expenses ought to be borne wholly or in part by someone other than the defendant, but due notice of the proceedings and an opportunity to be heard must be given to these other persons (sect. 17(6)).

Definitions

See page 15.

OVERFLOWING AND LEAKING CESSPOOLS

Reference

Public Health Act 1936 sect. 50 and Part 12.

Scope

The section does not apply in relation to the effluent from a properly constructed tank for the reception and treatment of sewage provided that the effluent is not prejudicial to health or a nuisance. Otherwise it applies to any leaking or overflowing cesspool even though the situation may not be prejudicial to health or a nuisance (sect. 50(1)).

Person responsible

Notices are served by the LA on the person by whose act, default or sufferance the soakage or overflow occurs or continues (sect. 50(1)).

Notices etc.

The provisions of Part 12 of the PHA 1936 apply to appeals against the enforcement of notices served under this procedure (page 13).

Appeals against notices

The grounds of appeal against notices requiring works are set out on page 15. Notices which require periodic emptying are not subject to the provisions of sect. 290 and no appeal is provided for. In defending any proceedings for penalty because of non-compliance with this form of notice, it is not open to the defendant to question the reasonableness of the requirements (sect. 50(3)).

FC4 Overflowing and leaking cesspools

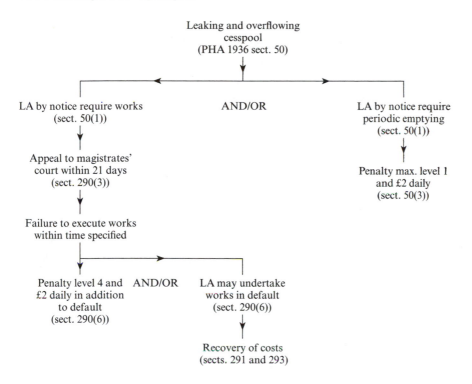

Leaking and overflowing
cesspool
(PHA 1936 sect. 50)

LA by notice require works
(sect. 50(1))

AND/OR

LA by notice require
periodic emptying
(sect. 50(1))

Appeal to magistrates'
court within 21 days
(sect. 290(3))

Penalty max. level 1
and £2 daily
(sect. 50(3))

Failure to execute works
within time specified

Penalty level 4 and
£2 daily in addition
to default
(sect. 290(6))

AND/OR

LA may undertake
works in default
(sect. 290(6))

Recovery of costs
(sects. 291 and 293)

Notes
1. There is no provision for appeal against notices not requiring works under this section (e.g. periodic emptying), but it is open to the defendant to question the reasonableness of the notice in subsequent proceedings.
2. The LA may at any time undertake any works which it has requested if asked to do so by the owner or occupier (PHA 1936 sect. 275).

Definition (also page 15)

Cesspool includes a settlement tank or other tank for the reception or disposal of foul matter from buildings (sect. 90(1)).

DEFECTIVE SANITARY CONVENIENCES

Reference

Public Health Act 1936 sect. 45 and Part 12.

Scope

This procedure deals with closets provided in a building which are in such a state as to be prejudicial to health or a nuisance and can, without reconstruction, be put into a satisfactory condition. This includes both repair and cleansing (PHA 1936 sect. 45(1)).

It does not apply to closets in a factory, workshop or workplace (sect. 45(4)). In such cases the provisions of the HASAWA 1974 can be used. (FC70)

Notices etc.

The provisions of Part 12 of the PHA 1936 apply to appeals against and the enforcement of notices requiring repairs but not to those requiring cleansing only (page 13).

Appeals

In relation to a notice not requiring works, i.e. requiring cleansing etc., the defendant in subsequent proceedings for penalty may question the reasonableness of the requirements and the decision to address the notice to him and not the owner or occupier as the case may be (sect. 45(3)).

Closets

It is suggested that the word includes the structure in which the appliance is housed as well as the appliance itself.

Definition (also page 15)

Closet includes privy (sect. 90(1)).

FC5 Defective sanitary conveniences

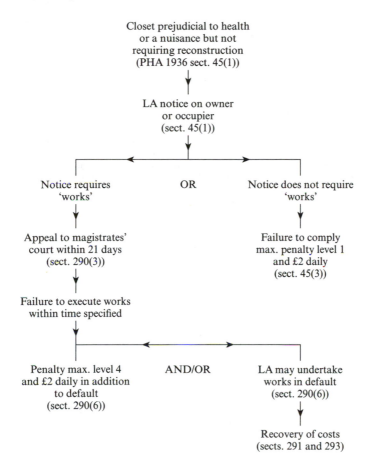

Closet prejudicial to health
or a nuisance but not
requiring reconstruction
(PHA 1936 sect. 45(1))

LA notice on owner
or occupier
(sect. 45(1))

Notice requires   OR   Notice does not require
'works'   'works'

Appeal to magistrates'   Failure to comply
court within 21 days   max. penalty level 1
(sect. 290(3))   and £2 daily
                                            (sect. 45(3))

Failure to execute works
within time specified

Penalty max. level 4   AND/OR   LA may undertake
and £2 daily in addition   works in default
to default   (sect. 290(6))
(sect. 290(6))

Recovery of costs
(sects. 291 and 293)

Notes
1. For closets prejudicial to health or a nuisance but requiring reconstruction, see BA 1984 sect. 64 (FC14).
2. At any time after service of the LA notice, the LA may undertake the required works at the request of the owner or occupier and charge the costs (sect. 275).

FC6 Closure or restriction of polluted water supply

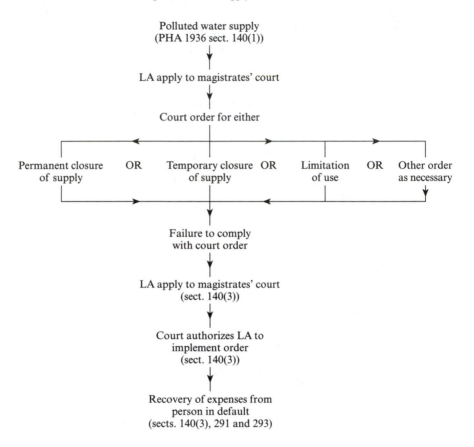

Polluted water supply
(PHA 1936 sect. 140(1))

↓

LA apply to magistrates' court

↓

Court order for either

Permanent closure
of supply

OR

Temporary closure
of supply

OR

Limitation
of use

OR

Other order
as necessary

Failure to comply
with court order

↓

LA apply to magistrates' court
(sect. 140(3))

↓

Court authorizes LA to
implement order
(sect. 140(3))

↓

Recovery of expenses from
person in default
(sects. 140(3), 291 and 293)

MISCELLANEOUS

CLOSURE OR RESTRICTION OF POLLUTED WATER SUPPLY

Reference

Public Health Act 1936 sect. 140.

Scope

The section applies to wells, tanks or any other source of water supply not vested in the local authority which are used, or likely to be used, for domestic purposes or the preparation of food or drink for human consumption and which are so polluted as to be prejudicial to health (sect. 140(1)).

It would seem that this procedure is available to a LA to deal with water supplied by statutory water undertakings operating under the WIA 1991 and therefore provides the only direct, statutory remedy available to LAs in respect of such public water supplies (Chapter 4).

Persons responsible

The court order following applications by the LA may be addressed to owners or occupiers of the premises to which the source of supply belongs or to any other person having control over it (sect. 140(1) and (2)).

Court orders

These may order any of the following:

(a) permanent closure or cutting off of the source; or
(b) temporary closure or cutting off of the source; or
(c) restriction of use for certain purposes only; or
(d) any other measure to prevent injury or damage to health of persons using water or consuming food or drink prepared from it (sect. 140).

LICENSING OF CAMPING SITES

Reference

Public Health Act 1936 sect. 269 (as amended by Caravan Sites and Control of Development Act 1960).

Scope

The procedure applies to all moveable dwellings (other than caravans) and licences are required for:

FC7 Licensing of camping sites

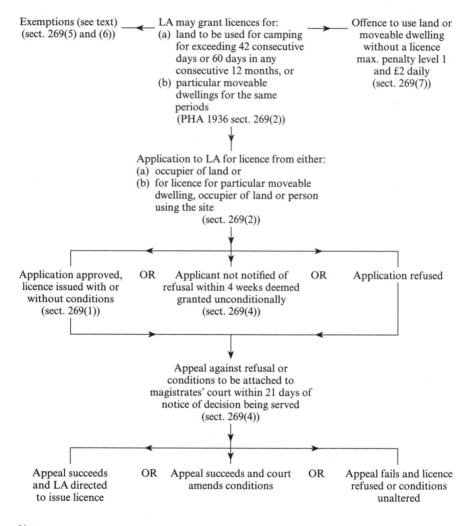

Exemptions (see text) ← LA may grant licences for: → Offence to use land or
(sect. 269(5) and (6))

LA may grant licences for:
(a) land to be used for camping
   for exceeding 42 consecutive
   days or 60 days in any
   consecutive 12 months, or
(b) particular moveable
   dwellings for the same
   periods
   (PHA 1936 sect. 269(2))

Offence to use land or
moveable dwelling
without a licence
max. penalty level 1
and £2 daily
(sect. 269(7))

Application to LA for licence from either:
(a) occupier of land or
(b) for licence for particular moveable
   dwelling, occupier of land or person
   using the site
   (sect. 269(2))

Application approved,   OR   Applicant not notified of   OR   Application refused
licence issued with or   refusal within 4 weeks deemed
without conditions   granted unconditionally
(sect. 269(1))   (sect. 269(4))

Appeal against refusal or
conditions to be attached to
magistrates' court within 21 days of
notice of decision being served
(sect. 269(4))

Appeal succeeds   OR   Appeal succeeds and court   OR   Appeal fails and licence
and LA directed   amends conditions   refused or conditions
to issue licence   unaltered

Notes
1. This procedure does not apply to caravans. For licensing of caravan sites, see FC105.
2. For removal of unauthorized campers under the CJPOA 1994, see FC117.

(a) the use of land for camping purposes on more than 42 consecutive days or more than 60 days in any 12 consecutive months; or
(b) the keeping of a moveable dwelling on any one site, or two or more sites in succession if any of those sites is within 100 yards of another of them, for more than the same periods as in (a).

In respect of (a), land which is in the occupation of the same person, and within 100 yards of a site on which a moveable dwelling is stationed during any part of a day, is regarded as being used for camping on that day (sect. 269(1)–(3)). The removal of a moveable dwelling for not more than 48 hours does not constitute an interruption of the 42 days' period (sect. 269(8)).

Exemptions

Apart from the use of land or moveable dwellings for periods less than those in (a) and (b) above, the following situations do not require licensing by the LA:

(a) moveable dwellings kept by the owner on land occupied by him in connection with his dwelling if used only by him or members of his household;
(b) moveable dwellings kept by the owner on agricultural land occupied by him and used for habitation only at certain times of the year and only by persons employed in farming operations on that land;
(c) moveable dwellings not in use for human habitation kept on premises by an occupier who does not permit moveable dwellings to be kept there for habitation;
(d) organizations granted a certificate of exemption by the Minister, being satisfied that:
  (i) the camping sites belonging to, provided by or used by the organization are properly managed and kept in good sanitary condition; and
  (ii) the moveable dwellings are used so as not to give rise to any nuisance (sect. 269(5) and (6)).

Applicants

Applications for licences may be made by:

(a) occupiers in respect of either the use of land or the stationing of moveable dwellings on that land; or
(b) a person intending to station a moveable dwelling on land (sect. 269(2) and (3)).

Conditions

LAs may attach to licences such conditions as they think fit with respect to:

   (a) for licences authorizing the use of land:
      (i) number and classes of moveable dwellings;
      (ii) the space between them;
      (iii) water supply; and
      (iv) for securing sanitary conditions.
   (b) for licences authorizing the use of a moveable dwelling for:
      (i) the use of that dwelling, including space to be kept free between dwellings;
      (ii) securing its removal at the end of a specified period; and
      (iii) for securing sanitary conditions (sect. 269(1)(i)(ii)).

Definitions

Caravan, see FC105.

   Moveable dwelling includes any tent, any van or other conveyance (other than a caravan) whether on wheels or not, and . . . any shed or similar structure, being a tent, conveyance or structure which is used either regularly or at seasons only, or intermittently, for human habitation but does not include a structure to which the building regulations apply (sect. 269(8)).

REMOVAL OF NOXIOUS MATTER

Reference

Public Health Act 1936 sect. 79.

   NB. Although provision was made in the CPA 1974 Sch. 4 for sect. 79 of the PHA 1936 to be repealed, this has not been implemented and this procedure remains available.

Scope

The procedure may be used where the proper officer of the LA considers that any accumulation of noxious matter ought to be removed. The word 'noxious' is not defined, but according to the Concise Oxford Dictionary means 'harmful, unwholesome' (sect. 79(1)).

Notice

The notice is served by the proper officer on either the owner or the occupier of the premises concerned requiring removal of the accumulation within 24 hours (sect. 79(1)). The notes on pages 12–13 relating to the form, authentication and service of the notices apply here, but the Part 12 provisions do not apply. There is, therefore, no provision for appeal etc.

FC8 Removal of noxious matter

Noxious matter on
any premises
(PHA 1936 sect. 79(1))

↓

Notice from proper officer to
owner or occupier requiring
removal within 24 hours
(sect. 79(1))

↓

If notice not complied with

↓

Proper officer may remove
accumulation
(sect. 79(1))

↓

LA may recover expenses
from owner or occupier
(sect. 79(2))

Notes
1. There is no penalty for non-compliance with the notice.
2. For accumulations constituting a statutory nuisance, see FC33a.
3. For removal of rubbish resulting from demolition, see FC20.
4. For removal of abandoned articles, see FC113.
5. For removal of abandoned cars, see FC112a.

Default

If the accumulation is not removed within 24 hours from the service of the notice, the proper officer may have it removed (sect. 79(1)).

Recovery of costs

The LA is enabled to recover from either the owner or occupier the expenses of any reasonable actions taken by its proper officer in acting in default (sect. 79(2)).

ACCUMULATIONS OF RUBBISH

Reference

Public Health Act 1961 sect. 34.

Scope

The procedure deals with any rubbish which is in the open air and which is seriously detrimental to the amenities of the neighbourhood (sect. 34(1)).

Notices

Notices must be in writing (PHA 1936 sect. 283), specify the steps which the LA proposes to take and indicate the provisions for counter-notice and appeals. The LA can take no action until 28 days from the service of the notice (sect. 34(2)).

In addition to district councils, London borough councils, the Common Council and county councils have concurrent powers and may serve notices under sect. 34 (LGA 1972 Sch. 14, para. 37).

Appeals

Appeal is to the magistrates' court within 28 days of service on the grounds that the LA was not justified in taking action under this section, or that the steps which it proposes to take are unreasonable (sect. 34(2)(b)).

Definition

Rubbish means rubble, waste paper, crockery and metal, and any other kind of refuse (including organic matter), but does not include any material accumulated for, or in the course of, any business or waste deposited in accordance with a waste management licence under the EPA 1990 (sect. 34(5)).

FC9 Accumulations of rubbish

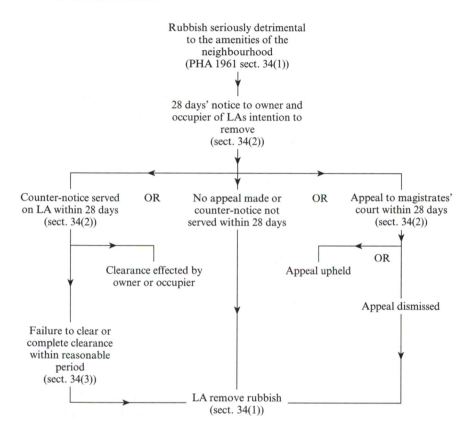

Rubbish seriously detrimental
to the amenities of the
neighbourhood
(PHA 1961 sect. 34(1))

28 days' notice to owner and
occupier of LAs intention to
remove
(sect. 34(2))

Counter-notice served
on LA within 28 days
(sect. 34(2))

OR

No appeal made or
counter-notice not
served within 28 days

OR

Appeal to magistrates'
court within 28 days
(sect. 34(2))

Clearance effected by
owner or occupier

Appeal upheld

OR

Appeal dismissed

Failure to clear or
complete clearance
within reasonable
period
(sect. 34(3))

LA remove rubbish
(sect. 34(1))

Notes

1. Where rubbish constitutes a statutory nuisance, see FC33a.
2. For proper officer's power relating to noxious matter, see PHA 1936, sect. 79, FC8.
3. For LA power relating to removal of manure, see PHA 1936 sect. 80.
4. For removal of rubbish resulting from demolition, see FC20.
5. There is no power for the LA to recover its costs.
6. The TCPA 1990 sect. 215 allows a LA to deal with a site which adversely affects the neighbourhood.

CLEANSING OF FILTHY OR VERMINOUS PREMISES

References

Public Health Act 1936 sect. 83 (as amended by PHA 1961 sect. 35).
Public Health Act 1961 sect. 36.

Scope

These provisions apply to any premises except those forming part of a mine or quarry but including ships, boats, tents, vans and sheds (sects. 83(4), 267(4) and 268(1)), and cover situations which are either:

(a) filthy or unwholesome so as to be prejudicial to health; or
(b) verminous.

Works required

The steps which are required to be taken must be specified in the notice and may include:

(a) removal of wallpaper or other wall covering;
(b) destruction or removal of vermin;
(c) interior surfaces of houses, shops or offices to be papered, painted or distempered and in all other cases painted, distempered or whitewashed (sect. 83(1)).

A special procedure is set out for situations where the LA wishes to use gas to destroy vermin, and this includes power to require vacation of both the affected premises and ones adjoining at a date specified in the notice, provided that temporary accommodation has been provided, whether it be dwelling accommodation or otherwise. The period of vacation must be specified in the notice (PHA 1961 sect. 36).

Notices

Notices under these provisions are not subject to sect. 290 Part 12 requirements, but must be in writing and the time allowed for compliance must be reasonable. In the case of notices indicating the LA's intention to gas, the period of notice should be not less than the appeal period, i.e. 7 days.

Appeals

There is no appeal against the service of notices under sect. 83 (except those relating to gassing), but in any subsequent proceedings by the LA for recovery of costs following work in default, the defendant can question the reasonableness of the requirements and the fact that the notice was served on him and not the owner or occupier as the case may be (sect. 83(2)).

FC10 Cleansing of filthy or verminous premises

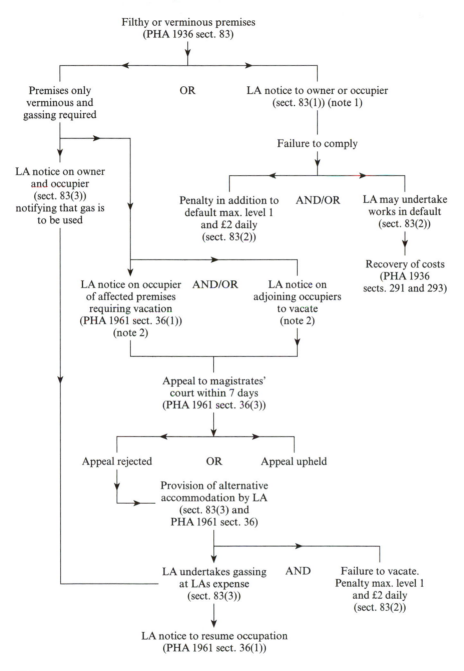

Notes

1. There is no appeal against this notice.
2. For cleansing of verminous articles, see PHA 1936 sect. 84, verminous persons and clothing sect. 85, and sale of verminous articles PHA 1961 sect. 37.

Definition

Vermin in its application to insects and parasites, includes their eggs, larvae and pupae, and the expression 'verminous' shall be constructed accordingly (sect. 90(1)).

Chapter 3

BUILDING ACT 1984

GENERAL PROCEDURAL PROVISIONS

Local authority

The LAs charged with the enforcement of these procedures are:

(a) district councils including metropolitan and other unitary authorities;
(b) London borough councils;
(c) the Common Council of the City of London;
(d) the Sub-Treasurer of the Inner Temple;
(e) the Under-Treasurer of the Middle Temple;
(f) in relation to Wales, councils of county or county borough councils (sect. 126).

These LAs are charged with the statutory duty of executing the provisions (sect. 91). This includes the unitary authorities in England.

Extent

The procedures in this chapter do not apply to Scotland or to Northern Ireland (sect. 135(2)).

Notices

(a) **Form.** All notices must be in writing (sect. 92(1)) and, although there is provision for the SoS to prescribe forms for procedures, none have so far been prescribed.
(b) **Authentication.** Notices must be signed (includes a facsimile signature) by either the proper officer or any officer authorized in writing to do so (sect. 93).
(c) **Service.** This may be effected by any of the following methods:
    (i) delivery to the person;
    (ii) in the case of service on an officer of the council, leaving it or sending it in a prepaid letter addressed to him at his office;

(iii) in the case of any other person, leaving it or sending it in a prepaid letter addressed to him at his usual or last known residence;

(iv) for an incorporated company or body, by delivering it to its secretary or clerk at its registered or principal office, or sending it in a prepaid letter addressed to him at that office;

(v) in the case of an owner by virtue of the fact that he receives the rackrent as agent for another person, by leaving it or sending it in a prepaid letter addressed to him at his place of business;

(vi) in the case of the owner or occupier of a premises, where it is not practicable after reasonable enquiries to ascertain his name and address or where the premises is unoccupied, by addressing it to the 'owner' or 'occupier' of the premises (naming them) to which the notice relates and delivering it to some person on that premises, or, if there is no one to whom the notice can be delivered, by affixing it, or a copy of it, to a conspicuous part of the premises (sect. 94).

Notices served in any of these ways will be held in later proceedings to have been properly served, but it is possible to serve in some other way and prove service in subsequent proceedings. In particular, the methods of service set out in sect. 233 LGA 1972 (page 8), are also available. Proof of receipt and/or service of notices should be obtained.

Notices requiring the execution of works

Where specifically indicated by the particular provisions relating to a procedure, the following apply:

(a) **Content.** The notice must indicate the nature of the works required and state the time within which they are to be executed. The time allowed should not be less than the period for appeal, i.e. 21 days (sect. 99(1)).

(b) **Appeals.** Appeal against these notices may be made to a magistrates' court within 21 days on any of the following grounds:
　(i) the notice is not justified;
　(ii) there has been some material informality, defect or error in, or in connection with, the notice;
　(iii) the LA has refused unreasonably to approve alternative works;
　(iv) the works required are unreasonable in character or extent or are unnecessary;
　(v) the time allowed for compliance is not reasonable;
　(vi) it would have been equitable for the notice to have been served on the occupier instead of the owner or vice versa;
　(vii) that some other person ought to contribute towards the cost of compliance (sects. 102 and 103).

(c) **Enforcement.** Following failure to comply with a notice to execute works within the time specified, subject to appeal, the LA may, at its discretion, carry out the work itself and recover expenses reasonably incurred from the person on whom the notice was served. In addition, the person concerned is liable to a fine not exceeding level 4 and a maximum daily penalty of £2 for a continuing offence (sect. 99(2)).

Recovery of costs

Costs incurred by LAs in carrying out works in default of an owner are recoverable either by proceedings in the county court or by becoming a charge on the property. They may be recovered from an owner by instalments and with interest (sects. 107(1) and 108).

Power of entry

An authorized officer of a council, producing if required a duly authenticated document showing his authority, has a right to enter any premises at all reasonable hours for the following purposes:

(a) ascertaining if there has been a contravention of building regulations;
(b) seeing if circumstances exist which would authorize or require the council to take any action or execute work;
(c) to take any action or execute any work authorized or required;
(d) generally for the council's performance of its functions.

For premises other than a factory, workshop or workplace at least 24 hours' notice must be given before entry may be demanded.
Where:

(a) admission has been refused; or
(b) refusal is apprehended; or
(c) the premises are unoccupied; or
(d) the occupier is temporarily absent; or
(e) the case is one of urgency; or
(f) application for admission would defeat the object of entry;

the LA may make application to a JP by way of sworn information in writing and the JP may authorize entry by an authorized officer, if necessary by force. The authorized officer may take with him any other person as may be necessary but must leave any unoccupied premises as secure as he found them (sects. 95 and 96).

The penalty for obstructing a person acting in the execution of this Act is a fine not exceeding level 1 on the standard scale (sect. 112).

Where an owner is prevented from executing any work which he has been required to undertake, he may apply to a magistrates' court for an order to the occupier to permit execution of those works (sect. 98).

Appeals

Any appeals to magistrates' courts are to be made by way of complaint for an order within 21 days from the date on which the council's requirement, notice, etc., was served (unless a longer period is specified). Notices and other documents must give the rights of appeal and state the time within which appeal may be made (sect. 103).

Persons aggrieved by decisions of the magistrates' court may appeal to the Crown Court except where there is a provision for arbitration (sect. 86).

Power of LA to execute work on behalf of owners or occupiers

A LA, by agreement with an owner or occupier of any premises, may undertake on his behalf works which it has required him to carry out under the Act or undertake sewer construction, alteration or repair which the owner or occupier is entitled to execute. The costs are chargeable to the owner or occupier (sect. 97).

DEFINITIONS

The following definitions are applicable to all procedures in this chapter unless otherwise indicated in the procedure.

Cesspool includes a settlement tank or other tank for the reception or disposal of foul matter from buildings.

Drain means a drain used for the drainage of one building or of any buildings or yards appurtenant to buildings within the same curtilage and includes any manholes, ventilating shafts, pumps or accessories belonging to the drain.

House means a dwelling-house, whether a private dwelling-house or not.

Owner means the person for the time being receiving the rackrent of the premises . . . whether on his own account or as agent or trustee for any other person, or who would so receive the same if those premises were let at a rackrent.

Prejudicial to health means injurious, or likely to cause injury, to health.

Premises includes buildings, lands, easements and hereditaments of any tenure.

Private sewer means a sewer which is not a public sewer (sect. 126).

Public sewer means a sewer for the time being vested in a SU in its capacity as such, whether vested in that undertaker by virtue of a scheme under Sch. 2 to the WA 1989 or Schedule 2 to this Act or under sect. 179 above or otherwise (WIA 1991 sect. 219).

Sewer does not include a drain . . . but . . . includes all sewers and drains used for the drainage of buildings and yards appurtenant to buildings and any manholes, ventilating shafts, pumps or other accessories belonging to the sewer (sect. 126).

DRAINAGE

DRAINAGE FOR NEW BUILDINGS

Reference

Building Act 1984 sects. 21 and 22.

Plans

The provisions apply only following the deposit of plans under the building regulations for either a new building or an extension to a building (sect. 21(1)).

Satisfactory provision for drainage

A drain will not satisfy the requirements of sect. 21 unless it discharges into either:

(a) a sewer;
(b) a cesspool; or
(c) some other place;

depending upon the requirement of the LA.
  The LA cannot require connection to a sewer unless:

(a) the sewer is within 100 ft of the site of the building; and
(b) the sewer is at a level which makes connection to it reasonably practicable; and
(c) if the sewer is not a public sewer, it is a sewer which the person constructing the drain is entitled to use; and
(d) the person concerned is entitled to construct the drain through the intervening land.

Provided that the other requirements are satisfied, connection with a sewer over 100 ft away can be required if the LA undertakes to pay for the additional costs of construction and subsequent maintenance and repair (sect. 21(5)).

Drainage of buildings in combination

The LA may require drains from two or more buildings to be combined for discharge into a sewer if it considers that this would be more economic or otherwise advantageous, provided that the requirement is made before or at the time of approving the plans. After that time, drainage in combination can be achieved only by agreement. At the time of its requirement, the LA must fix the apportionment of costs of both construction and subsequent

FC11 Drainage for new buildings

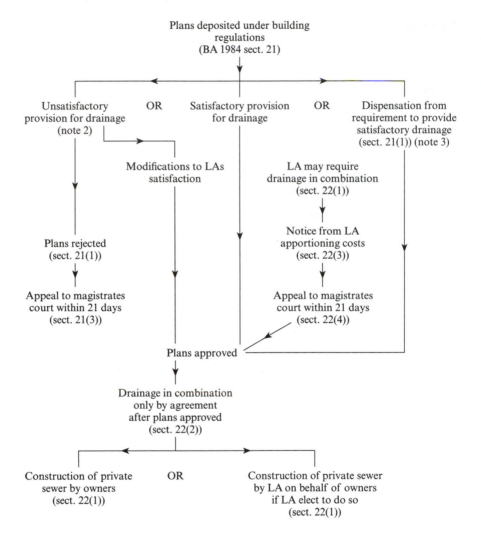

Plans deposited under building
regulations
(BA 1984 sect. 21)

Unsatisfactory   OR   Satisfactory provision   OR   Dispensation from
provision for drainage   for drainage   requirement to provide
(note 2)   satisfactory drainage
(sect. 21(1)) (note 3)

Modifications to LAs   LA may require
satisfaction   drainage in combination
(sect. 22(1))

Notice from LA
apportioning costs
(sect. 22(3))

Plans rejected
(sect. 21(1))

Appeal to magistrates   Appeal to magistrates
court within 21 days   court within 21 days
(sect. 21(3))   (sect. 22(4))

Plans approved

Drainage in combination
only by agreement
after plans approved
(sect. 22(2))

Construction of private   OR   Construction of private sewer
sewer by owners   by LA on behalf of owners
(sect. 22(1))   if LA elect to do so
(sect. 22(1))

Notes
1. Disputes or appeals arising from these provisions are determined by application to the magistrates' court except those relating to costs arising from a requirement of the LA to connect with a sewer in excess of 100 ft away where, alternatively, arbitration may be sought (sect. 21(3) and (6)).
2. For remedy for buildings without satisfactory provision for drainage, see FC12.
3. The LA may decide that it may dispense with the requirement for satisfactory provision of drainage in the case of particular buildings or extensions if it considers it proper to do so (sect. 21(1)).

maintenance/repair to be borne by the owners of buildings concerned and, where the existing sewer is over 100 ft away, the costs to be borne by the LA (sect. 22).

Definitions (also page 40)

Drainage includes the conveyance, by means of a sink and any other necessary appliance, of refuse water and the conveyance of rainwater from roofs (sect. 21(2)).

Plans A reference to the deposit of plans in accordance with building regulations is a reference to the deposit of plans in accordance with those regulations for the purposes of sect. 16 above unless the context otherwise requires (sect. 124) and 'plans' includes drawings of any other description and also specifications or other information in any form (sect. 126).

DEFECTIVE DRAINAGE TO EXISTING BUILDINGS

References

Building Act 1984 sects. 59 and 60.

Scope

The procedure is applied to buildings which have:

(a) unsatisfactory provision for drainage (sect. 59(1)(a));
(b) cesspools, private sewers, drains, soil pipes, rainwater pipes, spouts, sinks or other necessary appliances which are insufficient or, in the case of a private sewer or drain communicating with a public sewer, is so defective as to admit subsoil water (sect. 59(1)(b));
(c) cesspools etc. as detailed in (b) above, in such a condition as to be prejudicial to health or a nuisance and this also covers cesspools, private sewers and drains no longer in use (sect. 59(1)(c) and (d));
(d) rainwater pipes being used for foul waste, soil pipes from water closets not properly ventilated and surface water pipes acting as vents to foul drains or sewers (sect. 60).

In relation to (a), unsatisfactory drainage is not defined. For guidance see page 41 relating to sect. 21 BA 1984.

Notices etc.

The provisions on page 37 apply to appeals against and the enforcement of notices served under this procedure. Notices are to be served on the owner of the building concerned.

FC12 Defective drainage to existing buildings

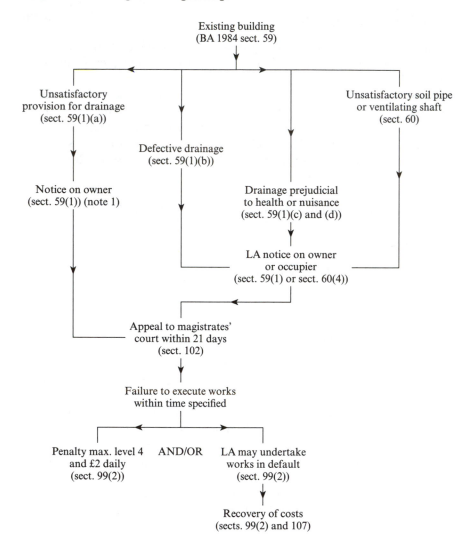

Existing building
(BA 1984 sect. 59)

Unsatisfactory
provision for drainage
(sect. 59(1)(a))

Unsatisfactory soil pipe
or ventilating shaft
(sect. 60)

Defective drainage
(sect. 59(1)(b))

Notice on owner
(sect. 59(1)) (note 1)

Drainage prejudicial
to health or nuisance
(sect. 59(1)(c) and (d))

LA notice on owner
or occupier
(sect. 59(1) or sect. 60(4))

Appeal to magistrates'
court within 21 days
(sect. 102)

Failure to execute works
within time specified

Penalty max. level 4
and £2 daily
(sect. 99(2))

AND/OR

LA may undertake
works in default
(sect. 99(2))

Recovery of costs
(sects. 99(2) and 107)

Notes

1. The provisions of sect. 21(4) and (5) apply in this situation, see page 41.
2. At the request of an owner or occupier, a LA may cleanse or repair drains, water closets, sinks or gullies and recover costs. This could be done by agreement before the expiration of any notice served under sect. 59 and whether or not a notice has been served (PHA 1961 sect. 22).
3. See also PHA 1961 sect. 17 as amended LG(MP)A 1982 (FC1).

Testing of drains etc.

LAs have the power to examine sanitary conveniences, drains, private sewers or cesspools and to apply tests, other than tests by water under pressure, where they suspect deficiencies under sect. 59. This includes exposing the pipes by opening the ground (sect. 48 PHA 1936).

Notice of intention to repair drains etc.

Before carrying out repairs to drains or private sewers the person undertaking the work must give at least 24 hours' notice to the LA except if the work is required in emergency, in which case he must not cover over the drain or sewer without notice (BA sect. 61).

Definitions (see also page 40)

Surface water includes water from roofs.

Water closet means a closet that has a separate fixed receptable connected to a drainage system and separate provision for flushing from a supply of clean water either by the operation of mechanism or by automatic action (sect. 126).

PAVING OF YARDS AND PASSAGES

Reference

Building Act 1984 sect. 84.

Courts, yards, passages

The section applies to any court or yard appurtenant to, or any passage giving access to, buildings as defined below, including those used in common by two or more houses provided it is not repairable by the inhabitants at large as a highway (sect. 84). It does not apply to garden paths.

The procedure applies to buildings which are houses or industrial and commercial buildings.

Paving and drainage

Courts, yards and passages must be so formed, flagged, asphalted or paved or provided with such works on, above or below its surface, as to allow satisfactory drainage of its surface or subsoil to a proper outfall (sect. 84(1)).

Notices

The provision relating to enforcement of notices and appeals on page 37 applies to this procedure.

FC13 Paving of yards and passages

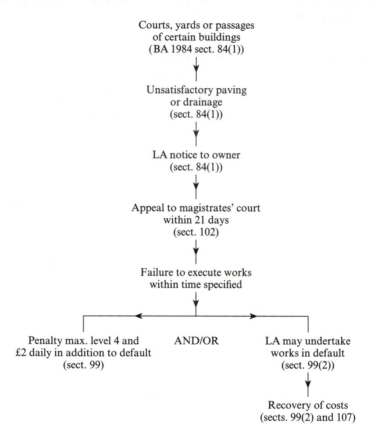

Courts, yards or passages
of certain buildings
(BA 1984 sect. 84(1))

Unsatisfactory paving
or drainage
(sect. 84(1))

LA notice to owner
(sect. 84(1))

Appeal to magistrates' court
within 21 days
(sect. 102)

Failure to execute works
within time specified

Penalty max. level 4 and
£2 daily in addition to default
(sect. 99)

AND/OR

LA may undertake
works in default
(sect. 99(2))

Recovery of costs
(sects. 99(2) and 107)

Notices, which are to be served on the owner, must indicate the nature of the works required and state the time within which they must be executed. The time allowed must be reasonable and not less than the period for appeal, i.e. 21 days (sects. 99 and 102).

Grounds of appeal

These are as in sect. 102 set out on page 38.

SANITARY APPLIANCES

SANITARY CONVENIENCES: PROVISION/REPLACEMENT

Reference

Building Act 1984 sects. 64 and 65.

Scope

The procedure applies to:

(a) **Insufficient or replacement accommodation in buildings** (other than workplaces) where the existing closet is prejudicial to health or a nuisance and requires reconstruction. The notices may require the provision of additional or replacement closets, but water closets may not be required unless a sufficient water supply and sewer are available (sect. 64):
  (i) sewer available – must be within 100 ft of the building at a level which makes connection reasonably practicable, is a sewer which the owner of the building is entitled to use and the intervening land is land through which he is entitled to construct a drain;
  (ii) water supply available – water is laid on or can be laid on from a point within 100 ft of the building and the intervening land is land through which the owner of the building is entitled to lay a pipe (sect. 125).
(b) **Sanitary conveniences in workplaces.** In deciding whether or not sanitary conveniences for buildings used as workshops are sufficient and satisfactory, regard must be paid to the number of persons employed in or in attendance at the building and the need to provide separate accommodation for the sexes, although the LA may waive this latter provision if it so wishes (sect. 65(1)).

Notices, appeals, etc.

The provisions of sects 99 and 102 apply to notices used in this procedure (pages 37 and 38).

FC14 Sanitary conveniences: provision/replacement

```
                    Closets                          Sanitary conveniences
                       │                                      │
   ┌───────────────◄───┤                                      ▼
   │                   │
Insufficient  OR   Prejudicial to
(BA 1984           health or nuisance                 Insufficient or
sect. 64(1))           │                              unsatisfactory in
(note 1)               ▼                                 workplaces
   │               Requiring                          (BA 1984 sect. 65)
   │               reconstruction                        (note 3)
   │               (BA 1984 sect. 64(1)(c))                 │
   ▼                  (note 2)                              │
   │                   │                                    ▼
   └─────────┬─────────┘                                    │
             ▼                                              ▼
      LA notice on owner                         LA notice on owner or
        (sect. 64(1))                                  occupier
             │                                        (sect. 65(2))
             ▼                                             │
             └──────────────────┬──────────────────────────┘
                                ▼
                      Appeal to magistrates'
                      court within 21 days
                      (BA 1984 sect. 102)
                                │
                                ▼
                      Failure to execute works
                      within time specified
                                │
             ┌─────────◄────────┼─────────►────────────┐
             ▼                                          ▼
     Penalty max. level 4        AND/OR          LA may undertake
     and £2 daily in addition                    works in default
        to default                                  (sect. 99)
        (sect. 99)                                      │
                                                        ▼
                                               Recovery of costs
                                               (sects. 99(2) and 107)
```

Notes

1. This relates to buildings, or parts of buildings occupied as a separate dwelling, which are without sufficient closet accommodation but does not include workplaces.
2. The procedure for dealing with existing closets of buildings which are prejudicial to health or a nuisance is split between those which can be put into satisfactory condition without construction (sect. 45 PHA 1936 FC5) and those which cannot (sect. 64 BA 1984).
3. This section covers all workplaces (does not include factories). Alternatively the situation could be dealt with by using the enforcement procedures of the HASAWA 1974, see FC70.
4. For sanitary conveniences in a place of public entertainment etc., see FC110.
5. For the power of a LA to require replacement of earth closets by water closets at the joint expense of the owner and LA, see FC15.

Definition (see also page 40)

Closet includes privy (sect. 126).

CONVERSION OF EARTH CLOSETS ETC. TO WATER CLOSETS

Reference

Building Act 1984 sect. 66.

Scope

The procedure allows a LA to secure the replacement of any closet, other than a water closet, by a water closet, even though the existing closet may be sufficient and not prejudicial to health or a nuisance, provided that the costs are equally shared between the LA and the owner of the building and that a water supply and sewer are available.

Water supply and sewer available

A building is not deemed to have a sewer available unless:

    (a) there is within 100 ft of the site of the building, and at a level which makes it reasonably practicable to construct a drain to communicate therewith, a public sewer or other sewer which the owner of the building is entitled to use and

    (b) the intervening land is land through which he is entitled to construct a drain.

A building is not deemed to have a sufficient water supply available unless it has a sufficient supply of water laid on, or unless such supply can be laid on to it from a point within 100 ft of the site of the building, and the intervening land is land through which the owner of the building or proposed building is entitled to lay a communication pipe. The limit of 100 ft does not apply if the LA undertakes so much of the expenses reasonably incurred in constructing, and in maintaining and repairing, a drain to communicate with a sewer or, as the case may be, in laying and in maintaining and repairing, a pipe for the purpose of obtaining a supply of water, as may be attributable to the fact that the distance of the sewer, or the point from which a supply of water can be laid on, exceeds 100 ft (sect. 125(2)).

The availability of a sewer or water supply in these terms is not necessary for the LA to contribute towards the costs of an owner converting by agreement.

Notices etc.

The provisions of sects 99 and 102 (pages 38–9) apply to appeals against, and the enforcement of notices served under, this procedure.

FC15 Conversion of earth closets etc. to water closets

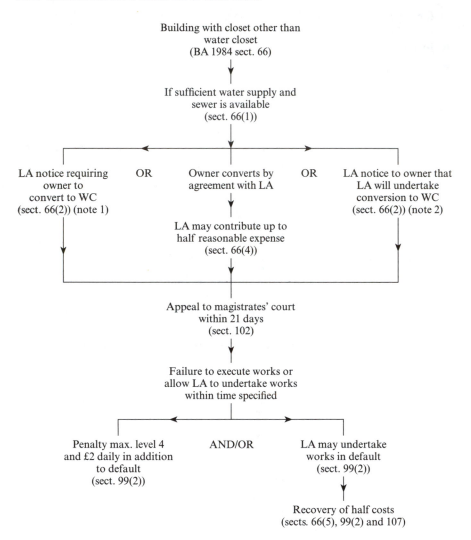

Building with closet other than
water closet
(BA 1984 sect. 66)

↓

If sufficient water supply and
sewer is available
(sect. 66(1))

| LA notice requiring owner to convert to WC (sect. 66(2)) (note 1) | OR | Owner converts by agreement with LA | OR | LA notice to owner that LA will undertake conversion to WC (sect. 66(2)) (note 2) |

LA may contribute up to
half reasonable expense
(sect. 66(4))

Appeal to magistrates' court
within 21 days
(sect. 102)

Failure to execute works or
allow LA to undertake works
within time specified

| Penalty max. level 4 and £2 daily in addition to default (sect. 99(2)) | AND/OR | LA may undertake works in default (sect. 99(2)) |

Recovery of half costs
(sects. 66(5), 99(2) and 107)

Notes

1. In this case the owner may recover half of the reasonable cost from the LA.
2. In this case the LA may recover half of the reasonable cost from the owner.

Appeals against notices

The grounds of appeal against notices under sect. 66 are as set out on page 38 except that there is no appeal that the works are unnecessary (sect. 66(5)).

Definitions (also page 40)

Closet includes privy (sect. 126).

Water closet means a closet which has a separate fixed receptable connected to a drainage system and separate provision for flushing from a supply of clean water, either by the operation of mechanism or by automatic action (sect. 126).

MISCELLANEOUS

WATER SUPPLY FOR NEW HOUSES

Reference

Building Act 1984 sect. 25.

Satisfactory water supply

Before approving plans submitted for approval under the building regulations, the LA must be satisfied that the water supply proposed will be sufficient and wholesome for domestic purposes and provided by:

(a) either connecting to a piped supply provided by the WU, or
(b) if (a) is not reasonable, by otherwise taking water into the house, by pipe, e.g. from a well supply; or
(c) if neither (a) nor (b) can be reasonably required, by providing a supply within a reasonable distance of the house (sect. 25(1)).

In relation to a judgement about whether or not water is 'wholesome', the standards of sect. 67 of the WIA 1991 and of any regulations made under that section (i.e. the Water Supplies (Water Quality) Regulations 1989 and 2000 and the Private Water Supplies Regulations 1991) apply here (sect. 25(7)).

Notices

Notices requiring prohibition of occupation in the event of the supply not being sufficient or wholesome, despite the passing of plans, must be in writing and served on the owner (sect. 92).

FC16 Water supply for new houses

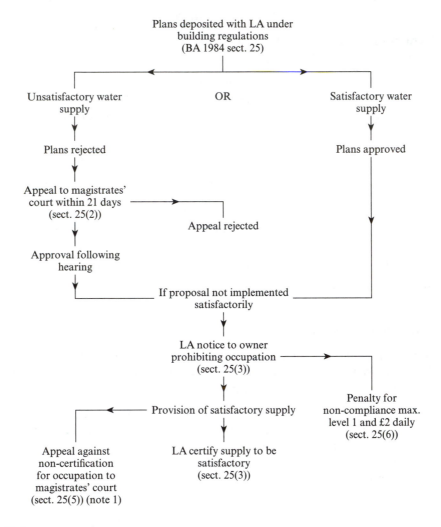

Notes
1. The court may either dismiss the appeal or authorize the occupation of the house, the latter having the same effect as certification by the LA.
2. For closure/restriction of polluted supplies, see FC6.

MEANS OF ESCAPE FROM CERTAIN HIGH BUILDINGS

Reference

Building Act 1984 sect. 72.

Scope

The section applies to existing or proposed buildings which are, or are to be:

(a) let as flats or tenement dwellings; or
(b) used as inns, hotels, boarding houses, hospitals, nursing homes, boarding schools, children's homes or similar institutions; or
(c) used as restaurants, shops, stores or warehouses which have sleeping accommodation on any upper floor for persons employed on the premises;

and which exceed two storeys in height and in which the floor of any upper storey is more than 20 ft above the surface of the street or ground on any side of the building (sect. 72(1) and (6)).

The section is, however, now only a residual section and does not cover premises subject to the Fire Precautions Act 1971 or houses in multiple occupation which are dealt with in the HA 1985 (see FCs 125 and 128).

Notices

The provisions of sects. 99 and 102 apply to appeals against, and the enforcement of, notices under this procedure where works are required (pages 38–9).

Appeals

No appeal is provided against notices which do not require works, but in any subsequent proceedings it is open to the defendant to question the reasonableness of the requirements (sect. 72(4)).

DEFECTIVE PREMISES

Reference

Building Act 1984 sect. 76.

Scope

This procedure is available to deal with any premises which are in such a state as to be prejudicial to health (i.e. 'defective state'), but where unreasonable delay in dealing with the situation would occur if the normal provisions

FC17 Means of escape from certain high buildings

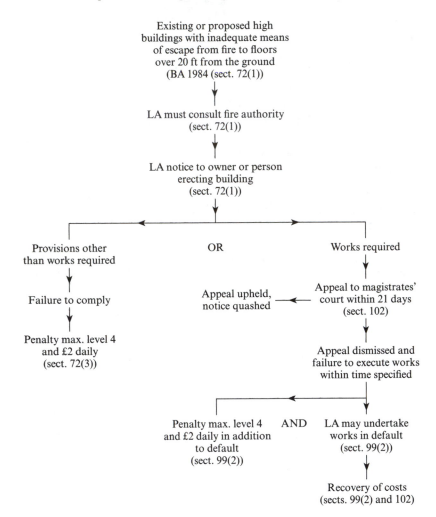

Existing or proposed high
buildings with inadequate means
of escape from fire to floors
over 20 ft from the ground
(BA 1984 (sect. 72(1))

LA must consult fire authority
(sect. 72(1))

LA notice to owner or person
erecting building
(sect. 72(1))

Provisions other   OR   Works required
than works required

Failure to comply   Appeal upheld,   Appeal to magistrates'
                           notice quashed   court within 21 days
                                                    (sect. 102)

Penalty max. level 4   Appeal dismissed and
and £2 daily   failure to execute works
(sect. 72(3))   within time specified

Penalty max. level 4   AND   LA may undertake
and £2 daily in addition   works in default
to default   (sect. 99(2))
(sect. 99(2))

                                 Recovery of costs
                                 (sects. 99(2) and 102)

Note
1. For means of escape from fire in houses in multiple occupation, see FC128.

FC18 Defective premises

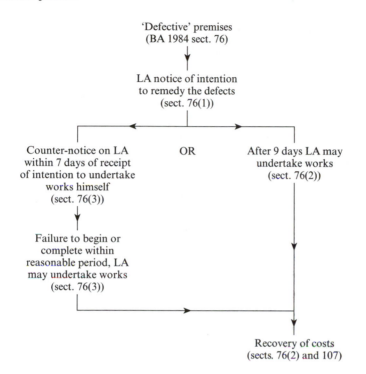

'Defective' premises
(BA 1984 sect. 76)

LA notice of intention
to remedy the defects
(sect. 76(1))

Counter-notice on LA
within 7 days of receipt
of intention to undertake
works himself
(sect. 76(3))

OR

After 9 days LA may
undertake works
(sect. 76(2))

Failure to begin or
complete within
reasonable period, LA
may undertake works
(sect. 76(3))

Recovery of costs
(sects. 76(2) and 107)

Notes

1. For normal statutory nuisance procedure relating to defective premises, see FC33a.
2. There is no provision for appeal against notices served under sect. 76 and there is no penalty for non-compliance with the notice.

for dealing with statutory nuisances in sect. 80 of the EPA 1990 were used, see FC33a (sect. 76(1)).

Person responsible

Notices are to be served either:

    (a) on the person responsible for the nuisance; or

    (b) where the person in (a) cannot be found or the nuisance has not yet occurred, on the owner or occupier of the premises on which the nuisance arises; or

    (c) where the nuisance arises from a structural defect, on the owner of the premises (sect. 27(1) and EPA 1990 sect. 80(2)).

Notices

The provisions of sect. 99 relating to the content and enforcement of notices do not apply here.

Notices must be in writing and, in addition to giving the LA's intention to remedy the defects, must specify the actual defects which it intends to remedy (sect. 76(1) and sect. 92). If a counter-notice is served by the recipient of the LA notice but works are not begun within a reasonable time or progress towards completion is not reasonable, the LA may execute the works itself. 'Reasonable' is not defined and the period allowed by the LA will therefore need to take account of all the relevant circumstances (sect. 76(2) and (3)).

Recovery of costs

In any proceedings by the LA to recover costs of carrying out works in default, the court must have regard to:

    (a) whether the LA was right in concluding that the premises were in such a state as to be prejudicial to health or a nuisance;

    (b) whether unreasonable delay would have occurred if the normal procedures for dealing with statutory nuisances had been followed;

    (c) following a counter-notice, whether the defendant had been given a reasonable time to begin or complete the works; and

    (d) whether any costs ought to be borne wholly or partly by someone other than the defendant (sect. 76(4)).

RUINOUS AND DILAPIDATED BUILDINGS

Reference

Building Act 1984 sect. 79.

FC19 Ruinous and dilapidated buildings

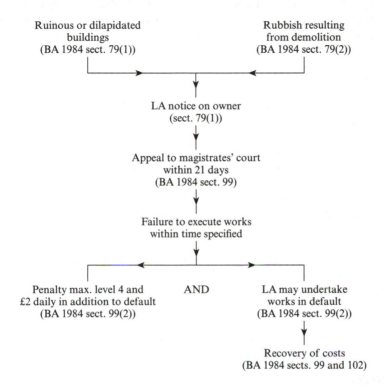

Ruinous or dilapidated
buildings
(BA 1984 sect. 79(1))

Rubbish resulting
from demolition
(BA 1984 sect. 79(2))

LA notice on owner
(sect. 79(1))

Appeal to magistrates' court
within 21 days
(BA 1984 sect. 99)

Failure to execute works
within time specified

Penalty max. level 4 and
£2 daily in addition to default
(BA 1984 sect. 99(2))

AND

LA may undertake
works in default
(BA 1984 sect. 99(2))

Recovery of costs
(BA 1984 sects. 99 and 102)

Note

1. The procedure cannot be applied to an advertisement structure as defined by sect. 290(1) of the TCPA 1990, but the provisions of that Act in relation to listed buildings etc. are still applicable to buildings dealt with under this procedure.

Ruinous and dilapidated buildings

The procedure is applicable to any buildings or structures which are seriously detrimental to the amenities of the neighbourhood because of their ruinous or dilapidated condition (sect. 79(1)), and to rubbish resulting from the demolition of a building which renders the site seriously detrimental to the amenities of the neighbourhood (sect. 79(2)).

Notices

Notices must be served in writing on the owner of the building or structure concerned and may require either:

(a) repair or restoration; or
(b) if the owner so elects, demolition of the building or structure or parts thereof and the removal of rubbish resulting from demolition (sect. 79(1)); or
(c) removal of the rubbish already existing (sect. 79(2)).

Notices must indicate both the nature of works of repair or restoration and the works of demolition and removal of rubbish or material (sect. 79(3)). The provisions of sects. 99 and 102 apply with regard to appeals against and the enforcement of these notices (pages 38–9).

DEMOLITION OF BUILDINGS

Reference

Building Act 1984 sects. 80–83 inclusive.

Scope

The notification procedure applies to the demolition of the whole or part of any building except:

(a) of an internal part of an occupied building which is to continue to be occupied;
(b) where the cubic content by external measurement does not exceed 1750 cubic feet;
(c) of greenhouses, conservatories, sheds or prefabricated garages forming part of a larger building;
(d) agricultural buildings unless it is contiguous to a non-agricultural building;
(e) demolition of a house subject to a demolition order under the HA 1985 (sect. 80(1)).

FC20 Demolition of buildings

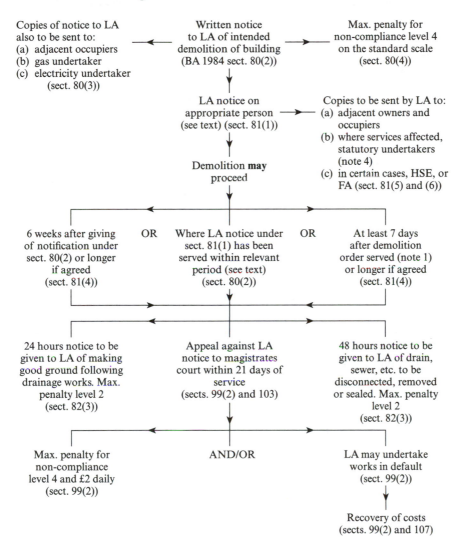

Copies of notice to LA
also to be sent to:
(a) adjacent occupiers
(b) gas undertaker
(c) electricity undertaker
    (sect. 80(3))

Written notice
to LA of intended
demolition of building
(BA 1984 sect. 80(2))

Max. penalty for
non-compliance level 4
on the standard scale
(sect. 80(4))

LA notice on
appropriate person
(see text) (sect. 81(1))

Copies to be sent by LA to:
(a) adjacent owners and
    occupiers
(b) where services affected,
    statutory undertakers
    (note 4)
(c) in certain cases, HSE, or
    FA (sect. 81(5) and (6))

Demolition **may**
proceed

6 weeks after giving
of notification under
sect. 80(2) or longer
if agreed
(sect. 81(4))

OR

Where LA notice under
sect. 81(1) has been
served within relevant
period (see text)
(sect. 80(2))

OR

At least 7 days
after demolition
order served (note 1)
or longer if agreed
(sect. 81(4))

24 hours notice to be
given to LA of making
good ground following
drainage works. Max.
penalty level 2
(sect. 82(3))

Appeal against LA
notice to magistrates
court within 21 days of
service
(sects. 99(2) and 103)

48 hours notice to be
given to LA of drain,
sewer, etc. to be
disconnected, removed
or sealed. Max. penalty
level 2
(sect. 82(3))

Max. penalty for
non-compliance
level 4 and £2 daily
(sect. 99(2))

AND/OR

LA may undertake
works in default
(sect. 99(2))

Recovery of costs
(sects. 99(2) and 107)

Notes
1. For demolition order, see FC119b.
2. For further procedure relating to rubbish following demolition, see FC19.
3. The sect. 99 procedure applies here for appeals against and the enforcement of LA notices (pages 38–9).
4. The statutory undertakers include the WUs.

Notification to a LA

The person undertaking the demolition of any building within the scope of this procedure must give prior notification to the LA in writing indicating the building concerned and the works of demolition which are intended. Copies of this notice must be sent or given to:

(a) occupiers of adjacent buildings;
(b) the gas undertaker;
(c) the electricity undertaker.

Demolition must not commence until the LA has served its notice under sect. 81 or the 'relevant period' (below) has expired (sect. 80(2) and (3)).

LA notices

In addition to those situations where notification must be made to the LA notices may also be served by the LA in the situations identified in (i), (ii) and (iii) under (a) below.

(a) **Person.** These notices are to be served on either:
  (i) a person upon whom a demolition order under the HA 1985 has been served; or
  (ii) a person intending to comply with an order under sect. 77 BA 1984 (dangerous structure etc.) or a notice under sect. 79 BA 1984 (ruinous and dilapidated buildings, etc.); or
  (iii) a person who is intending, or has begun, a demolition to which sect. 80 BA 1984 applies ('Scope' above).
(b) **Relevant period.** This is the period within which the LA is required to serve its notice and this is:
  (i) where a person has served a notification on the LA under sect. 80, within 6 weeks from the giving of the notice to the LA, or such longer period as may be agreed by the notifier in writing; and
  (ii) in relation to a demolition order under the HA 1985, not less than 7 days after service of the order, or such longer period as the person on whom the order was served may allow in writing (sect. 81(4)).
(c) **Contents.** The LA notice may require all or any of the following:
  (i) shoring up of the adjacent buildings;
  (ii) weather-proofing of exposed surfaces of adjacent buildings and repairing or making good damage;
  (iii) removal of material from site;
  (iv) disconnection, sealing and removal of drains and sewers;
  (v) making good ground surface following disconnection or removal of drains and sewers;
  (vi) arrangements with the statutory undertakers (including WUs) for the disconnection of gas, electricity and water supplies;

(vii) arrangements for the burning of structures and materials on site with the HSE and the FA (where a fire certificate is required) or, in all other cases, the FA;

(viii) any other steps considered necessary for the protection of the public and the preservation of public amenity.

Any requirement under any of the relevant statutory provisions of the HASAWA 1974 takes precedence over any requirement of these notices (sect. 81(2)).

(d) **Copies.** The LA is required to send copies of its notices to:
   (i) owners **and** occupiers of adjacent buildings;
   (ii) if the notice requires disconnection of gas, electricity or water supplies, the statutory undertaker (includes WUs) concerned;
   (iii) if the notice relates to the burning of structures or materials on site, the FA (if a fire certificate is in force to the HSE as well) (sect. 80(5) and (6)).

(e) **Enforcement of LA notices.** These notices are subject to sect. 99 procedures (page 38) and the LA is, therefore, able to undertake works in default and recover costs as well as prosecuting for non-compliance (sect. 82(6)).

(f) **Appeals.** Appeal against LA notices served under sect. 81 may be made under the grounds set out in BA 1984 sect. 102 (page 38) and in addition on the following grounds:
   (i) in relation to the shoring up of an adjacent building, that the owner is not entitled to support by the building to be demolished and he, therefore, should pay for, or contribute towards, the costs; and
   (ii) in relation to the weather proofing of an adjacent building, that the owner should pay for, or contribute towards the costs (sect. 83(2)).

In both cases the appellant must serve copies of his notices of appeal on the other persons concerned (sect. 83(3)).

DANGEROUS BUILDINGS

Reference

Building Act 1984 sects. 77 and 78.

Scope

The procedure may be used where a building or structure, or part thereof, is:

(a) in such a state; or
(b) is used to carry such loads;

as to be dangerous (sects. 77(1) and 78(1)).

FC21 Dangerous buildings

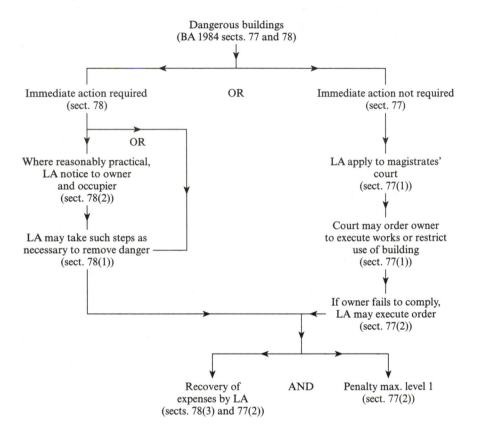

Note
1. This procedure is subject to the provisions of the Planning (Listed Buildings and Conservation Areas) Act 1990 in relation to listed buildings etc.

Emergency measures

These may be taken where immediate action is required in the opinion of the LA to remove the danger (sect. 78(1)).

The measures which may be taken are those necessary to remove the danger (sect. 78(1)). Where practicable, prior notice must be given to the owner and occupier (sect. 78(2)).

Court order

Where there is no call for immediate action, the LA may apply to a magistrates' court and the court may order the owner to:

(a) execute necessary works to remove the danger; or
(b) if the owner elects, to demolish the building or structure and remove resulting rubble or rubbish; or
(c) where the danger arises from the overloading of the building or structure, restrict the use until the court withdraws or modifies that restriction following works having been carried out (sect. 77(1)).

Recovery of expenses

The LA's expenses in executing a court order are recoverable from the owner (sect. 77(2)). Where the emergency provisions have been used, the LA may recover its costs, but in any dispute as to the recovery of these costs, the court may consider whether the LA was justified in taking emergency action as opposed to an application to the court for an order under sect. 77 (sect. 78(5)).

Recovery of expenses connected with the fencing off of a building or structure or arranging for it to be watched are not recoverable after:

(a) the danger has been removed by emergency action by the LA; or
(b) after works required under sect 77 have been carried out (sect. 78(4)).

Chapter 4

WATER INDUSTRY ACT 1991

WATER SUPPLY PROVISIONS

GENERAL PROCEDURAL PROVISIONS

The following provisions are applicable to the use by LAs of all procedures dealt with in this chapter.

Local authority

The LAs for the purposes of these procedures are:

(a) district councils;
(b) London borough councils;
(c) the Common Council of the City of London;
(d) in Wales, county and county borough councils (unitary authorities) (sect. 219).

This definition includes unitary authorities in England, i.e. district councils where there is no county and county councils where there is no district (Local Government Changes for England Regulations 1994).

Extent

These procedures apply in England and Wales only (sect. 223(3)).

Notices

The service of notices on any person may be undertaken by either:

(a) delivery to him or leaving it at his proper address; or
(b) sending it by post.

If the person is in partnership, service is on a partner or person having the control or management of the partnership business and in the case of a body corporate on the secretary or clerk.

Notices must be sent to the last known address except for:

(i) a body corporate – to be sent to the secretary or clerk at the registered or principal office:

(ii) partnerships – to the principal office.

Where, after reasonable inquiry, the name and address of any owner or occupier cannot be obtained or, in the case of an occupier the premises are unoccupied, the notice may be served by giving it to a person resident or working on the premises or by leaving it conspicuously fixed to some building or object on the land (sect. 216).

The following provisions relate only to the procedures dealing with water supplies.

Information

A LA may by notice require any person to provide such information as it may reasonably require to fulfil its powers and duties with respect to both public and private water supplies. The maximum penalty for non-compliance with such notices is level 5 (sect. 85).

Power of entry etc.

A person designated in writing by the LA may:

(a) enter premises supplied with private or public water to see if any action is required of the LA;

(b) carry out such inspections, measurements and tests, including the taking of samples of water, land and articles, as is considered appropriate.

Unless the situation is one of emergency, power of entry cannot be required unless it is requested at a reasonable time and after having given at least 24 hours notice to the occupier in relation to non-business premises. Persons convicted of obstructing an LA officer are subject to penalties not exceeding level 5 (sect. 84(3) and Schedule 6).

Where the LA shows that entry has been refused, or is anticipated, or the premises are unoccupied or the case is one of emergency, a JP may issue a warrant to authorize entry, if need be by force (Schedule 6 para. 2).

DEFINITIONS

Owner in relation to any premises, means the person who:

(a) is for the time being receiving the rackrent of the premises whether on his own account or as the agent or trustee for another person; or

(b) would receive the rackrent if the premises were let at a rackrent (sect. 219(1)).

Private supply means:

'(1) subject to subsection (2) below, a supply of water provided otherwise than by a water undertaker (including a supply provided for the purposes of the bottling of water), and cognate expressions shall be construed accordingly;

(2) for the purposes of any reference . . . to a private supply, or to supplying water by means of a private supply, water shall be treated as supplied to any premises not only where it is supplied from outside those premises, but also where it is abstracted, for the purpose of being used or consumed on those premises, from a source which is situated on the premises themselves; and for the purposes of this subsection water shall be treated as used on any premises where it is bottled on those premises for use or consumption elsewhere' (sect. 93(1) and (2)).

Water undertaker. A company may be appointed:

(a) by the SoS; or
(b) with the consent of or in accordance with a general authorization given by the SoS, by the Director,

to be the water undertaker . . . for any area of England and Wales (sect. 6(1)).

WATER SUPPLY

CONTROL OVER PUBLIC WATER SUPPLIES

References

Water Industry Act 1991 sects. 77 and 78.
Water Supply (Water Quality) Regulations 1989 (as amended) and 2000 (as amended).
Circular 20/89 DoE, The Water Act 1989 (consolidated in the WIA 1991).

Scope

Public supplies are those provided by an appointed WU, and each WU is placed under a statutory duty to supply water for domestic or food production purposes which is wholesome and to ensure that, in general, there is no deterioration in the quality of water from any of its sources (sect. 68(1)).

In this connection, domestic purposes include drinking, washing, cooking, central heating and sanitary purposes (sect. 218(1)).

'Food production purposes' means the manufacturing, processing, preserving or marketing purposes with respect to food or drink for which water supplied to food production premises may be used; and 'food production premises' means premises used for the purposes of a business of preparing food or drink for consumption otherwise than on premises (sect. 93).

FC22 Control over public water supplies

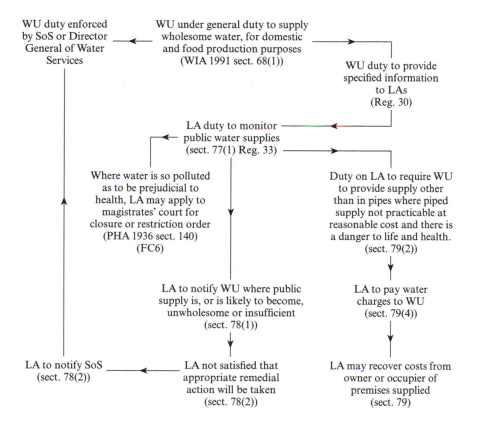

Notes

1. Regulation numbers refer to the Water Supply (Water Quality) Regulations 1989 (as amended).
2. For procedure for closure or restriction of polluted supply see FC6.

Enforcement

Those duties placed on WUs are enforceable by the SoS and the Director General of Water Services through enforcement orders (sect. 18).

Wholesomeness

Water is to be judged as being wholesome if it satisfies the requirements of reg. 2 of the Water Supply (Water Quality) Regulations 2000 or reg. 3 of the Private Water Supplies Regulations 1991 as appropriate (sect. 67(1)).

LA role

LAs are placed under a statutory duty to take all such steps as they consider appropriate for keeping themselves informed about the wholesomeness and sufficiency of all public water supplies (sect. 77(1)).

In this connection, LAs:

(a) must make arrangements with the WU to be notified of events giving rise to significant risks to health; and

(b) may take and have analysed by designated officers such public water samples as they may reasonably require (Reg. 33 1989 Regs.).

In order to assist LAs in undertaking their surveillance role, WUs are required to pass on to LAs the following information:

(a) an annual statement before 30 June in the following year in a form set out in Schedule 4 of the regulations concerning the general quality of the water supplied in each LA area;

(NB. The form and content of this statement was substituted by the Water Supply (Water Quality) Regulations 1991.)

(b) the samples taken in respect of each parameter;

(c) the extent to which the WU has complied with the requirements of the regulations regarding wholesomeness and relaxations;

(d) any action taken to comply with any undertakings given to the SoS in relation to any relaxations; and

(e) the occurrence of any event which gives rise, or may give rise, to a significant risk to health (this information is also to be supplied directly to the health authority) (Reg. 30(4) and (5) 1989 Regs.).

Powers of the LA

There are three situations which the LA deals with in relation to public supplies:

1. Where a LA believes that the water:
   (a) is, or is likely to become, unwholesome;
   (b) is insufficient for domestic purposes;

(c) is, or is likely to cause, danger to life or health because of unwhole-someness or insufficiency; or

(d) that the quality of sources of water are deteriorating;

the LA is required to inform the WU. Where the LA is not satisfied that all appropriate remedial action is being taken by the WU, it must notify the SoS who may then use his powers of enforcement (sect. 78).

2. The LA is required to notify the WU wherever it (the WU) is satisfied that it is not reasonably practicable at reasonable cost to provide, or to maintain a piped supply of wholesome water that is sufficient for domestic purposes, and to request the WU to provide a supply of water to particular premises by means other than pipes where:

(a) this is practicable at reasonable cost; and

(b) the unwholesomeness or insufficiency is such as to cause danger to life and health.

In these circumstances, the LA is required to pay to the WU the appropriate water charges, but may recover these from the owners or occupiers of the premises benefiting from the supply. The WU is under a duty to comply with a LA request which fulfils the criteria outlined above (sect. 79).

3. Where the LA is of the opinion that a supply is so polluted as to be prejudicial to health, it may apply to a magistrates' court for an order closing or restricting that supply. This procedure is detailed in FC6 (PHA 1936 sect. 140).

CONTROL OVER PRIVATE WATER SUPPLIES

References

Water Industry Act 1991 sects. 80–83.
Water Supply (Water Quality) Regulations 1989 (as amended).
The Private Water Supplies Regulations 1991.
Circular 24/91 DoE, Private Water Supplies.

Scope

A private water supply is a supply of water provided otherwise than by a WU and includes a supply provided for the purposes of the bottling of water. This definition is extended to cover water sources abstracted both outside and inside of premises, and water bottling in premises for use or consumption elsewhere (sect. 93(1) and (2)). The FSA 1990 extended the definition to cover water supplied for food production (page 66 also).

LAs are under the same general duty to monitor the wholesomeness and sufficiency of private supplies as they are in relation to public supplies (sect. 77(1)), but in this case they have direct and sole enforcement powers.

FC23 Control over private water supplies

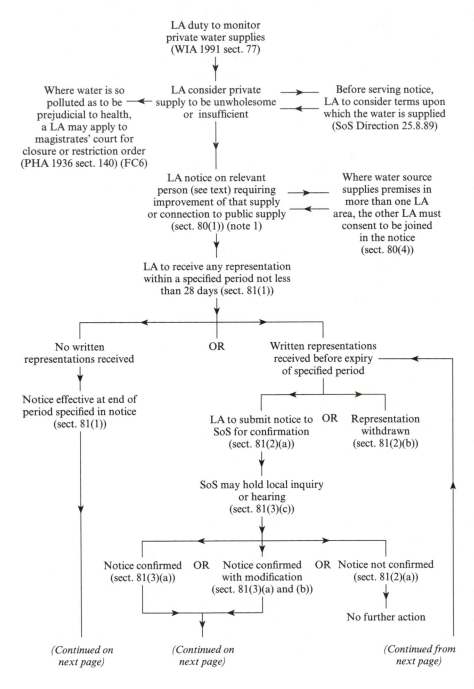

LA duty to monitor
private water supplies
(WIA 1991 sect. 77)

Where water is so
polluted as to be
prejudicial to health,
a LA may apply to
magistrates' court for
closure or restriction order
(PHA 1936 sect. 140) (FC6)

LA consider private
supply to be unwholesome
or insufficient

Before serving notice,
LA to consider terms upon
which the water is supplied
(SoS Direction 25.8.89)

LA notice on relevant
person (see text) requiring
improvement of that supply
or connection to public supply
(sect. 80(1)) (note 1)

Where water source
supplies premises in
more than one LA
area, the other LA must
consent to be joined
in the notice
(sect. 80(4))

LA to receive any representation
within a specified period not less
than 28 days (sect. 81(1))

No written
representations received

OR

Written representations
received before expiry
of specified period

Notice effective at end of
period specified in notice
(sect. 81(1))

LA to submit notice to
SoS for confirmation
(sect. 81(2)(a))

OR

Representation
withdrawn
(sect. 81(2)(b))

SoS may hold local inquiry
or hearing
(sect. 81(3)(c))

Notice confirmed
(sect. 81(3)(a))

OR

Notice confirmed
with modification
(sect. 81(3)(a) and (b))

OR

Notice not confirmed
(sect. 81(2)(a))

No further action

*(Continued on
next page)*

*(Continued on
next page)*

*(Continued from
next page)*

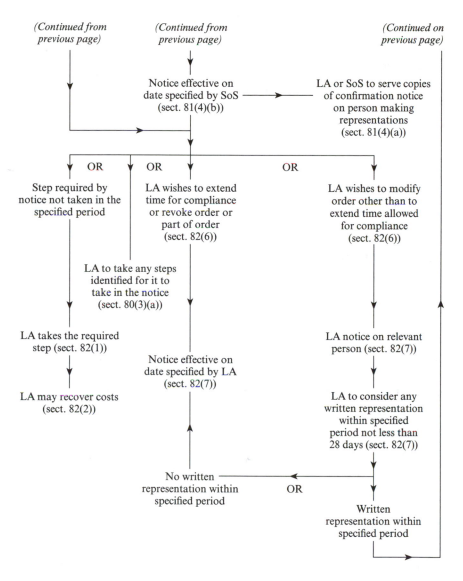

Notes
1. Notices under sect. 80 are to be registered as a local land charge and bind successive owners or occupiers unless revoked (sect. 82(5)).
2. There are no penalties for non-compliance with any requirements of sect. 80 notices. The LAs' only remedy is to undertake the requirements themselves and recover their costs.

Reg. 8 of the Private Water Supplies Regulations 1991 also makes it a duty of LAs to take and analyse samples of certain private supplies, i.e. categories 1 and 2 as defined in the regulations. The regulations also lay down a sampling regime for private supplies.

Information

The LA may serve a notice on any person requiring him to provide any information it may reasonably require in order to fulfil its responsibilities for private supplies (sect. 85(1)).

Monitoring

The classification of private supplies, sampling requirements and charging for sampling and analysis is dealt with by the Private Water Supplies Regulations 1991.

Improvement notices

Where a LA is satisfied that a private water supply is not wholesome or sufficient for domestic or food production purposes and that a relaxation of standards would not be appropriate (FC24a), it may serve an improvement notice on the relevant person (sect. 80(1)).

By reason of a direction issued by the SoS on 25 August 1989, before serving any notice LAs must ascertain the terms of any agreement, contract, licence or other document relating to the terms upon which the water is supplied, and have regard to this in determining the content of any notice.

Standards

Water from private supplies is to be judged as being wholesome for domestic purposes if it satisfies the requirements of the Private Water Supplies Regulations 1991 (sect. 67 and Reg. 3).

Relevant person

The persons upon whom the notice should be served are the owners **and** occupiers of premises served by the supply, the owner **and** occupiers of premises where the source is situated **and** other persons who exercise powers of management or control of that source (sect. 80(7)).

Requirements

A notice must:

   (a) identify the reasons why the supply is considered to be unwholesome and/or insufficient;

(b) specify the steps required to provide a wholesome and sufficient supply;

(c) allow a period of not less than 28 days for representation or objections to be received;

(d) indicate the need for referral to the SoS if representations or objections are made within the specified period (sect. 80(2)).

The notice may specify one or more of the following:

(a) the steps to be taken by the LA themselves;

(b) the steps to be taken by the person specified in the notice within a specified period;

(c) the payments to be made to another relevant person or the LA for expenses in meeting the LA requirements;

(d) the payments to be made by the LA to a person for the costs of meeting the notice requirements (sect. 80(3));

and may require:

(a) a supply of water to be provided by the WU or another person; and

(b) steps to be taken to ensure that this is done (sect. 80(6)).

Operation of notices

Where no representation or objections are made to the LA within the period specified in the notice, the notice becomes effective at the end of the period allowed for such representation to be made (sect. 81(1)).

Where representations etc. are made and not withdrawn, then the LA must refer the matter to the SoS who may then either:

(a) confirm the notice with or without modification; or

(b) not confirm the notice.

Where the notice is confirmed, it becomes effective on a date as specified by the SoS in his confirmation (sect. 81(2), (3) and (4)).

Enforcement

The LA's powers of enforcement where an improvement notice is not complied with are restricted to the undertaking of required works and the recovery of related costs (sect. 82).

RELAXATIONS OF QUALITY STANDARDS FOR WATER SUPPLIES

References

Water Industry Act 1991 sect. 67(2)(e) and (f).
Water Supply (Water Quality) Regulations 1989 Part 3 (as amended 1991).

FC24a Relaxation of standards for public water supplies by SoS

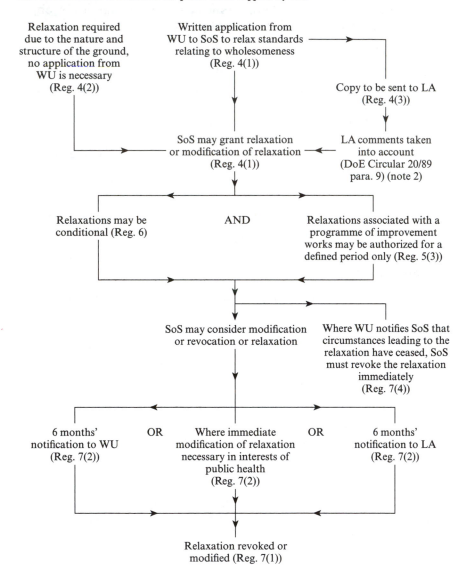

Relaxation required due to the nature and structure of the ground, no application from WU is necessary (Reg. 4(2))

Written application from WU to SoS to relax standards relating to wholesomeness (Reg. 4(1))

Copy to be sent to LA (Reg. 4(3))

SoS may grant relaxation or modification of relaxation (Reg. 4(1))

LA comments taken into account (DoE Circular 20/89 para. 9) (note 2)

Relaxations may be conditional (Reg. 6)

AND

Relaxations associated with a programme of improvement works may be authorized for a defined period only (Reg. 5(3))

SoS may consider modification or revocation or relaxation

Where WU notifies SoS that circumstances leading to the relaxation have ceased, SoS must revoke the relaxation immediately (Reg. 7(4))

6 months' notification to WU (Reg. 7(2))

OR

Where immediate modification of relaxation necessary in interests of public health (Reg. 7(2))

OR

6 months' notification to LA (Reg. 7(2))

Relaxation revoked or modified (Reg. 7(1))

Notes
1. Regulation numbers refer to the Water Supply (Water Quality) Regulations 1989.
2. In relation to application by the WU made because of the nature and structure of the ground, the LA may make formal representations to the SoS within 6 weeks (Reg. 4(3)).

FC24b Relaxations of standards for private supplies by LA

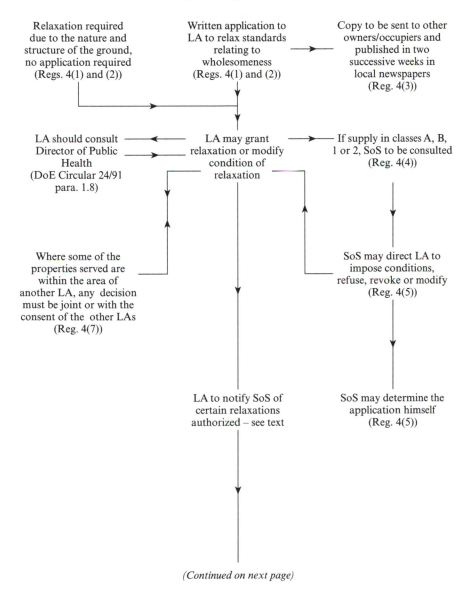

Relaxation required
due to the nature and
structure of the ground,
no application required
(Regs. 4(1) and (2))

Written application to
LA to relax standards
relating to
wholesomeness
(Regs. 4(1) and (2))

Copy to be sent to other
owners/occupiers and
published in two
successive weeks in
local newspapers
(Reg. 4(3))

LA should consult
Director of Public
Health
(DoE Circular 24/91
para. 1.8)

LA may grant
relaxation or modify
condition of
relaxation

If supply in classes A, B,
1 or 2, SoS to be consulted
(Reg. 4(4))

Where some of the
properties served are
within the area of
another LA, any decision
must be joint or with the
consent of the other LAs
(Reg. 4(7))

SoS may direct LA to
impose conditions,
refuse, revoke or modify
(Reg. 4(5))

LA to notify SoS of
certain relaxations
authorized – see text

SoS may determine the
application himself
(Reg. 4(5))

(Continued on next page)

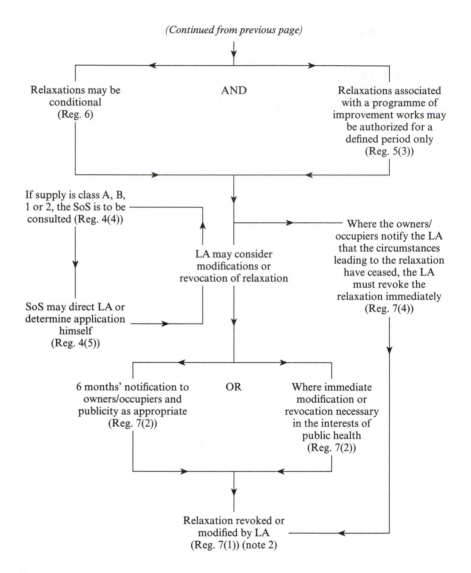

(Continued from previous page)

Relaxations may be
conditional
(Reg. 6)

AND

Relaxations associated
with a programme of
improvement works may
be authorized for a
defined period only
(Reg. 5(3))

If supply is class A, B,
1 or 2, the SoS is to be
consulted (Reg. 4(4))

LA may consider
modifications or
revocation of relaxation

Where the owners/
occupiers notify the LA
that the circumstances
leading to the relaxation
have ceased, the LA
must revoke the
relaxation immediately
(Reg. 7(4))

SoS may direct LA or
determine application
himself
(Reg. 4(5))

6 months' notification to
owners/occupiers and
publicity as appropriate
(Reg. 7(2))

OR

Where immediate
modification or
revocation necessary
in the interests of
public health
(Reg. 7(2))

Relaxation revoked or
modified by LA
(Reg. 7(1)) (note 2)

Notes
1. Regulation numbers refer to the Private Water Supplies Regulations 1991.
2. LA to notify SoS of revocation or modification of certain relaxations (text).
3. This procedure does not cover relaxations of standards for water used for food production
   purposes – for this, see Reg. 4(1)(d) Private Water Supplies Regulations 1991.

Private Water Supplies Regulations 1991.
Circulars 20/89 and 24/91 DoE.

Scope

(a) **Public supplies.** The SoS is able to relax any of the standards of wholesomeness prescribed by the 1989 regulations in relation to the supply of water for domestic or food production purposes by any appointed WU providing he is satisfied that the authorization is:
   (i) necessary as an emergency measure to maintain a supply of water for human consumption; or
   (ii) called for by reason of exceptional meteorological conditions; or
   (iii) called for by reason of the nature and structure of the ground in the area from which the supply emanates or that the supply is to be used; or
   (iv) solely for food production purposes (1989 Reg. 4 (1))
(b) **Private supplies.** Relaxations of standards of wholesomeness for other than public supplies are dealt with by LAs (page 64 for definition) with the same criteria as in (i)–(iv) in (a) above applying (1991 Reg. 4(1)). For definition of 'private supplies' see page 66 and also under 'Scope' on page 69.

    LAs may authorize a relaxation from prescribed standards in respect of private supplies for food production where the quality of the final food product will not be affected (Private Water Supplies Regulation 1991 Reg. 4(1)(d)). This is by way of a separate procedure from that dealt with in this section.

Restrictions

Both the SoS and any LA are restricted from approving relaxations which either:

(a) if being considered as being necessary as an emergency measure, would give rise to a risk to public health; or
(b) if being considered in relation to exceptional meteorological conditions or because of the nature and structure of the ground, would lead to a public health hazard in relation to parameters in Tables B or C or item 7 of Table D of both Regulation; or
(c) affect the fitness for human consumption of the food or drink in its final form (1989 and 1991 Reg. 5).

Applications

Both the SoS and a LA may proceed with a relaxation being considered as called for by reason of the nature and structure of the ground without having received an application, but otherwise a written application is necessary either from the WU to the SoS for public supplies or from the owner or occupier of premises where the source is situated, from any other person having control over the source or from the owner of any premises supplied to the LA in respect of any private supply (1989 Reg. 4(1) and (2) and 1991 Reg. 4(2)).

Consultations

In respect of public supplies, the WU must copy the application to the LA and, while not required by the Regulations, paragraph 9 of DoE Circular 20/89 indicates that the SoS will take account of any comments made to him by the LA. The amendments in the 1991 Regulations require that, in respect of applications made by reason of the nature and structure of the ground etc., the WU must inform the LA that it may make representations to the SoS within 6 weeks (1989 Reg. 4(3)).

In respect of private supplies, the LA is required to consult the SoS in relation to all relaxations relating to supplies classified as A, B, 1 or 2. The SoS is empowered to:

(a) call in the authorization for determination himself; or
(b) impose conditions; or
(c) order refusal; or
(d) order revocation or modification (1991 Reg 4(5)).

Where a private source supplies premises in the area of another LA, the LA dealing with the relaxation must either obtain the consent of that other LA, or exercise its powers jointly with it (1991) Reg. 4(7)).

Conditions

Relaxations made by either the SoS or a LA may be limited to either:

(a) particular sources or types of source; or
(b) particular water zones or parts of such zones or zones of particular descriptions; or
(c) particular food production purposes.

Specific conditions may be attached relating to:

(a) the quality of the water;
(b) the steps to be taken to improve the quality;
(c) the monitoring of the quality;
(d) the giving of information.

Relaxation for emergency reasons or because of exceptional meteorological conditions must be time related; those dealt with because of the nature and structure of the land may be so (1989 Reg. 6 and 1991 Reg. 5(3) and 6).

Revocation and modifications

Both the SoS and LAs are given powers to revoke or modify relaxations made by them after giving 6 months' notice, unless an immediate revocation or modification is required in the interests of public health (1989 Reg. 7(2) and 1991 Reg. 7(2)).

Notifications by LA to SoS

The LA must send a copy to the SoS of any authorization issued by it in relation to:

(a) emergency measures to maintain a supply of water for human consumption;
(b) relaxations in respect of exceptional meteorological conditions or by reason of the nature and structure of the ground in respect of a class A or 1 supply; and
(c) water used solely for food production purposes,

and of revocations or modifications of relaxations relating to any of these (1991 Regs. 4(5) and 7(1)).

Chapter 5

PUBLIC HEALTH (CONTROL OF DISEASE) ACT 1984

GENERAL PROCEDURAL PROVISIONS

The following provisions are applicable to procedures under the Public Health (Control of Disease) Act 1984.

Local authority

The LAs responsible for the execution of this Act and, unless otherwise stated, for the enforcement of regulations made under it are:

(a) district councils, metropolitan and unitary authorities;
(b) London borough councils;
(c) the Common Council of the City of London;
(d) the Sub-Treasurer of the Inner Temple and the Under-Treasurer of the Middle Temple;
(e) in Wales, county and county borough councils (unitary authorities) (sect. 1).

Port health authorities may have the powers assigned to them and in these cases the LAs within the port health district cease to discharge those functions (sect. 3).

Extent

This Act does not apply in Scotland or Northern Ireland (sect. 79(3)).

Notices

(a) **Form.** All notices must be in writing and the SoS may prescribe the form of any notice etc. by regulation (sect. 58).
(b) **Authentication.** Notices must be signed (includes a facsimile signature) by the proper officer or any other officer so authorized in writing (sect. 59).
(c) **Service.** This may be effected by one of the following methods:
    (i) delivery to the person;
    (ii) in the case of the proper officer of a LA, by leaving it or sending it in a prepaid letter addressed to him at either his residence

or his office, and in the case of any other officer of the LA, by leaving it or sending it in a prepaid letter addressed to him at his office;

(iii) in the case of any other person, leaving it or sending it in a prepaid letter addressed to him at his normal or last known residence;

(iv) for an incorporated company or body, by delivering it to its secretary or clerk at its registered or principal office, or by sending it in a prepaid letter addressed to him at that office;

 (v) in the case of the owner of any premises by virtue of the fact that he receives the rackrent of the premises for another person, by leaving it or sending it in a prepaid letter addressed to him at his place of business;

(vi) in the case of an owner or occupier of any premises, if it is not practicable after reasonable enquiry to ascertain the name and address or, if the premises are unoccupied, by addressing it to the person concerned by the description of 'owner' or 'occupier' of the premises (naming them) and delivering it to some person on those premises or, if there is no one to whom the notice can be delivered, by affixing it, or a copy of it, to some conspicuous part of the premises (sect. 60).

Power of entry

An authorized officer of a council, producing if required a duly authenticated document showing his authority, has a right to enter any premises at all reasonable hours for the following purposes:

(a) ascertaining if there has been a contravention of the Act or bye-laws made under it;

(b) seeing if circumstances exist which would authorize or require the council to take any action or execute work;

(c) to take any action or execute any work authorized or required;

(d) generally for the council's performance of its functions.

For premises other than a factory, workshop or workplace, at least 24 hours' notice must be given before entry may be demanded.

Where there is reasonable ground for entry under (a)–(d) above and:

(a) admission has been refused; or

(b) refusal is apprehended; or

(c) the premises are unoccupied; or

(d) the occupier is temporarily absent; or

(e) the case is one of urgency; or

(f) application for admission would defeat the object of entry;

the LA may make application to a JP by way of sworn information in writing, and the JP may authorize entry by an authorized officer, if necessary by force. Unless the case falls as one of (c)–(f) above, notice to apply for the warrant must be given to the occupier (sect. 61). The authorized officer may take with him any other persons as may be necessary but must leave any unoccupied premises as secure as he found them (sect. 62).

The penalty for obstructing any person acting in the execution of these Acts, Regulations and orders is a maximum of level 1 on the standard scale (sect. 63).

DEFINITIONS

Common lodging house means a house (other than a public assistance institution) provided for the purpose of accommodating by night poor persons, not being members of the same family, who resort to it and are allowed to occupy one common room for the purpose of sleeping or eating, and, where part only of a house is so used, includes the part so used.

Hospital includes any premises for the reception of the sick.

House means a dwelling-house, whether a private dwelling-house or not.

Owner means the person for the time being receiving the rackrent of the premises in connection with which the word is used, whether on his own account or as agent or trustee for any other person, or who would so receive the rackrent if those premises were let at a rackrent.

Premises includes buildings, lands, easements and hereditaments of any tenure.

Proper officer means, in relation to a purpose and to an authority, an officer appointed for that purpose by that authority (page 85).

Vessel has the same meaning as in the Merchant Shipping Act 1995, except that it includes a hovercraft within the meaning of the Hovercraft Act 1968, and 'master' shall be construed accordingly (sect. 74).

KEY TO POWERS AVAILABLE FOR
THE CONTROL OF DISEASE

The following powers are available to LAs for the control of communicable disease but only to the extent that each is specifically applied by the Act or by Regulations. Their application to each disease is shown in the list on page 84.

1. Sect. 17 PH(CoD)A 1984   – Penalty on exposure of persons and articles.
2. Sect. 18 PH(CoD)A 1984   – Information from occupier.
3. Sect. 19 PH(CoD)A 1984   – Persons not to carry out occupation to danger of others.
4. Sect. 20 PH(CoD)A 1984   – Stopping employment. (This power includes all occupations and could be used, for example, in relation to work

	at a slaughterhouse where food safety controls generally are not enforced by the LA.)
5. Sect. 21 PH(CoD)A 1984	– Child may be ordered not to attend school.
6. Sect. 22 PH(CoD)A 1984	– List of day scholars at school.
7. Sect. 23 PH(CoD)A 1984	– Exclusion of children from places of entertainment or assembly.
8. Sect. 24 PH(CoD)A 1984	– Sending infected articles to laundry etc.
9. Sect. 25 PH(CoD)A 1984	– Controls over library books.
10. Sect. 26 PH(CoD)A 1984	– Matter not to be placed in dustbins.
11. Sect. 28 PH(CoD)A 1984	– Prohibition of homework.
12. Sect. 29 PH(CoD)A 1984	– Letting of houses, rooms, etc.
13. Sect. 30 PH(CoD)A 1984	– Cessation of occupation of house.
14. Sect. 33 PH(CoD)A 1984	– Use of public conveyance.
15. Sect. 34 PH(CoD)A 1984	– Duties of owners of public conveyances.
16. Sect. 35 PH(CoD)A 1984	– Power of justice to order medical examination.
17. Sect. 36 PH(CoD)A 1984	– Medical examination of group of persons.
18. Sect. 37 PH(CoD)A 1984	– Removal of persons to hospital.
19. Sect. 38 PH(CoD)A 1984	– Detention in hospital.
20. Sect. 40–42 PH(CoD)A 1984	– Common lodging houses.
21. Sect. 43 PH(CoD)A 1984	– Restrictions on removal of bodies.
22. Sect. 44 PH(CoD)A 1984	– Avoidance of contact with bodies.
23. Sect. 45 PH(CoD)A 1984	– Prohibition of wakes.
23A. Sect. 48 PH(CoD)A 1984	– Removal of body to mortuary or for immediate burial.
24. Schedule 3 PH(ID) Regs. 1988	– Destruction of lice.
25. Schedule 4 PH(ID) Regs. 1988	– Prevention of spread of infection.

In addition, the following powers apply to all infectious diseases:

26. Sect. 31 PH(CoD)A 1984	– Disinfection of premises (FC26).
27. Sect. 32 PH(CoD)A 1984	– Removal of persons from infected houses.
28. Sect. 39 PH(CoD)A 1984	– Notification in common lodging house.
29. Sect. 51 PH(CoD)A 1984	– Canal boats.
30. Reg. 20 M&D (G) Regs. 1959	– Infected milk (FC92).
31. The Food Safety (Fishery Products and Live Shellfish) Regulations 1998	– Shellfish – temporary prohibition orders (FC89).

Note

Special arrangements exist for the control of infectious diseases in ships and aircraft in the Public Health (Aircraft) Regulations 1979 and the Public Health (Ships) Regulations 1979 and in respect of the Channel Tunnel the Public Health (International Trains) Regulations 1994.

POWERS AVAILABLE FOR THE CONTROL OF SPECIFIC DISEASES

In addition to those powers which relate to all infectious diseases (nos. 26–31 in the Key), there are powers available to the LA or its proper officer relating to specified diseases. The procedures involved are simple in most cases and, therefore, most are not included in chart form in this book. However, those which are have the page reference indicated. It will be seen that the powers vary for the different diseases.

Acquired immune deficiency syndrome	– 16, 18, 19 (modified), 21, 22.
Acute encephalitis	– 1–3, 5–8, 10–16, 18, 19, 22, 23.
Acute poliomyelitis	– 1–3, 5–8, 10–16, 18, 19, 22, 23.
Anthrax	– 1–3, 5, 6, 8, 10–16, 18, 19, 21–23.
Cholera	– 1–23 (except 4).
Diphtheria	– 1–8, 10–19, 22, 23.
Dysentery (amoebic and bacillary)	– 1–8, 10–19, 22, 23, 25.
Food poisoning	– 2, 25.
Leprosy	– 1, 3, 5, 11–13, 16, 18, 19, 22.
Leptospirosis	– 1–3, 5, 6, 8, 10–16, 18, 19, 22, 23.
Malaria	– 2, 16.
Measles	– 1–3, 5, 6, 8, 10–16, 18, 19, 22, 23.
Meningitis	– 1–3, 5–8, 10–16, 18, 19, 22, 23.
Meningococcal septicaemia	– 1–3, 5–8, 10–16, 18, 19, 22, 23.
Mumps	– 1–6, 8, 10–16, 18, 19, 22, 23.
Ophthalmia neonatorum	– 1, 8, 10.
Plague	– 1–23.
Paratyphoid fever	– 1–8, 10–19, 22, 23, 25.
Rabies	– 1–3, 5–17, 18, 19.
Relapsing fever	– 1–24 (except 4).
Rubella	– 1–6, 8, 10–16, 18, 19, 22, 23.
Salmonella infections	– 25, 27.
Scarlet fever	– 1–6, 8, 10–19, 22, 23, 27.
Smallpox	– 1–23 (except 4).
Staphylococcal infections likely to cause food poisoning	– 25.
Tetanus	– 2, 16.

Tuberculosis   – 1–3, 5–8, 10–13, 16, 22, 23 but 9, 18, 19, 26, 27 apply where tuberculosis of the respiratory tract is in an infectious state.

Typhoid fever   – 1–8, 10–19, 22, 23, 25.

Typhus   – 1–24 (except 4).

Viral haemorrhagic fever   – 1–3, 5–17, 18, 19, 21, 22, 23, 23A.

Viral hepatitis   – 1–3, 5–8, 10–19, 22, 23.

Whooping cough   – 1–3, 5, 6, 8, 10–16, 18, 19, 22, 23.

Yellow fever   – 2, 16.

PROPER OFFICER

All of the powers and procedures for the notification and control of communicative diseases identified in this Chapter are enforced by the LA (for definition, page 80). LAs are, however, advised that the majority of such powers should be delegated to the Consultant in Communicative Disease Control (CCDC) appointed by each Health Authority (Annex B to NHS Management Executive Guidance HSG (93)56). This will require a formal resolution of the LA detailing each of the powers being delegated.

LAs need to provide appropriate professional and other staff support to the CCDC and ensure that proper arrangements have been made for the receipt of notifications (FC25a).

By detailing such powers to the CCDC in this way, the LA does not shed its responsibilities as the enforcing authority under the 1984 Act and the CCDC (and not the Health Authority) is clearly responsible to them for any powers delegated.

NOTIFICATION OF DISEASES

References

Public Health (Control of Disease) Act 1984 sect. 11.
Public Health (Infectious Diseases) Regulations 1988.

Notification*

The diseases (confirmed or suspected) which are required to be notified by this procedure are:

Cholera	Relapsing fever
Food poisoning	Smallpox
Plague	Typhus

(Public Health (Control of Disease) Act 1984 sects. 10 and 11)

* For the notification of occupation related diseases see pages 297–308.

Acute encephalitis	Ophthalmia neonatorum
Acute poliomyelitis	Paratyphoid fever
Anthrax	Rabies
Diphtheria	Rubella
Dysentery (amoebic or bacillary)	Scarlet fever
Leprosy	Tetanus
Leptospirosis	Tuberculosis
Malaria	Typhoid fever
Measles	Viral haemorrhagic fever
Meningitis	Viral hepatitis
Meningococcal septicaemia	Whooping cough
(without meningitis)	Yellow fever
Mumps	

(Public Health (Infectious Diseases) Regulations 1988)

An individual LA may declare a particular disease to be notifiable within its own district either permanently or temporarily by the procedure set out on FC25b (sect. 16).

Duty to notify

The person responsible for notifying the proper officer of the LA is any registered medical practitioner who becomes aware or suspects that a patient whom he is attending is suffering from a notifiable disease or food poisoning. The only defence against prosecution for not doing so is if he believed that some other registered medical practitioner had already made the required notification (sect. 11(1)). Certificates for notification must be supplied free of charge by the HA (sect. 11(2)).

This procedure is applicable to vessels lying in any inland or coastal water (sect. 9) and to tents, vans, sheds and similar structures used for human habitation (sect. 56).

Details to be notified

The form of certificate to be used (or similar) is set out in Schedule 2 of the PH(ID) Regs. 1988 and notification must include:

(a) the name, age and sex of patient;
(b) the address where patient is;
(c) the disease or suspected disease from which patient is suffering;
(d) the date of onset;
(e) if the patient is in hospital, the day of admission, address where the patient came from and an opinion by the medical practitioner as to whether or not the disease was contracted in the hospital (sect. 11(1)).

FC25a Notification of diseases: notification

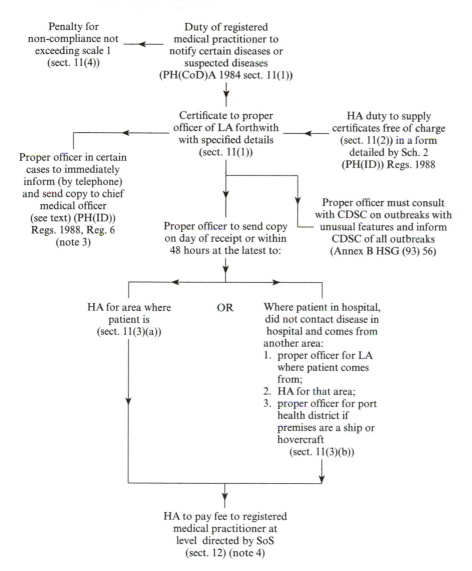

Penalty for non-compliance not exceeding scale 1 (sect. 11(4))

Duty of registered medical practitioner to notify certain diseases or suspected diseases (PH(CoD)A 1984 sect. 11(1))

Certificate to proper officer of LA forthwith with specified details (sect. 11(1))

HA duty to supply certificates free of charge (sect. 11(2)) in a form detailed by Sch. 2 (PH(ID)) Regs. 1988

Proper officer in certain cases to immediately inform (by telephone) and send copy to chief medical officer (see text) (PH(ID)) Regs. 1988, Reg. 6 (note 3)

Proper officer to send copy on day of receipt or within 48 hours at the latest to:

Proper officer must consult with CDSC on outbreaks with unusual features and inform CDSC of all outbreaks (Annex B HSG (93) 56)

HA for area where patient is (sect. 11(3)(a))

OR

Where patient in hospital, did not contact disease in hospital and comes from another area:
1. proper officer for LA where patient comes from;
2. HA for that area;
3. proper officer for port health district if premises are a ship or hovercraft (sect. 11(3)(b))

HA to pay fee to registered medical practitioner at level directed by SoS (sect. 12) (note 4)

Notes
1. This procedure does not apply to Scotland or Northern Ireland.
2. For special procedures relating to the notification of infectious diseases on board ships or aircraft, see the Public Health (Ships) Regulations 1979 and the Public Health (Aircraft) Regulations 1979.
3. DoH should also be informed of any case or outbreak of infectious disease which may have wider than local significance (Annex B HSG (93) 56).
4. Whilst the notification is to be made to the proper officer of the LA, it is the Health Authority which pays the fee.

FC25b Notification of diseases: declaration by LA

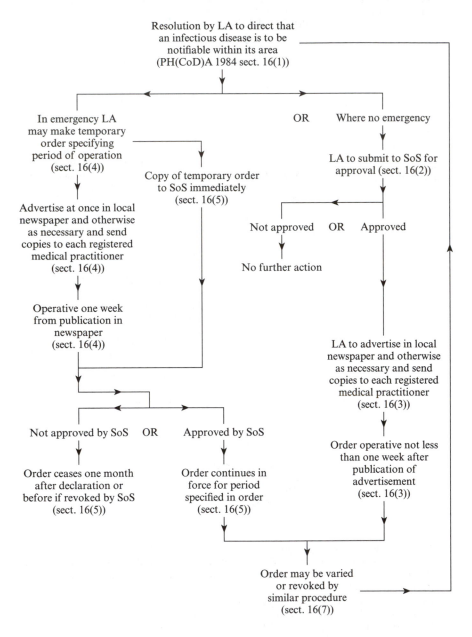

Resolution by LA to direct that an infectious disease is to be notifiable within its area (PH(CoD)A 1984 sect. 16(1))

In emergency LA may make temporary order specifying period of operation (sect. 16(4))

OR   Where no emergency

LA to submit to SoS for approval (sect. 16(2))

Copy of temporary order to SoS immediately (sect. 16(5))

Advertise at once in local newspaper and otherwise as necessary and send copies to each registered medical practitioner (sect. 16(4))

Not approved   OR   Approved

No further action

Operative one week from publication in newspaper (sect. 16(4))

LA to advertise in local newspaper and otherwise as necessary and send copies to each registered medical practitioner (sect. 16(3))

Not approved by SoS   OR   Approved by SoS

Order ceases one month after declaration or before if revoked by SoS (sect. 16(5))

Order continues in force for period specified in order (sect. 16(5))

Order operative not less than one week after publication of advertisement (sect. 16(3))

Order may be varied or revoked by similar procedure (sect. 16(7))

Fees

The health authority (and not the LA receiving the notification) is required to pay to the medical practitioner making a notification a fee at such a level as directed by the SoS (sect. 12(11)).

Notification etc. by a proper officer

Following notification of certain diseases or suspected diseases, the proper officer is required to immediately inform (by telephone) the chief medical officer at the DoH. These diseases are:

1. cholera (including cholera due to *el tor vibrio*);
2. plague;
3. smallpox (including variola minor, alastrim);
4. yellow fever;
5. any serious outbreak of disease including food poisoning.

In the following cases, the proper officer is required to send a copy of the notification certificate to the chief medical officer:

1. cholera (including cholera due to *el tor vibrio*);
2. plague;
3. smallpox (including variola minor, alastrim)
4. yellow fever;
5. leprosy;
6. malaria ⎫
7. rabies ⎬ if contracted in Britain
8. viral haemorrhagic fever.

(Public Health (Infectious Diseases) Regulations 1988, Reg. 6.)
In addition, proper officers should inform the DoH of any case or outbreak of infectious disease which may have wider than local significance – this is to allow consideration of action at national level, e.g. withdrawal of a food for sale. Proper officers also have a responsibility to consult as appropriate with the CDSC, specifically with cases of outbreaks which present unusual features (HSG (93)56 Annex B).

The proper officer must make weekly and quarterly returns of notifications to the Registrar General and these are collated by the CDSC for weekly publication in the Communicable Disease Report.

CLEANSING AND DISINFECTION OF PREMISES AND ARTICLES

Reference

Public Health (Control of Diseases) Act 1984 sect. 31.

FC26 Cleansing and disinfection of premises and articles

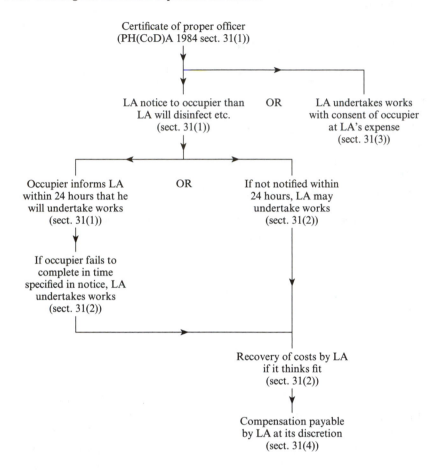

Certificate of proper officer
(PH(CoD)A 1984 sect. 31(1))

LA notice to occupier than
LA will disinfect etc.
(sect. 31(1))

OR

LA undertakes works
with consent of occupier
at LA's expense
(sect. 31(3))

Occupier informs LA
within 24 hours that he
will undertake works
(sect. 31(1))

OR

If not notified within
24 hours, LA may
undertake works
(sect. 31(2))

If occupier fails to
complete in time
specified in notice, LA
undertakes works
(sect. 31(2))

Recovery of costs by LA
if it thinks fit
(sect. 31(2))

Compensation payable
by LA at its discretion
(sect. 31(4))

Note

1. Where necessary on certification by the proper officer, the LA may remove people from an infected house before disinfection, either by agreement or by application to a JP, at no cost to the person removed (sect. 32 PH(CoD)A 1984).

Scope

This procedure deals with any premises where cleansing and disinfection of the premises, or disinfection or destruction of articles within those premises, is required to prevent spread of an infectious disease and first requires a certificate from the proper officer that the action is required. It applies in relation to any infectious disease and not only to 'notifiable' diseases (sect. 31(1)).

Notices

Notices are to be served on occupiers except where the premises are empty, in which case the owner is made responsible (sect. 31(5)). Notices must be in writing (sect. 58). There is no provision for appeal against these notices.

Compensation

At the discretion of the LA, compensation may be paid to any person who has suffered damage by its action in carrying out disinfection or destruction (sect. 31(4)).

Chapter 6

ENVIRONMENTAL PROTECTION ACT 1990 (with Noise Act 1996 and the Dogs (Fouling of Land) Act 1996)

GENERAL PROCEDURAL PROVISIONS (FOR ENVIRONMENTAL PROTECTION ACT 1990)

There are relatively few general provisions in this Act which are applicable to all the procedures. Each set of procedures contains its own definitions, powers of entry, etc., and these are included in the text of each section. The following are, however, applicable to all EPA procedures in this chapter when indicated.

Extent

The Act applies to England, Wales and Scotland (with modifications) but not to Northern Ireland except to the minor extent which does not involve procedures in this chapter (sect. 164).

Service of notices

Notices to be served by a LA may be so served by:

(a) giving it to the person; or
(b) leaving it at his last known address; or
(c) sending it by post to that address.

In the case of a body corporate, the notice is to be served or given to the secretary or clerk at the registered or principal office. For a partnership, it is to be addressed to a partner or a person having control or management of the partnership business at the principal office of the partnership (where the principal office of the company or partnership is outside of the UK, the notice may be sent to its principal office within the UK) (sect. 160).

It should be noted that there is no provision similar to that of the PHA 1936 sect. 285 (page 13) dealing with situations where the name and address of the owner or occupier cannot be obtained after reasonable enquiry. However, there is provision in sect. 233 of the LGA 1972 (page 8) which can be used in these circumstances.

Crown premises

The Act generally is applicable to Crown premises but does not create a criminal liability. The powers of entry given to LA officers in respect of the various procedures may, however, be curtailed by the SoS in the interests of national security in relation to specific premises and powers to be specified in a certificate (sect. 159).

Powers of entry

The general powers of entry for authorized officers provided under sect. 69 of the EPA have been repealed. Such powers for procedures under this chapter are either provided especially or are contained in new powers relating to 'pollution control functions' in sect. 108 of the EA 1995, page 220. The resulting position is indicated in each section of the chapter.

AIR POLLUTION CONTROL

N.B. The procedures in this section, constituting part 1 of the EPA 1990, are to be progressively repealed and replaced by the Pollution Prevention and Control (England and Wales) Regulations 2000 which were made under the Pollution Prevention and Control Act 1999. This process will be phased in over a period up to 2007 (see table 7.1 in chapter 7) and up to that time the procedures in this section and those of the new regime, which are dealt with in chapter 7 will run concurrently.

GENERAL PROCEDURAL PROVISIONS

Definitions

Environment consists of all, or any, of the following media, namely, the air, water and land: and the medium of air includes the air within buildings and the air within other natural or man-made structures above or below ground.

Pollution of the environment means pollution of the environment due to the release (into any environmental medium) from any process of substances which are capable of causing harm to man or any other living organisms supported by the environment.

Harm means harm to the health of living organisms or other interference with the ecological systems of which they form part; and, in the case of man, includes offence caused to any of his senses or harm to his property; and 'harmless' has a corresponding meaning.

Process means any activities carried on in Great Britain, whether on premises or by means of mobile plant, which are capable of causing pollution of the environment, and 'prescribed process' means a process prescribed under sect. 2(1). For the purposes of this subsection:

(a) 'activities' means industrial or commercial activities or activities of any other nature whatsoever (including, with or without other activities, the keeping of a substance);

(b) 'Great Britain' includes so much of the adjacent territorial sea as is, or is treated as, relevant territorial waters for the purposes of Part III of the WRA 1991, or, as respects Scotland, Part II of the CPA 1974; and

(c) 'mobile plant' means plant which is designed to move or to be moved whether on roads or otherwise (sect. 1(1)–(6)).

Authorization means an authorization for a process (whether on premises or by means of mobile plant) granted under sect. 6; and a reference to the conditions of an authorization is a reference to the conditions subject to which at any time the authorization has effect (sect. 1(9)).

Enforcing authority. The enforcing authority in relation to England and Wales is the Environment Agency or the local authority by which, under sect. 4, the functions conferred or imposed by this Part otherwise than on the SoS are for the time being exercisable in relation respectively to the release of substances into the environment or into the air; and 'local enforcing authority' means any such local authority. The 'enforcing authority' in relation to Scotland is the Scottish Environment Protection Agency.

Local authority means, subject to the paragraph below:

(a) in Greater London, a London borough council, the Common Council of the City of London, the Sub-Treasurer of the Inner Temple and the Under-Treasurer of the Middle Temple;

(b) in England and Wales outside Greater London, a district council and the Council of the Isles of Scilly.

District council includes the unitary authorities in England and Wales.

Where, by an order under sect. 2 of the PH(CoD)A 1984, a port health authority has been constituted for any port health district, the port health authority shall have by virtue of this subsection, as respects its district, the functions conferred or imposed by this Part and no such order shall be made assigning those functions; and 'local authority' and 'area' shall be construed accordingly (sect. 4(11) and (12)).

Powers of entry etc.

These powers are contained in the EA 1995 as part of provisions which deal out all local enforcing authority pollution control functions (page 220).

Obtaining of information

The LA may require by notice in writing any person to provide such information as it may reasonably require to discharge its functions under Part 1 of the EPA 1990 either within a specified period or at a specified time (sect. 19(2)).

The penalty for non-compliance with such a notice is a fine not exceeding the statutory maximum or, on indictment, an unlimited fine and/or up to 2 years' imprisonment (sect. 23(3)).

AUTHORIZATION OF PRESCRIBED PROCESSES BY LAs

References

Environmental Protection Act 1990 sects. 6–12 (as amended by the Environment Act 1995).
Environmental Protection (Applications, Appeals and Registers) Regulations 1991 (as amended 1991, 1994 and 1996).
Environmental Protection (Prescribed Processes and Substances) Regulations 1991 (as amended 1991, 1992, 1993, 1994, 1995 and 1996).
Disposal of Controlled Waste (Exceptions) Regulations 1991.
Environmental Protection (Authorization of Processes) (Determination Periods) Order 1991 (amended 1994).
Environmental Protection (Amendment of Regulations) Regulations 1991.
The Local Enforcing Authorities Air Pollution Fees and Charges Scheme (England and Wales) – revised annually.
Pollution Prevention and Control (England and Wales) Regulations 2000.

DoE General Guidance Notes:
  GG1(91): Introduction to Part 1 of the Act
  GG2(91): Authorizations
  GG3(91): Applications and Registers
  GG4(91): Interpretation of terms used in process guidance notes
  GG5(91): Appeals
DoE Updating Guidance Notes (UG Series)
DoE Additional Guidance Notes* (AQ Series) – see table 6.2

Scope

No prescribed process (below) may be operated without an authorization from the enforcing authority (sect. 6(1)).

---

* These notes supplement and clarify the PG series. They can be accessed from the NETCEN website found at www.aeat.co.uk/netcen/airqual/info/labrief.html and also at www.enviroment-agency.gov.uk/business/lapc/Aqmindex

FC27 Authorization of prescribed processes by LAs

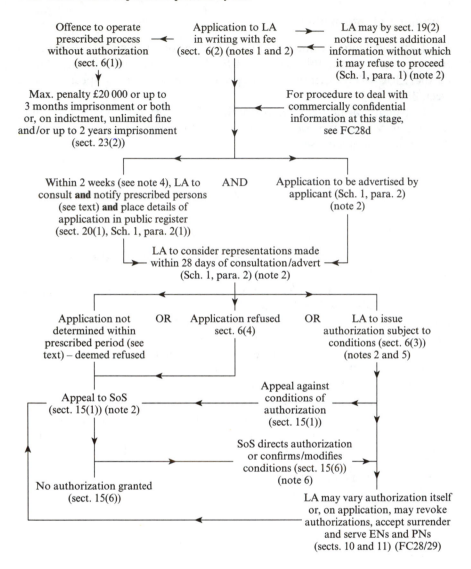

Offence to operate prescribed process without authorization (sect. 6(1))

Application to LA in writing with fee (sect. 6(2) (notes 1 and 2)

LA may by sect. 19(2) notice request additional information without which it may refuse to proceed (Sch. 1, para. 1) (note 2)

Max. penalty £20 000 or up to 3 months imprisonment or both or, on indictment, unlimited fine and/or up to 2 years imprisonment (sect. 23(2))

For procedure to deal with commercially confidential information at this stage, see FC28d

Within 2 weeks (see note 4), LA to consult **and** notify prescribed persons (see text) **and** place details of application in public register (sect. 20(1), Sch. 1, para. 2(1))

AND

Application to be advertised by applicant (Sch. 1, para. 2) (note 2)

LA to consider representations made within 28 days of consultation/advert (Sch. 1, para. 2) (note 2)

Application not determined within prescribed period (see text) – deemed refused

OR

Application refused sect. 6(4)

OR

LA to issue authorization subject to conditions (sect. 6(3)) (notes 2 and 5)

Appeal to SoS (sect. 15(1)) (note 2)

Appeal against conditions of authorization (sect. 15(1))

No authorization granted (sect. 15(6))

SoS directs authorization or confirms/modifies conditions (sect. 15(6)) (note 6)

LA may vary authorization itself or, on application, may revoke authorizations, accept surrender and serve ENs and PNs (sects. 10 and 11) (FC28/29)

Notes
1. A specimen application form is included as Appendix 1 to GG3(91).
2. Details of these events (with certain exceptions, see text) to be entered in the public register.
3. For the use of enforcement and prohibition notices relating to authorized prescribed processes, see FC29.
4. Where the applicant has successfully applied to the LA for the exclusion of commercially confidential information from the register, the LA must consult as shown within 4 weeks of the letter of agreement to the exclusion being sent.
5. It is considered that, other than authorizations for waste oil burners under 0.4 MW, LAs should send drafts of authorizations to operators for comment before formal issue (AQ16(93)).
6. The SoS may hold a hearing (in public or private) or may hold a public inquiry before determining the appeal (see. 15(5)).
7. This procedure is to be incrementally repealed and replaced by a permit system under the Pollution Prevention and Control Act 1999 – see chapter 7.

Reg. 4(2) of the Environmental Protection (Prescribed Processes and Substances) Regulations 1991 (as amended) – the so-called 'triviality' provisions – allows operation without authorization of Part B processes which result in the release of substances at or below specified quantities. LAs should confirm acceptance of such exemption in writing in accordance with AQ1 (94).

Air pollution control by LAs (LAAPC)

Part 1 of the EPA 1990 gives air pollution control powers to LAs in relation to 'medium-polluting' processes prescribed in Part B of Schedule 1 of the Environmental Protection (Prescribed Processes and Substances) Regulations 1991 (as amended). These processes are listed in Table 6.1 which also gives an indication of the relevant process guidance note.

In exercising its functions under Part 1 a LA is required to have regard to the National Air Quality Strategy published by the Government under sect. 80 of the EA 1995 (sect. 4A).

Prescribed processes not allocated to the LA for air pollution control purposes are dealt with by the Environment Agency (EA) on the basis of integrated pollution control.

Table 6.1 Process allocated to local authority enforcement

Process guidance note reference	Environmental Protection Act 1990, Part I Process
*PG1/1(95)	Waste Oil Burners, less than 0.4 MW net rated thermal input
*PG1/2(95)	Waste Oil Burners or Recovered Oil Burners, less than 3 MW net rated thermal input
*PG1/3(95)	Boilers and Furnaces, 20–50 MW net rated thermal input
*PG1/4(95)	Gas Turbines, 20–50 MW net rated thermal input (also see AQ1(97))
*PG1/5(95)	Compression Ignition Engines, 20–50 MW net rated thermal input
*PG1/10(92)	Waste Derived Fuel Burning Processes less than 3 MW net rated thermal input
PG1/11(96)	Reheat and heat treatment furnaces, 20–54 MW net rated thermal input
PG1/12(95)	Combustion of fuel manufactured from or comprised of solid waste in appliances between 0.4 and 3 MW net rated thermal input
PG1/13(96)	Processes for storage, loading and unloading of petrol at terminals – also see AQ6(97) and AQ10(99)
PG1/14(96)	Unloading of petrol into storage at service stations – also see AQ6(97) and AQ10(99)
PG1/15(97)	Odorising natural gas and liquified petroleum gas

* Also see UG-1(92)

Table 6.1 (*cont'd*)

Process guidance note reference	Environmental Protection Act 1990, Part I Process
PG2/1(96)	Furnaces for Extraction of Non-Ferrous Metal from scrap
PG2/2(96)	Hot Dip Galvanizing Processes
PG2/3(96)	Electrical and Rotary Furnaces
PG2/4(96)	Iron, Steel and Non-ferrous Metal Foundry Processes
PG2/5(96)	Hot and Cold Blast Cupolas
PG2/6(96)	Aluminium and Aluminium Alloy Processes
PG2/7(96)	Zinc and Zinc Alloy Processes
PG2/8(96)	Copper and Copper Alloy Processes
PG2/9(96)	Metal Decontamination Processes
PG3/1(95)	Blending, Packing, Loading and Use of Bulk Cement
PG3/2(95)	Manufacture of Heavy Clay Goods and Refractory Goods
PG3/3(95)	Glass (excluding lead glass) Manufacturing Processes
PG3/4(95)	Lead Glass Manufacturing Processes
PG3/5(95)	Coal, Coke and Coal Product Processes
PG3/6(95)	Polishing or Etching of Glass or Glass Products using Hydrofluoric Acid
PG3/7(95)	Exfoliation of Vermiculite and Expansion of Perlite
PG3/8(96)	Quarry Processes Including Roadstone Plants and the Size of Reduction of Bricks, Tiles and Concrete
PG3/9(91)	Sand Drying and Cooling
PG3/12(95)	Plaster Processes
PG3/13(95)	Asbestos Processes – see also AQ3(96)
PG3/14(95)	Lime Processes
PG3/15(96)	Mineral drying and roadstone coating processes
PG3/16(96)	Mobile crushing and screening processes
PG3/17(95)	China and ball clay processes including spray drying of ceramics
PG4/1(94)	Processes for the Surface Treatment of Metals
PG4/2(96)	Processes for the manufacture of fibre reinforced plastics – also see AQ1(97)
PG5/1(95)	Clinical Waste Incineration Processes under 1 tonne an hour
PG5/2(95)	Crematoria
PG5/3(95)	Animal Carcase Incineration Processes under 1 tonne an hour
PG5/4(95)	General Waste Incineration Processes under 1 tonne an hour
PG5/5(91)	Sewage Sludge Incineration Processes under 1 tonne an hour
PG6/1(99)	Animal by-product Rendering
PG6/2(95)	Manufacture of Timber and Wood Based Products

Table 6.1 (*cont'd*)

Process guidance note reference	Environmental Protection Act 1990, Part I Process
PG6/3(99)	Chemical Treatment of Timber and Wood Based Products
PG6/4(95)	Processes for Manufacture of Particleboard and Fibreboard
PG6/5(95)	Maggot Breeding Processes – also see AQ1(97)
PG6/6(91)	Fur Breeding Processes (withdrawn)
PG6/7(97)	Printing and Coating of Metal Packaging
PG6/8(97)	Textile and Fabric Coating and Finishing Processes
PG6/9(96)	Manufacture of Coating Powder
PG6/10(97)	Coating Manufacturing Processes
PG6/11(97)	Manufacture of Printing Ink
PG6/12(91)	Production of Natural Sausage Casings, Tripe, Chitterlings and Other Boiled Green Offal Products
PG6/13(97)	Coil Coating Processes
PG6/14(97)	Film Coating Processes
PG6/15(97)	Coating in Drum Manufacturing and Reconditioning Processes
PG6/16(97)	Printworks
PG6/17(97)	Printing of Flexible Packaging
PG6/18(97)	Paper Coating Processes
PG6/19(97)	Fish Meal and Fish Oil Processes
PG6/20(97)	Paint Application in Vehicle Manufacturing – also see AQ4(98)
PG6/21(96)	Hide and Skin Processes
PG6/22(97)	Leather Finishing Processes
PG6/23(97)	Coating of Metal and Plastic
PG6/24(96)	Pet Food Manufacturing Processes
PG6/25(97)	Vegetable Oil Extraction and Fat and Oil Refining Processes
PG6/26(96)	Animal Feed Compounding Processes – also see AQ1(97)
PG6/27(96)	Vegetable Matter Drying Processes
PG6/28(97)	Rubber Processes
PG6/29(97)	Di-isocyanate Processes – also see AQ3(97)
PG6/30(97)	Production of Compost for Mushrooms
PG6/31(96)	Powder Coating Processes, Including Sheradizing
PG6/32(97)	Adhesive Coating Processes
PG6/33(97)	Wood Coating Processes
PG6/34(97)	Respraying of Road Vehicles – also see AQ4(98)
PG6/35(96)	Metal and Other Thermal Spraying Processes
PG6/36(97)	Tobacco Processing
PG6/37(92)	Knackers Yards (withdrawn)
PG6/38(92)	Blood Processing
PG6/39(92)	Animal By-Product Dealers
PG6/40(94)	Coating and Recoating of Aircraft and Aircraft Components
PG6/41(94)	Coating and Recoating of Rail Vehicles
PG6/42(94)	Bitumen and tar processes

Table 6.2 Additional Guidance Notes

AQ5(91)	Registers – Rehabilitation of Offenders Act.
AQ7(91)	Meaning of 'existing process'.
AQ8(91)	Obtaining further information on specimen notices.
AQ9(91)	VAT on charges.
AQ3(92)	Radioactive substances.
AQ5(92)	Foundries-triviality.
AQ7(92)	Amendments to applications.
AQ8(92)	Rubber processes-meaning of 'if carbon black is used'.
AQ9(92)	Mobile Plant.
AQ14(92)	Cement Processes.
AQ1(93)	EPA Part 1: advising operators on their rights of appeal. See also AQ7(94) and AQ8(96).
AQ3(93)	Triviality and zinc die-casters.
AQ4(93)	Transfer of authorizations under section 9 EPA 1990.
AQ7(93)	Small coal mines-triviality.
AQ8(93)	Mobile Plant-necessity for consultation with English Nature & Countryside Council for Wales.
AQ9(93)	Variation Notices-consolidation of authorizations.
AQ10(93)	Notification of Technical Guidance Note *Guidelines on Discharge Stack Heights for Polluting Emissions* – see also AQ18(93).
AQ12(93)	Obtaining additional information.
AQ14(93)	Service of notices.
AQ15(93)	Commercial confidentiality.
AQ16(93)	Submission of draft authorizations to operators.
AQ18(93)	Errors in Technical Guidance D1 – see also AQ10(93).
AQ1(94)	Notification to operators of triviality exemption.
AQ2(94)	Solvent substitution [of 1,1,1 trichloroethane or methyl chloroform to another solvent] in vapour degreasing.
AQ3(94)	Animal by-product rendering [PG6/1(91)] amendment.
AQ4(94)	HMIP Technical Guidance Note M2 *Monitoring of pollutants at source*.
AQ5(94)	Content of public registers.
AQ6(94)	HMIP Technical Guidance Note A2 *Pollution Abatement Technology for Reduction of Solvent Vapour Emissions*.
AQ7(94)	Appeals arrangements.
AQ9(94)	HMIP Technical Guidance Note A3 *Pollution Abatement Technology for Particulate and Trace Gas Removal*.
AQ11(94)	Waste Management Licensing Regulations 1994 & implications for LAAPC.
AQ12(94)	Environmental Technology Best Practice Programme.
AQ15(94)	Explicit advice to operators on appeals.
AQ17(94)	Recirculation of workroom air and triviality exemption from LAAPC.
AQ18(94)	Triviality.

Table 6.2 (*cont'd*)

AQ2(95)	Time limits for processing applications and upgrades.
AQ3(95)	Refractory materials definitions.
AQ4(95)	Inspection frequency, basic principles.
AQ5(95)	Four yearly reviews of authorizations.
AQ7(95)	Commercially confidential material on public registers. See also AQ1(93), 15(93) and AQ15(94).
AQ10(95)	PG5/5(91) not being updated.
AQ11(95)	Amendments to LAAPC by Environment Act 1995 and note re. Air quality provisions.
AQ14(95)	Technical Guidance on standards for IPC.
AQ16(95)	Authorization conditions on odours.
AQ17(95)	Summary of recent LAAPC appeals decisions – see also AQ10(96) and AQ7 (98).
AQ18(95)	Amended sections 10 and 13 EPA 1990.
AQ20(95)	Monitoring of processes and emissions by local authorities.
AQ1(96)	Chimney heights for process heaters.
AQ2(96)	Amendments to LAAPC by Environment Act 1995.
AQ3(96)	Asbestos processes re. PG3/13(95).
AQ8(96)	Amendment to specimen notes on appeal 11/92 and AQ1(93).
AQ10(96)	Summary of recent LAAPC appeal decisions.
AQ1(97)	Corrections to PG1/4(95), 4/2(96), 6/5(95) and 6/26(96).
AQ3(97)	Corrections to PG6/29(97).
AQ4(97)	Categories of process operated by different local authorities.
AQ5(97)	Guidance on appeals procedures.
AQ6(97)	Petroleum processes PG1/13(96) and 1/14(96).
AQ7(97)	Categories of process operated by different local authorities.
AQ1(98)	Information about LAAPC on Internet at www.aeat.co.uk/netcen/airqual/info/labrief.html
AQ2(98)	Address for AEQ Division and Local Authority Unit.
AQ3(98)	List of current and out dated AQ notes at July 1998.
AQ4(98)	Corrections to PG6/18(97), 6/20(97), 6/34(97).
AQ5(98)	Chimney height calculation advice from Stanger Science.
AQ7(98)	Summary of recent appeals.
AQ8(98)	Proposed changes to statistical returns.
AQ9(98)	BATNEEC for delivery of petrol to retail stations.
AQ10(98)	Tank connections at petrol stations.
AQ11(98)	Guidance in response to LAAPC action plan.
AQ12(98)	Application of amine emission limits in PG2/4(96).
AQ1(99)	PG notes and the printed circuit board industry.
AQ2(99)	Transfer efficiency of spray coatings.
AQ3(99)	List of current and out dated AQ notes at March 1999.
AQ4(99)	1999/00 increased fees and charges.

Table 6.2 (*cont'd*)

AQ6(99)	Heavy clay and refractory goods PG3/2(95).
AQ7(99)	Transfer of Local Authority Unit to Environment Agency.
AQ8(99)	Local Authority Unit.
AQ9(99)	Cancelled and replaced by AQ10(99).
AQ10(99)	Use of dipsticks on mobile containers *see* PG1/13 and 1/14(96).
AQ1(00)	Categories of process regulated by different local authorities [updates AQ4 (97)]
AQ2(00)	Summary of High Court judgement Dudley MBC v Henley Foundries.
AQ3(00)	Amendments to PG6/1(00) Processing of animal remains and by-products.
AQ4(00)	Additional guidance on preparation of court cases.
AQ5(00)	Update of previous notes on MACC2.
AQ6(00)	Lists of current outdated AQ notes and updates AQ3(99).
AQ7(00)	Up to date list of all extant link authorities.
AQ8(00)	Advice re. enquiries about 'installation' under new IPPC regime: implications for climate change levy [CCL].
AQ1(01)	Solvents Directive Guidance: implementation of 1999/13/EC limitation of VOCs.
AQ2(01)	Guidance for LAPC cost accounting.
AQ3(01)	General information about DEFRA & new contact.
AQ4(01)	Use of crematoria for organ disposal.

Applications

Applications are to be made to the LA by the person carrying on the pre-scribed process, and must be accompanied by the appropriate fee as deter-mined by the SoS (sect. 6(2)). These fees are set in the Local Enforcing Authorities Air Pollution Fees and Charges Scheme (England and Wales), which is amended from time to time. The fee structure includes a payment to accompany the application and an annual charge where an authorization has been granted.

The details to be included in applications are specified in the Environ-mental Protection (Applications, Appeals and Registers) Regulations 1991 (as amended 1996) and are expanded in GGN3(91). The LA is empowered to ask for any additional information which it may require in order to deter-mine the application and, if this is not provided within the period specified by the LA, the LA may refuse to proceed (sect. 19(2) and Reg. 2).

ISSUE OF VARIATION NOTICES BY LAs

FC28a Issue of variation notices by LAs

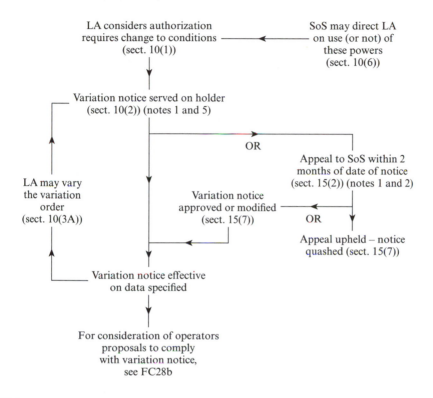

Notes
1. Details must be included in the public register.
2. The notice is not suspended pending determination of the appeal (sect. 15(9)).
3. Specimen variation notices have been issued by the DoE (November 1992) but LAs are not obliged to use them (also AQ14(93)).
4. LAs are to use the variation notice procedure to incorporate upgrading programmes into the authorization details.
5. It is suggested that where the variation notice will significantly amend an existing authorization a draft should first be sent to the operator for comment (AQ16(93)).

CONSIDERATION OF OPERATOR'S PROPOSALS TO COMPLY WITH VARIATION NOTICES

FC28b Consideration of operator's proposals to comply with variation notice

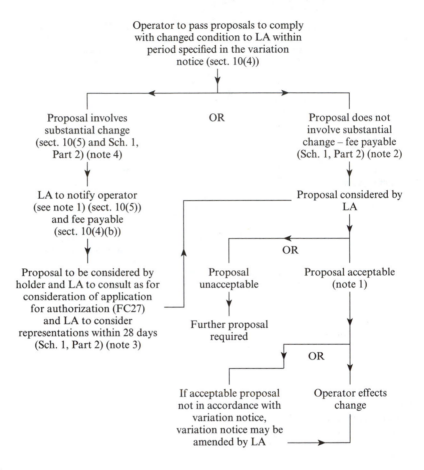

Notes

1. Details to be entered into public register.
2. Proposal to be dealt with by LA as soon as possible.
3. The timescales applicable are as for the consideration of applications for authorization (FC27).
4. For definition of 'substantial change', see page 112.

APPLICATION BY HOLDERS FOR VARIATIONS

FC28c Application by holders for variations

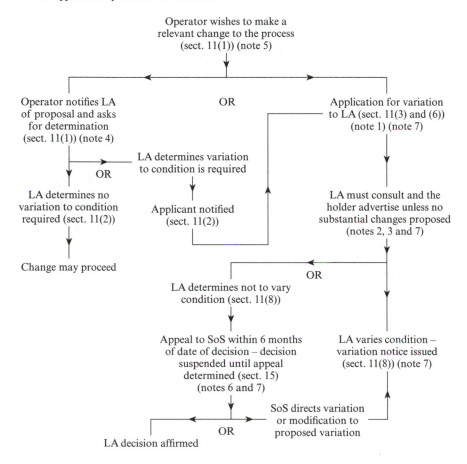

Notes
1. The application is to be made with the prescribed particulars and the required fee.
2. The consultation and advertisement procedure is as for applications for authorization (page 108).
3. For the procedure to deal with commercially confidential information at this stage, see FC28d.
4. The application must be in the prescribed form and contain the prescribed information (page 102).
5. For a definition of 'relevant change' see page 112.
6. The SoS may hold a hearing (in public or private) or hold a public inquiry before determining the appeal (sect. 15(5)).
7. Details to be entered in public register (Reg. 15).

HANDLING COMMERCIALLY CONFIDENTIAL INFORMATION

FC28d Handling commercially confidential information

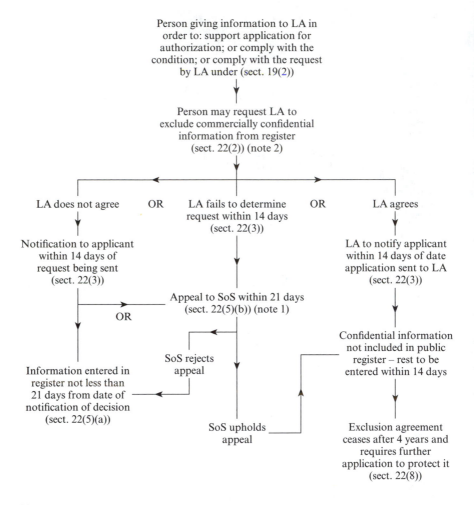

Notes
1. The information concerned may not be entered in the register until 7 days after the appeal is finally determined or withdrawn (sect. 22(5)).
2. Operators are advised to consult the LA before any information is furnished in order to ensure that they are aware of the considerations which will be taken into account by the LA and that they should not include unnecessary material in their submission (AQ15(93)).

REVOCATION OF AUTHORIZATIONS BY LAs

FC28e Revocation of authorizations by LAs

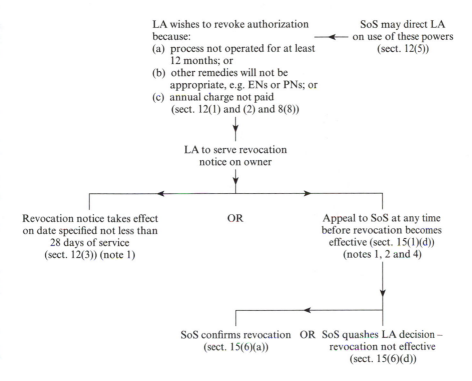

LA wishes to revoke authorization
because:
(a) process not operated for at least
12 months; or
(b) other remedies will not be
appropriate, e.g. ENs or PNs; or
(c) annual charge not paid
(sect. 12(1) and (2) and 8(8))

SoS may direct LA
on use of these powers
(sect. 12(5))

LA to serve revocation
notice on owner

Revocation notice takes effect
on date specified not less than
28 days of service
(sect. 12(3)) (note 1)

OR

Appeal to SoS at any time
before revocation becomes
effective (sect. 15(1)(d))
(notes 1, 2 and 4)

SoS confirms revocation OR SoS quashes LA decision –
(sect. 15(6)(a))   revocation not effective
(sect. 15(6)(d))

Notes
1. At any time before the revocation becomes effective, the LA may withdraw the notice or amend the date on which it becomes effective (sect. 12(4)).
2. The revocation is suspended pending determination of the appeal.
3. Specimen revocation notices were issued by the DoE in November 1992 although LAs are not obliged to use them (also AQ14(93)).
4. The SoS may hold a hearing (in public or private) or may hold a public inquiry before determining the appeal (sect. 15(5)).

Consultation

The LA is required:

(a) to place details of the application in a public register, except that which is commercially confidential; and
(b) consult the following bodies within 14 days of receipt of the application:
    (i) DEFRA (or the SoS for Wales or Scotland as appropriate);
    (ii) where there is to be a release of any substance into a sewer, the SU (or in Scotland the Sewerage authority);
    (iii) the Nature Conservancy Council for England (or the Scottish Natural Heritage or Countryside Council for Wales) where the process may affect a SSI site;
    (iv) where there may be a release into a harbour, the harbour authority;
    (v) where there may be a release into territorial waters or coastal waters, the Sea Fisheries Committee;
    (vi) the HSE (except where HSE have already been notified by the LA under sect. 10(5) that a substantial change is to be carried out).

The applicant must also advertise the application in accordance with the detailed requirements of the Environmental Protection (Applications, Appeals and Registers) Regulations 1991 (as amended 1996).

The LA must consider any representations made within 28 days of notifying the consultee or of the advertisement being placed (Sch. 1, para. 2).

Confidential information

Information which is commercially confidential is not to be included within the public register providing that the LA agrees that it can be justifiably withheld. In the event of disagreement, the SoS decides (sect. 22).

Operators may also apply to the SoS for information to be excluded from the register because its inclusion would be contrary to the interests of national security. The SoS may direct the LA on the matter (sect. 21).

The procedure for dealing with commercially confidential information is shown on FC28d.

AQ15(93) indicates that in considering appeals the SoS will have as his main concerns:

(a) whether inclusion of the information would prejudice to an unreasonable degree an applicant's commercial interests;
(b) in assessing prejudice, to take account of both the extent of any damage which might be caused or the likelihood of such damage looking at the balance of probabilities rather than demanding conclusive proof;
(c) to balance against prejudice to an appellant's commercial interests any benefit to the public interest which would arise from including the information in the register; and

(d) assessing directly the importance to the public of the information, e.g. better informed on the likely environmental impact of the process.

It is recommended that LAs take account of the same issues when considering applications made to them under sect. 22(2).

Determination

An authorization cannot be issued by the LA unless it considers that the operator will be able to carry on the process so as to comply with the conditions which would be included in the authorization (sect. 6(4)).

An authorization must contain:

(a) conditions which the LA consider appropriate to meet the objectives set out below;
(b) any conditions which the SoS directs the LAs to include (the direction may also relate to conditions not to be included);
(c) any other conditions which the LA considers appropriate (sect. 7(1)).

The objectives referred to in (a) above are:

1. To use Best Available Techniques Not Entailing Excessive Cost (BATNEEC) to prevent or minimize the release of substances prescribed for release to air in regulations made under sect. 2(5) of the Act, and to render harmless all substances, whether or not prescribed, which may be released into the air.
2. To comply with any directions the SoS issues to local authorities for the purposes of implementing EC treaty or international law obligations which relate to environmental protection.
3. To comply with any limits or requirements and to achieve any quality standards or quality objectives which the SoS has set down in regulations made under sect. 2 of the CAA 1968 (now CAA 1993) sect. 2 of the European Communities Act 1972, Part I of the HASAWA 1974 (in relation to air pollution control), Parts II–IV of the CPA 1974, WRA 1991, sect. 3 of the EPA 1990 and sect. 87 of the EA 1995. (Not all these limits etc. will be applicable to the single-medium air pollution control regime – notably any in regulations made under Part III of the WRA 1991.)
4. To comply with any relevant requirement in a plan made by the SoS under sect. 3(5) of the Act (sect. 7(2)).

Further general advice on the meaning of BATNEEC is contained in GG1(91). The series of process and technical guidance notes provides technical advice on the application of these objectives, both generally and to the different categories of prescribed process. Sect. 7(11) of the Act places a duty on local authorities to have regard to any guidance issued to them by the SoS on appropriate techniques relating to the BATNEEC objective in sect. 7(2) and (10).

A specimen authorization and conditions are contained in Appendices 1 and 2 of GG2(91). Conditions must be reviewed by the LA no less frequently than every 4 years, although the SoS may direct a different period (sect. 6(6) and (7)). These reviews are now limited by the transitional arrangements relating to existing installations into the IPPC regime (sects. 6(6A) and 6(B)) as inserted by the Pollution Prevention and Control (England and Wales) Regulations 2000. The IPPC process will replace these reviews where they fall within 2 years of the beginning of the relevant period for that industry sector as set out in Sch. 3 of the PPC Regulations. Guidance on reviews under sect. 6(6) was issued by the DoE in April 1995.

Periods for determination

Where decisions have not been made on applications within the following timescale, the applicant is able to appeal to the SoS:

(a) new waste oil burning appliances rated 0.4 megawatts or less – 2 weeks;
(b) all new processes other than waste oil burners in (a) above – 4 months (Sch. 1, para. 5 and the Environmental Protection (Authorization of Processes) (Determination Periods) Order 1991) (as amended).

Amendments contained in the Order of 1994 specify a 9 month determination period for applications for newly allocated processes for local control as detailed in Table 6.1.

Refusal of authorization

Where the LA is not satisfied that the operator will be able to operate the plant in compliance with the conditions which would be included in an authorization, the application must be refused (sect. 6(4)).

Transfers

Authorizations are transferable to another operator provided the new holder of the authorization gives the LA written notice that the authorization has been transferred to it not later than 21 days from the date of transfer (sect. 9). Details of transfer should be included in the public register.

Variations (FC28a–c)

Proposals to vary an existing authorization may be made by:

(a) the operator by notice in writing to the LA in order to accommodate a relevant change to the process (applications are subject to a fee as set out in the Local Enforcing Authorities Air Pollution Fees and Charges Scheme (England and Wales) and are updated from time to time); or
(b) the LA through a variation notice served on the holder of the authorization.

Unless there is no substantial change, applications from the holder must be advertised in accordance with the regulations, and the LA must consult in the same way as for new applications.

LAs should use the variation notice procedure to translate upgrading programmes into authorization conditions (sect. 10).

Amendments introduced by the EA 1995 allow the LA to vary any variation notice served by serving a further notice which specifies the variations and the dates on which they are to take effect. The procedure is the same as that for the service of the variation notice (sect. 10(3A)).

Revocations

Subject to appeal, LAs may revoke authorizations at any time by notice in writing to the holder, giving a minimum of 28 days' notice before the operative date including in circumstances where the process has not been carried on for a period of 12 months or more (sect. 12). Revocation is also available to the LAs if the holder has not paid the required fee (sect. 8(8)). It was recommended by the DoE that revocation notices should be accompanied by a letter giving the reasons for the revocation.

Appeal against the revocation notice has the effect of suspending the notice until the appeal is determined or withdrawn by the LA (sect. 15(8)).

Appeals

An appeal procedure to the SoS is available against decisions of the LA:

(a) to refuse authorization or variation or against conditions included in an authorization – within 6 months of the date of decision;
(b) in relation to a deemed refusal in an application not determined by the LA within the prescribed period (page 110) – within 6 months of the period having elapsed;
(c) to serve a variation notice – 2 months from the date of the notice;
(d) to revoke the authorization – before the date upon which the revocation takes place.

The SoS has the power to allow a longer period for each appeal (sect. 15).

The detailed provisions relating to appeals are given within The Environmental Protection (Applications, Appeals and Registers) Regulations 1991 (as amended) and guidance to LAs is provided in GG5(91): Appeals.*

Public registers

The LA (other than port health authorities) must keep a register (which should comprise a separate set of papers, not the working file) giving information on the following issues relating to authorizations:

* Also see AQ1(93), AQ5(97), AQ7(94) and AQ8(96).

(a) applications for authorizations and variations and of advertisements related to them;
(b) LA notices requiring further information regarding an application and details of the response;
(c) representations made (statutory consultees and the public);
(d) authorizations and transfers of authorizations;
(e) notice of variation, enforcement, prohibition or revocation;
(f) appeals;
(g) court cases/convictions;
(h) monitoring data (for four years);
(i) published reports;
(j) directions by the SoS;
(k) information given to the LA in compliance with an authorization, variation notice, enforcement notice or prohibition notice or under sect. 19(2).

There are exclusions for certain commercially confidential information and for information affecting national security.

Registers must be available for public inspection at all reasonable times and the LA may charge for copies.

Similar information must be kept by the LA in respect of processes dealt with by the Environment Agency (sects. 20(2) and the Environmental Protection (Applications, Appeals and Registers) Regulations 1991 as amended).

Advice on what information should be included in the public register is given in GG3 and AQ5(94). Generally, this information should be in accord with information identified in Reg. 15 of the Environmental Protection (Application, Appeals and Registers) Regulations 1991 (as amended) and should not contain any additional information.

Offences

Offences relating to operating an unauthorized business are punishable by a fine not exceeding £20 000 or imprisonment for up to 3 months (or both) or, on indictment, an unlimited fine and/or up to 2 years' imprisonment.

Offences relating to a failure to notify transfer of an authorized process are punishable by a fine not exceeding the statutory maximum or, on indictment, an unlimited fine and/or up to 2 years' imprisonment (sect. 23).

Definitions

Relevant change is a change in the manner of carrying on the process which is capable of altering the substances released from the process, or of affecting the amount or any other characteristic of any substance so released (sect. 11(11)).

Substantial change means a substantial change in the substances released from the process or in the amount or any other characteristic of any substance so released; and the SoS may give directions to the enforcing authorities as to what does or does not constitute a substantial change in relation to processes generally, any description or process or any particular process (sect. 10(7)).

ENFORCEMENT AND PROHIBITION NOTICES FOR PRESCRIBED PROCESSES

References

Environmental Protection Act 1990 sects. 13–15.
The Environmental Protection (Applications, Appeals and Registers) Regulations 1991 (as amended 1996).
DoE General Guidance Notes GG1(91) and GG5(91).

Scope

These procedures may be used by LAs in respect of any prescribed processes which have been authorized by them.

Enforcement notices

Instead of a prosecution an EN **may** be served where the LA considers that a condition of the authorization (page 109) is being, or is likely to be, contravened (sect. 13(1)).
The notice must:

(a) state the LA's opinion on the contravention;
(b) specify details of the contravention and the steps required to remedy it; and
(c) specify the period allowed for compliance (sect. 13(2)).

An EN may be withdrawn by the LA at any time by notice in writing to the person in receipt of the original notice (sect. 13(4)).

Prohibition notices

The LA is **required** to serve a PN where, in respect of a process authorized by it, there is an imminent risk of serious pollution of the environment. This may involve a contravention of a condition of the authorization, but may be used where this is not the case (sect. 14(1) and (2)).
The notice must:

(a) state the LAs opinion as to the risk of serious pollution of the environment;
(b) specify the risk;
(c) specify the steps required and the period allowed; and
(d) direct that the authorization cease to have effect wholly or to the extent indicated until the prohibition notice is withdrawn (sect. 14(3)).

FC29 Enforcement and prohibition notices for prescribed processes

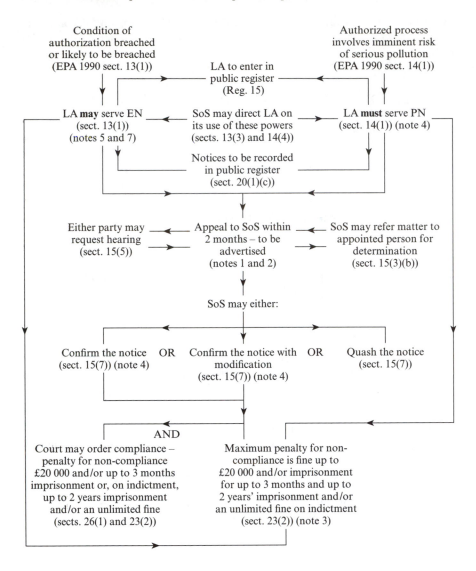

Condition of authorization breached or likely to be breached (EPA 1990 sect. 13(1))

LA to enter in public register (Reg. 15)

Authorized process involves imminent risk of serious pollution (EPA 1990 sect. 14(1))

LA **may** serve EN (sect. 13(1)) (notes 5 and 7)

SoS may direct LA on its use of these powers (sects. 13(3) and 14(4))

LA **must** serve PN (sect. 14(1)) (note 4)

Notices to be recorded in public register (sect. 20(1)(c))

Either party may request hearing (sect. 15(5))

Appeal to SoS within 2 months – to be advertised (notes 1 and 2)

SoS may refer matter to appointed person for determination (sect. 15(3)(b))

SoS may either:

Confirm the notice (sect. 15(7)) (note 4)

OR

Confirm the notice with modification (sect. 15(7)) (note 4)

OR

Quash the notice (sect. 15(7))

AND

Court may order compliance – penalty for non-compliance £20 000 and/or up to 3 months imprisonment or, on indictment, up to 2 years imprisonment and/or an unlimited fine (sects. 26(1) and 23(2))

Maximum penalty for non-compliance is fine up to £20 000 and/or imprisonment for up to 3 months and up to 2 years' imprisonment and/or an unlimited fine on indictment (sect. 23(2)) (note 3)

Notes
1. Detailed provisions dealing with appeals are specified in the Environmental Protection (Applications, Appeals and Registers) Regulations 1991 (as amended 1996).
2. The notices are not suspended pending determination of the appeal (sect. 15(9)).
3. Where proceedings would afford an ineffectual remedy, the LA may take proceedings in the High Court (sect. 24).
4. The LA is required to give notice in writing to the person receiving the PN of its withdrawal when it is satisfied that the steps required by the notice have been taken (sect. 14(5)).
5. LAs may instead of serving an EN prosecute for a failure to comply with the conditions of the authorization.
6. Specimen ENs and PNs have been issued by the DoE (November 1992) but LAs do not have to use them (also AQ14(93)).
7. An EN may be withdrawn by the service of a further notice to that effect (sect. 13(5)).

Once the steps required have been taken, the LA must give written notification to the person carrying on the process and withdraw the notice (sect. 14(5)).

Responsible person

Both ENs and PNs are to be served on the person carrying out the prescribed process under the authorization (sects. 13(1) and 14(1)).

Directions by the Secretary of State

The SoS is empowered to direct a LA on the use of these powers and on the matters or steps to be included in the notices (sects. 13(3) and 14(4)).

Appeals

The person on whom the notice has been served may appeal to the SoS within 2 months of the date of the notice (sect. 15(2) and Reg. 10(1)(d)).

Notices are **not** suspended pending determination of the appeal (sect. 15(9)).

Detailed arrangements and requirements for the furtherance of appeals are provided for in the Environmental Protection (Applications, Appeals and Registers) Regulations 1991 (as amended), and guidance is given in GG5(91): Appeals.

Public register

Subject to certain restrictions relating to national security and confidential information, details of ENs and PNs served by the LA are required to be kept in a register, details of which are prescribed in regulations 15–17 inc. of the Environmental Protection (Applications, Appeals and Registers) Regulations 1991 (amended 1996). Guidance on registers is also given in GG3(91): Applications and Registers.

Offences

Failure to comply with any requirement of an EN or PN is an offence for which the penalty on conviction is:

(a) for summary action, a maximum fine of £20 000 and/or imprisonment for up to 3 months; and

(b) on indictment, an unlimited fine or imprisonment for up to 2 years, or both (sect. 23(1)(c) and (2)).

High Court action

Where the LA considers that proceedings for non-compliance with an EN or PN in a magistrates' court would afford an ineffectual remedy, it may seek a High Court injunction to secure compliance (sect. 24).

Powers of the magistrates' court

Where the operator of a prescribed process is found guilty of non-compliance with an EN or PN, the court may in addition to imposing a penalty, order specified steps to be taken to secure compliance within a specified period (sect. 26).

WASTE ON LAND

DEFINITIONS

General

Environment consists of all, or any, of the following media, namely land, water and the air.

Pollution of the environment means pollution of the environment due to the release or escape (into any environmental medium) from:

(a) the land on which controlled waste is treated;

(b) the land on which controlled waste is kept;

(c) the land in or on which controlled waste is deposited;

(d) fixed plant by means of which controlled waste is treated, kept or disposed of;

of substances or articles constituting or resulting from the waste, and capable (by reason of the quantity or concentrations involved) of causing harm to man or any other living organisms supported by the environment.

This applies in relation to mobile plant by means of which controlled waste is treated or disposed of as it applies to plant on land by means of which controlled waste is treated or disposed of.

For the purposes of the above **harm** means harm to the health of living organisms or other interference with the ecological systems of which they form part, and in the case of humans includes offence to any of his senses or harm to his property; and **harmless** has a corresponding meaning.

Disposal of waste includes its disposal by way of deposit in or on land and, subject to the paragraph below, waste is '**treated**' when it is subjected to any process, including making it re-usable or reclaiming substances from it, and **recycle** (and cognate expressions) shall be construed accordingly.

Land includes land covered by waters where the land is above the low water mark of ordinary spring tides and references to land on which controlled waste is treated, kept or deposited are references to the surface of the land (including any structure set into the surface).

Substance means any natural or artificial substance, whether in solid or liquid form or in the form of a gas or vapour (sect. 29).

The definition of waste

Waste means

(a) any substance or object in the categories set out in Schedule 2B of the EPA 1990 which the holder discards or intends or is required to discard;
(b) any substance or article which requires to be disposed of as being broken, worn out, contaminated or otherwise spoiled;

but does not include an explosive.

Controlled waste means household, industrial and commercial waste or any such waste.

Household waste means waste from:

(a) domestic property, that is to say, a building or self-contained part of a building which is used wholly for the purposes of living accommodation;
(b) a caravan (as defined in sect. 29(1) of the CSCDA 1960) which usually and for the time being is situated on a caravan site (within the meaning of that Act);
(c) a residential home;
(d) premises forming part of a university or school or other educational establishment;
(e) premises forming part of a hospital or nursing home.

Industrial waste means waste from any of the following premises:

(a) any factory (within the meaning of the Factories Act 1961);
(b) any premises used for the purposes of, or in connection with, the provision to the public of transport services by land, water or air;
(c) any premises used for the purposes of, or in connection with, the supply to the public of gas, water or electricity or the provision of sewerage services; or
(d) any premises used for the purposes of, or in connection with, the provision to the public of postal or telecommunications services.

Commercial waste means waste from premises used wholly or mainly for the purposes of a trade or business or the purposes of sport, recreation or entertainment excluding:

(a) household waste;
(b) industrial waste;
(c) waste from any mine or quarry and waste from premises used for agriculture within the meaning of the Agriculture Act 1947 or, in Scotland, the Agriculture (Scotland) Act 1948; and
(d) waste of any other description prescribed by regulations made by the SoS for the purposes of this paragraph. (sect. 75)

The Controlled Waste Regulations 1992 (as amended) and DoE Circular 14/92 clarify the definitions of household, commercial and industrial waste in relation to the application of the duty of care, the obligations of collection authorities and the charging regime.

A detailed explanation of the definition of waste is given in Annex 2 of DoE Circular 11/94.

Local authorities

The **waste collection authorities** are:

(a) for any district in England not within Greater London, the council of the district;

(b) in Greater London, the following:
   (i) for any London borough, the council of the borough;
   (ii) for the City of London, the Common Council;
   (iii) for the Temples, the Sub-Treasurer of the Inner Temple and the Under-Treasurer of the Middle Temple respectively;

(c) for any county or county borough in Wales, the council of the county or county borough;

(d) in Scotland, a council constituted under the Local Government (Scotland) Act 1994 (sect. 30(3)).

POWERS OF ENTRY

The powers of entry previously provided for authorized officers under sect. 69(9) of the EPA 1990 have been repealed. Although similar powers have been provided for officers of the Environment Agency this has not happened for LA officers who appear to be without such powers for those procedures within this section which are to be enforced by them.

RECEPTACLES FOR HOUSEHOLD WASTE

Reference

Environmental Protection Act 1990 sect. 46.

Scope

This procedure may be used by a WCA to require the occupier of any premises from which it has the duty to collect household waste ('the Definition of Waste' on page 117) to place the waste for collection in receptacles of the kind and number specified in the notice (sect. 46(1)).

FC30 Receptacles for household waste

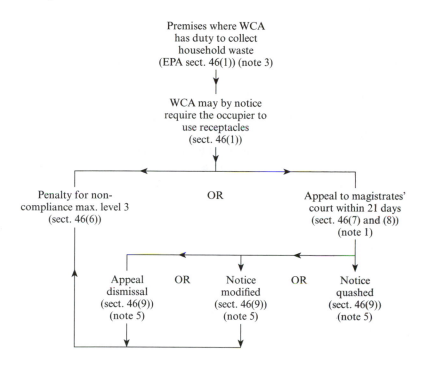

Premises where WCA
has duty to collect
household waste
(EPA sect. 46(1)) (note 3)

WCA may by notice
require the occupier to
use receptacles
(sect. 46(1))

Penalty for non-
compliance max. level 3
(sect. 46(6))

OR

Appeal to magistrates'
court within 21 days
(sect. 46(7) and (8))
(note 1)

Appeal
dismissal
(sect. 46(9))
(note 5)

OR

Notice
modified
(sect. 46(9))
(note 5)

OR

Notice
quashed
(sect. 46(9))
(note 5)

Notes
1. The notice is suspended pending determination of the appeal (sect. 46(9)(a)).
2. For the provision of receptacles for industrial and commercial waste, see FC31.
3. For definition of waste, see page 117.
4. For service of notices, see page 92.
5. Either party may appeal to the Crown Court against the decision of the magistrates' court (sect. 73(1)).

WCAs have a duty to collect household waste. No charge may be made except in prescribed cases detailed in the Controlled Waste Regulations 1992 (sect. 45).

Content of notice

The kind and number of receptacles to be used must be reasonable, but separate receptacles can be required for waste which is to be recycled (sect. 46(2)).

The notice may include requirements dealing with:

(a) the size, construction and maintenance of receptacles;
(b) the placing of them to facilitate emptying and access for that purpose;
(c) the placing of receptacles on the highway (with the consent of the highway authority and with an arrangement for damage liability);
(d) limitation of the waste which may or may not be put into the receptacles and the precautions to be taken with particular substances or articles; and
(e) the steps to be taken by occupiers to facilitate the collection of the waste (sect. 46(4) and (5)).

Provision of receptacles

The WCA has a choice of the way in which the necessary receptacles are to be provided:

(a) by the WCA free of charge;
(b) with the occupier's agreement, by the WCA on payment by the occupier (single purchase or periodic payments);
(c) by the occupier, including where the occupier refuses an agreement under (b) (sect. 46(3)).

Appeal

The occupier may appeal to the magistrates' court within 21 days of the notice being served or, where required to enter into an agreement under (b) above, within 21 days of the expiry of the period allowed by the WCA to enter into the agreement.

The grounds of appeal are:

(a) that any requirement is unreasonable; and
(b) the receptacles already used are adequate (sect. 46(7) and (8)).

Upon appeal, the notice is suspended until the appeal is determined and in any subsequent proceedings for non-compliance with the notice, no question regarding the reasonability of any requirement can be entertained (sect. 46(9)).

Definition (also page 116)

Receptacle includes a holder for receptacles (sect. 46(10)).

RECEPTACLES FOR COMMERCIAL AND INDUSTRIAL WASTE

Reference

Environmental Protection Act 1990 sect. 47.

Power of WCA to provide receptacles

A WCA has power to supply receptacles for commercial and industrial waste ('the Definition of Waste' on page 117) which the WCA has been requested to collect. A charge for a receptacle for industrial waste is mandatory, but the charge for a receptacle for commercial waste is discretionary (sect. 47(1)). The charge may be on a purchase or rental basis.

WCAs are under a duty to collect commercial waste when asked to do so by an occupier but have discretion on the collection of industrial waste. Where collections are made the WCA must make a charge (sect. 45).

Scope

This procedure may be applied where:

   (a) commercial or industrial waste is stored at a premises; and
   (b) if not stored in receptacles of a particular kind, the waste is likely to cause nuisance or be detrimental to the amenities of the neighbourhood (sect. 47(2)).

Contents of notice

The notice may include requirements covering:

   (a) size, construction and maintenance of the receptacle (the kind and number must be reasonable);
   (b) the siting of the receptacle for emptying and access for that purpose;
   (c) placing of the receptacle on the highway, with the consent of the highway authority and in accordance with arrangements for damage liability;
   (d) the substances which may or may not be put into the receptacles and the precautions to be taken;
   (e) the steps to be taken by occupiers to facilitate collection (sect. 47(3) and (4)).

The time allowed for compliance (where appropriate) must be reasonable, and should not be less than the period within which an appeal may be made, i.e. 21 days.

FC31 Receptacles for commercial and industrial waste

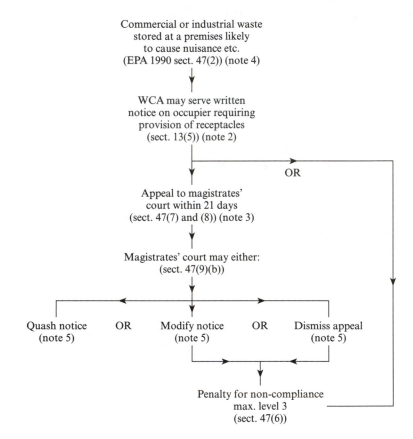

Commercial or industrial waste
stored at a premises likely
to cause nuisance etc.
(EPA 1990 sect. 47(2)) (note 4)

WCA may serve written
notice on occupier requiring
provision of receptacles
(sect. 13(5)) (note 2)

OR

Appeal to magistrates'
court within 21 days
(sect. 47(7) and (8)) (note 3)

Magistrates' court may either:
(sect. 47(9)(b))

Quash notice   OR   Modify notice   OR   Dismiss appeal
(note 5)   (note 5)   (note 5)

Penalty for non-compliance
max. level 3
(sect. 47(6))

Notes
1. For provision of receptacles for household waste, see FC30.
2. For service of notices, see page 92.
3. The notice is suspended pending determination of the appeal (sect. 47(9)(1)).
4. For definition of industrial and commercial waste, see page 117.
5. Either party may appeal to the Crown Court against the decision of the magistrates court (sect. 73(1)).

Appeals

Appeal may be made to the magistrates' court within 21 days of receipt of the notice on the following grounds:

(a) any requirement is unreasonable;
(b) the waste is not likely to cause nuisance or be detrimental to the amenities of the neighbourhood (sect. 47(7) and (8)).

Upon appeal, the notice is suspended until the appeal is determined and, in any subsequent proceedings, no question concerning the reasonability of any requirement can be entertained (sect. 47(9)(c)).

Definition (also page 116)

Receptacle includes a holder for receptacles (sect. 47(10)).

REMOVAL OF CONTROLLED WASTE ON LAND

Reference

Environmental Protection Act 1990 sect. 59.

Scope

These provisions apply whenever controlled waste ('Definition of waste' on page 117) has been deposited on land without the authority of a waste management licence under sect. 33, and may be operated by either a WCA or the Environment Agency as the WRA (sect. 59(1)).

Removal etc. by a WCA or WRA

This may be effected, without recourse to the notice procedure, where:

(a) there is no occupier of the land affected; or
(b) the occupier neither made nor knowingly permitted the deposit; or
(c) removal, and/or steps to eliminate or reduce the consequences of the deposit is necessary forthwith to remove or prevent pollution of land, water or air or harm to human health – see definitions, page 93 (sect. 59(7)).

Recovery of costs of both removing the waste and of taking steps to eliminate or reduce the consequences of the deposit is possible from any person who caused, or knowingly permitted, the deposit (sect. 59(8)).

Removal by notice

The WCA or WRA may serve written notice on the occupier of the land requiring:

FC32 Removal of controlled waste on land

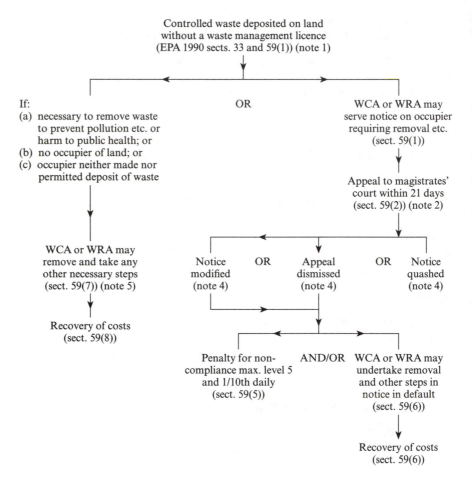

Controlled waste deposited on land
without a waste management licence
(EPA 1990 sects. 33 and 59(1)) (note 1)

If:
(a) necessary to remove waste to prevent pollution etc. or harm to public health; or
(b) no occupier of land; or
(c) occupier neither made nor permitted deposit of waste

OR

WCA or WRA may serve notice on occupier requiring removal etc. (sect. 59(1))

Appeal to magistrates' court within 21 days (sect. 59(2)) (note 2)

WCA or WRA may remove and take any other necessary steps (sect. 59(7)) (note 5)

Notice modified (note 4)   OR   Appeal dismissed (note 4)   OR   Notice quashed (note 4)

Recovery of costs (sect. 59(8))

Penalty for non-compliance max. level 5 and 1/10th daily (sect. 59(5))   AND/OR   WCA or WRA may undertake removal and other steps in notice in default (sect. 59(6))

Recovery of costs (sect. 59(6))

Notes
1. For definition of controlled waste, see page 117.
2. The notice is suspended pending determination of the appeal (sect. 59(4)).
3. For procedure for removing abandoned vehicles, see FC112a.
4. Either party may appeal to the Crown Court against the decision of the magistrates' court (sect. 73(1)).
5. Once removed, the waste belongs to the WCA or WRA and may be dealt with accordingly (sect. 59(9)).
6. For service of notices, see page 92.

(a) removal of the waste within a specified period of not less than 21 days; and/or

(b) the taking of specified steps to eliminate or reduce the consequences of the deposit (sect. 59(1)).

Appeals

Appeal is to the magistrates' court within 21 days of service and no particular grounds of appeal are specified. The court is required to quash the notice if either:

(a) the appellant neither deposited nor knowingly caused or permitted the deposit; or

(b) there is a material defect in the notice (sect. 59(3)).

STATUTORY NUISANCES (INCLUDING THE NOISE ACT 1996)

References

Environmental Protection Act 1990 sects. 79–82 and Schedule 3 (as amended by the Noise and Statutory Nuisance Act 1993, the Environment Act 1995 and the Pollution Prevention and Control Act 1999).
The Statutory Nuisance (Appeals) Regulations 1995.
The Noise Act 1996.
DoE Circular 8/97. The Noise Act 1996.
DoE Circular 9/97. The Noise Act and Statutory Nuisance Act 1993.

Extent

The 1990 Act applies in England and Wales and, with certain amendments, in Scotland. It does not apply in Northern Ireland. The 1996 Act applies in England, Wales and Northern Ireland.

Anti-social behaviour orders (ASBOs)

The use of the procedure for making ASBOs under the Crime and Disorder Act 1998 rather than the statutory nuisance provisions dealt with here may be preferable in some circumstances. These are detailed in FC118.

Statutory nuisances

These are nuisances to which the abatement procedures of Part 3 of the EPA 1990 have been applied and are:

FC33a Statutory nuisances: LA action

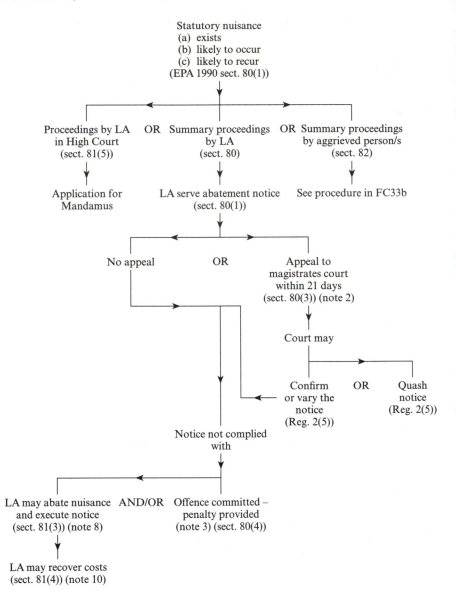

Statutory nuisance
(a) exists
(b) likely to occur
(c) likely to recur
(EPA 1990 sect. 80(1))

Proceedings by LA in High Court (sect. 81(5)) **OR** Summary proceedings by LA (sect. 80) **OR** Summary proceedings by aggrieved person/s (sect. 82)

Application for Mandamus

LA serve abatement notice (sect. 80(1))

See procedure in FC33b

No appeal **OR** Appeal to magistrates court within 21 days (sect. 80(3)) (note 2)

Court may

Confirm or vary the notice (Reg. 2(5)) **OR** Quash notice (Reg. 2(5))

Notice not complied with

LA may abate nuisance and execute notice (sect. 81(3)) (note 8) **AND/OR** Offence committed – penalty provided (note 3) (sect. 80(4))

LA may recover costs (sect. 81(4)) (note 10)

Notes
1. Regulation numbers refer to the Statutory Nuisance (Appeals) Regulations 1995.
2. For the suspension of some notices pending hearing of the appeal, see page 134.
3. Details of penalties are given on page 135.
4. For expedited procedure to deal with the defective premises, see FC18.
5. For service of notices under EPA 1990, see page 92.
6. No account may be taken under this procedure of radioactivity (RSA 1993 sect. 40).
7. For a special procedure dealing with certain noise nuisance in streets, see FC33c.
8. For seizure of equipment causing noise nuisance, see FC33e.
9. For powers to deal with night-noise from dwellings, see FC34a and b.
10. Or LA may put a charge on the property (sect. 81A).

FC33b Statutory nuisances: action by aggrieved persons

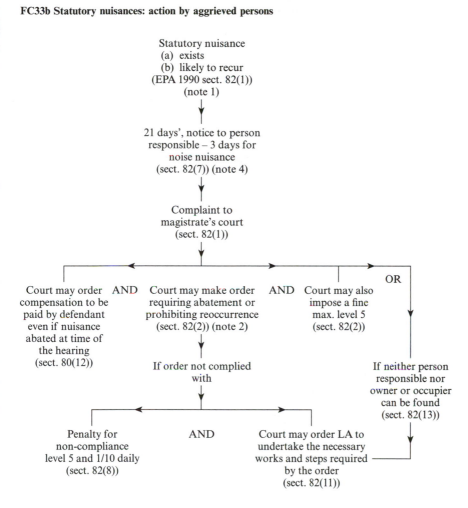

Statutory nuisance
(a) exists
(b) likely to recur
(EPA 1990 sect. 82(1))
(note 1)

↓

21 days', notice to person
responsible – 3 days for
noise nuisance
(sect. 82(7)) (note 4)

↓

Complaint to
magistrate's court
(sect. 82(1))

↓

OR

Court may order   AND   Court may make order   AND   Court may also
compensation to be   requiring abatement or   impose a fine
paid by defendant   prohibiting reoccurrence   max. level 5
even if nuisance   (sect. 82(2)) (note 2)   (sect. 82(2))
abated at time of
the hearing
(sect. 80(12))

If order not complied
with

If neither person
responsible nor
owner or occupier
can be found
(sect. 82(13))

Penalty for   AND   Court may order LA to
non-compliance   undertake the necessary
level 5 and 1/10 daily   works and steps required
(sect. 82(8))   by the order
                                 (sect. 82(11))

Notes

1. Use of this procedure is not possible to take action to prohibit an anticipated nuisance which does not exist or has not already occurred.
2. The court may also order the closure of a premises rendered unfit for human habitation by the nuisance until they are rendered fit (sect. 82(3)).
3. No account may be taken under this procedure of radioactivity (RSA 1993 sect. 40).
4. This includes noise nuisances for vehicles, machinery or plant in streets.

FC33c Statutory nuisances: LA action; special procedure for certain noise in streets

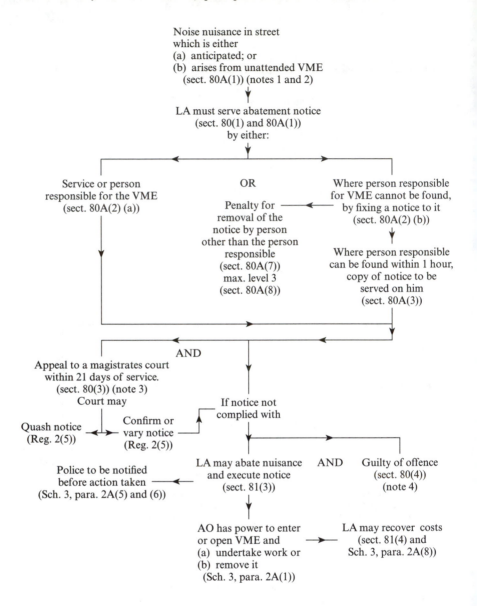

Noise nuisance in street
which is either
(a) anticipated; or
(b) arises from unattended VME
(sect. 80A(1)) (notes 1 and 2)

LA must serve abatement notice
(sect. 80(1) and 80A(1))
by either:

Service or person responsible for the VME (sect. 80A(2) (a))

OR

Penalty for removal of the notice by person other than the person responsible (sect. 80A(7)) max. level 3 (sect. 80A(8))

Where person responsible for VME cannot be found, by fixing a notice to it (sect. 80A(2) (b))

Where person responsible can be found within 1 hour, copy of notice to be served on him (sect. 80A(3))

AND

Appeal to a magistrates court within 21 days of service. (sect. 80(3)) (note 3) Court may

Quash notice (Reg. 2(5))

Confirm or vary notice (Reg. 2(5))

If notice not complied with

Police to be notified before action taken (Sch. 3, para. 2A(5) and (6))

LA may abate nuisance and execute notice (sect. 81(3))

AND

Guilty of offence (sect. 80(4)) (note 4)

AO has power to enter or open VME and (a) undertake work or (b) remove it (Sch. 3, para. 2A(1))

LA may recover costs (sect. 81(4) and Sch. 3, para. 2A(8))

Notes
1. For other statutory noise nuisances in streets see page 136.
2. VME – vehicle, machinery or equipment.
3. The LA will have indicated in the notice that it is to be suspended pending the hearing of any appeal (page 134).
4. Details of penalties are given on page 135.
5. For the control over loudspeakers in streets see FC69.
6. A similar procedure for use in Scotland is set out in Schedule 1 of the NSNA 1993 in conjunction with sect. 58 of the CPA 1974.
7. Regulation numbers refer to the Statutory Nuisance (Appeals) Regulations 1995.
8. See DoETR circ. 9/97 for guidance on the operation of this procedure.

FC33d Statutory nuisances: control over noisy parties

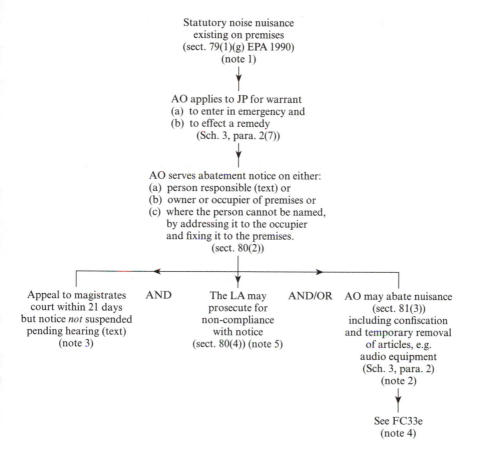

Statutory noise nuisance
existing on premises
(sect. 79(1)(g) EPA 1990)
(note 1)

↓

AO applies to JP for warrant
(a) to enter in emergency and
(b) to effect a remedy
(Sch. 3, para. 2(7))

↓

AO serves abatement notice on either:
(a) person responsible (text) or
(b) owner or occupier of premises or
(c) where the person cannot be named,
by addressing it to the occupier
and fixing it to the premises.
(sect. 80(2))

| Appeal to magistrates court within 21 days but notice *not* suspended pending hearing (text) (note 3) | **AND** | The LA may prosecute for non-compliance with notice (sect. 80(4)) (note 5) | **AND/OR** | AO may abate nuisance (sect. 81(3)) including confiscation and temporary removal of articles, e.g. audio equipment (Sch. 3, para. 2) (note 2) |

↓

See FC33e
(note 4)

Notes

1. Anticipated noise nuisances can also be dealt with (page 132).
2. These powers must be exercised reasonably (sect. 81(3)).
3. The court may quash or vary the notice or dismiss the appeal (Reg. 2(5)).
4. The procedure for dealing with the seizure of equipment is provided for in sect. 10 and the schedule to the Noise Act 1996 – see FC33e.
5. For the adoptive procedure dealing with night noise from dwellings, see FC34a.

1. Any premises in such a state as to be prejudicial to health or a nuisance (sect. 79(1)(a)). (For an expedited procedure to deal with the defective premises, see BA 1984 sect. 76, FC18).
2. Smoke emitted from premises so as to be prejudicial to health or a nuisance; but this does not apply to:
   (a) premises occupied by the Crown for military or Ministry of Defence purposes;
   (b) smoke emitted from the chimney of a house within a smoke control area;
   (c) dark smoke from the chimney of a building or of a furnace attached to a building or installed on any land;
   (d) smoke from a railway locomotive steam engine;
   (e) dark smoke from any industrial or trade premises (sect. 79(1)(b), (2) and (3)).
3. Fumes or gases emitted from private dwellings so as to be prejudicial to health or a nuisance (sect. 79(1)(c) and (4)).
4. Any dust, steam (other than from a railway locomotive engine), smell or other effluvia arising on industrial, trade or business premises and being prejudicial to health or a nuisance (sect. 79(1)(d) and (5)).
5. Any accumulation or deposit which is prejudicial to health or a nuisance (sect. 79(1)(e)).
6. Any animal kept in such a place or manner as to be prejudicial to health or a nuisance (sect. 79(1)(f)).
7. Noise (except that from aircraft other than model aircraft) emitted from premises so as to be prejudicial to health or a nuisance; but this does not apply to Crown premises used for military or Ministry of Defence purposes (sect. 79(1)(ga), (2) and (6) as amended).
8. Noise that is prejudicial to health or a nuisance and is emitted from or caused by a vehicle, machinery or equipment in a street (other than noise made by traffic, by any military force or by political demonstration or a demonstration supporting or opposing a cause or campaign) (sect. 79(1)(ga) and (6A)).
9. Any other matter declared by any enactment to be a statutory nuisance (sect. 79(1)(h)) and these include:
   (a) any well, tank, cistern or water butt used for the supply of water for domestic purposes which is so placed, constructed or kept as to render the water therein liable to contamination prejudicial to health (PHA 1936 sect. 141);
   (b) any pond, pool, ditch, gutter or watercourse which is so foul or in such a state as to be prejudicial to health or a nuisance (PHA 1936 sect. 259(1)(a));
   (c) any part of a watercourse, not being a part ordinarily navigated by vessels employed in the carriage of goods by water, which is so choked or silted up as to obstruct or impede the proper flow of water and thereby to cause a nuisance or give rise to conditions prejudicial to health (PHA 1936 sect. 259(1)(b));

(d) a tent, van, shed or similar structure used for human habitation,
  (i) which is in such a state, or so overcrowded, as to be prejudicial to the health of the inmates; or
  (ii) the use of which, by reason of the absence of proper sanitary accommodation, or otherwise, gives rise, whether on the site or on other land, to a nuisance or to conditions prejudicial to health (PHA 1936 sect. 268(2));

(e) a shaft or outlet of certain abandoned or disused mines where:
  (i) it is not provided with a property maintained device designed and constructed to prevent persons from accidentally falling down the shaft or accidentally entering the outlet; or
  (ii) by reason of its accessibility from a highway or a place of public resort, it constitutes a danger to the public (M and QA 1954 sect. 151);

(f) a quarry which is not provided with an efficient and properly maintained barrier so designed and constructed as to prevent persons from accidentally falling into it and which, by reason of its accessibility from a highway or place or public resort, constitutes a danger to the public (M and QA 1954 sect. 151).

A local authority may not take summary proceedings for statutory nuisance falling within paras. 2, 4 and 5 above without the consent of the Secretary of State where proceedings might be instituted under part 1 of the EPA 1990 dealing with the authorization of processes or under regulations made under sect. 2 of the Pollution Prevention and Control Act 1999 – see FCs 47–56.

In implementing the statutory nuisance procedures, LAs may not have regard to the radioactive state of any substance, article or premises. These are matters to be dealt with under the Radioactive Substances Act 1993 (RSA 1993 sect. 40).

Unless the SoS has granted consent, a LA may not bring summary proceedings for causing a statutory nuisance where proceedings may be brought under the IPPC regime under the PP and CA 1999 – see FC47 (sect. 79 (10)).

However, activities that are not covered by IPPC, even though they are on the same site as an IPPC installation, may be regulated by the statutory nuisance procedures. The restriction also does not affect the right of members of the public to use private proceedings under sect. 82 of the EPA – see page 137 – or prevent a LA from taking proceedings in the High Court – see page 137.

Contaminated land

These statutory nuisance provisions do not apply where the nuisance consists of or is caused by land being in a 'contaminated state', i.e. land where there are substances in, on or under that land such that harm is being caused, or there is a possibility of its being caused, or the pollution of controlled waters is being, or is likely to be, caused (sect. 79(1A) and (1B) as inserted by the EA 1995.

Contaminated land is dealt with by special procedures, see FCs 45–46.

Abatement notices

Where the LA (definition page 138) is satisfied that a statutory nuisance:

(a) exists; or
(b) is likely to occur; or
(c) is likely to recur;

it is required to serve an abatement notice:

(a) requiring the abatement of the nuisance or prohibiting or restricting its occurrence;
(b) requiring the execution of such works or steps as necessary for those purposes;
(c) specifying the time or times within which the notice is to be complied with (sect. 80(1)).

The notice must also indicate the rights for and times of appeal (Sch. 3, para. 6). The period allowed for compliance must be reasonable ((d) under 'Appeals' below) but may nevertheless be short, e.g. to deal with noise from a party. Where the period allowed is less than the time allowed for appeal, the notice is suspended in certain circumstances ('Suspension of notices' below). For the service of notices, see page 92.

The notice may be served by a LA to deal with a nuisance which exists or has occurred from outside its area (sect. 81(2)).

Persons responsible

The abatement notice is to be served:

(a) except as in (b) and (c) below, on the person/s responsible for the nuisance;
(b) on the owner where the nuisance arises from any defect of a structural character;
(c) where the person responsible cannot be found or the nuisance has not yet occurred, on the owner or occupier of the premises (sect. 80(2)).

Where more than one person is responsible, the notice must be served on each (sect. 81(1)).

For noise nuisances from vehicles, machinery or plant in streets, the notice is to be served:

(a) on the person responsible for that vehicle, etc.; or
(b) where that person cannot be found, by fixing it to the vehicle, etc. (sect. 80A(2)).

Appeals

Any person served with an abatement notice may appeal to a magistrates' court within 21 days of the date of service (sect. 80(3), Sch. 3, paras. 1–4) on one or more of the following grounds:

(a) that the abatement notice is not justified by sect. 80 of the 1990 Act;

(b) that there has been some material informality, defect or error in, or in connection with, the abatement notice or in, or in connection with, any copy of the abatement notice served under sect. 80A(3) (which relates to notices in respect of vehicles, machinery or equipment);

(c) that the authority has refused unreasonably to accept compliance with alternative requirements, or that the requirements of the abatement notice are otherwise unreasonable in character or extent, or are unnecessary;

(d) that the time, or, where more than one time is specified, any of the times, within which the requirements of the abatement notice are to be complied with is not reasonably sufficient for the purpose;

(e) where the nuisance to which the notice relates,

   (i) is a nuisance falling within (1) premises, (4) dust etc., (5) accumulations etc., (6) animals, (7) noise or (8) noise in streets in 'Statutory nuisances' above, and arises on industrial, trade or business premises or in the case of a vehicle, equipment or machinery is being used for industrial, trade or business purposes; or

   (ii) is a nuisance falling within (2) smoke etc. of the same section above and the smoke is emitted from a chimney;

that the best practicable means were used to prevent, or to counteract the effects of, the nuisance;

(f) that, in the case of a nuisance under (7) noise or (8) noise in streets, the requirements imposed by the abatement notice are more onerous than the requirements for the time being in force, in relation to the noise to which the notice relates, of:

   (i) any notice served under sect. 60 or 66 of the CPA 1974; or

   (ii) any consent given under sect. 61 or 65 of the CPA 1974; or

   (iii) any determination made under sect. 67 of the CPA 1974; or

   (iv) in relation to noise in streets, of any condition of consent given under para. 1 of Sch. 2 of the EPA 1990;

(g) that the abatement notice should have been served on some person instead of the appellant, being:

   (i) the person responsible for the nuisance; or

   (ii) the person responsible for the vehicle, machinery or equipment; or

   (iii) in the case of a nuisance arising from any defect of a structural character, the owner of the premises; or

   (iv) in the case where the person responsible for the nuisance cannot be found or the nuisance has not yet occurred, the owner or occupier of the premises;

(h) that the abatement notice might lawfully have been served on some person instead of the appellant being:
  (i) in the case where the appellant is the owner of the premises, the occupier of the premises; or
  (ii) in the case where the appellant is the occupier of the premises, the owner of the premises;
and that it would have been equitable for it to have been so served (copy of the appeal notice to be served on the person/s to be implicated);

(i) that the abatement notice might lawfully have been served on some person in addition to the appellant, being:
  (i) a person responsible for the nuisance;
  (ii) a person who is also an owner of the premises; or
  (iii) a person who is also an occupier of the premises;
  (iv) a person who is also the person responsible for the vehicle, machinery or equipment (copy of appeal notice to be served on the person/s to be implicated);
and that it would have been equitable for it to have been so served (copy of the appeal notice to be served on the person/s to be implicated) (Reg. 2 of the Statutory Nuisances (Appeals) Regulations 1995).

Any party may appeal to the County Court against any decision of the magistrates' court (Sch. 3, para. 1(3)).

The court may either

(a) quash the notice, or
(b) vary it, or
(c) dismiss the appeal and may make orders about the responsibility of persons for costs of workers.

Suspension of notices

Where the nuisance:

(a) is injurious to health; or
(b) is likely to be of limited duration; or
(c) any expenditure incurred in compliance with the notice would not be disproportionate to the public benefit expected; **and**
(d) the notice has specified the existence of these circumstances;

the abatement notice is **not** suspended pending the hearing of any appeal. In any other case where:

(a) expenditure is involved in attaining compliance; or
(b) in the case of noise nuisance, the noise is caused in the course of the performance of a duty imposed by law;

the notice is suspended pending the determination of the appeal (Reg. 3).

Offences

Upon conviction for non-compliance with the requirements of an abatement notice, a person is liable on summary conviction to a fine not exceeding level 5 and one-tenth of that level daily for a continuation. Where the offence is committed on industrial, trade or business premises, the maximum fine is £20 000 (sect. 80(5) and (6)).

Defences

It will be a defence to prove that the best practicable means have been used to prevent or counteract the effects of the nuisance **except**:

(a) where the nuisance arises on **other than** industrial, trade or business premises **and** involves nuisance categories (1) premises, (4) dust etc., (5) accumulations etc., (6) animals, (7) noise or (8) noise in streets;

(b) in the case of a category (2) smoke nuisance except where it arises from smoke emitted from a chimney; and

(c) in relation to category (3) fumes etc. and (8) statutory nuisances declared by Acts other than the EPA 1990 (sects 80(7) and (8)).

In relation to noise nuisances only, including noise in streets, it will be a defence to prove that:

(a) the situation was covered by a notice under sect. 60 or a consent under sects. 61 or 65 of the CPA 1974 relating to construction sites; or

(b) where a sect. 66 CPA 1974 noise reduction notice was in force, the level of noise was below that specified in the notice; or

(c) although a sect. 66 notice was not in force, there was a sect. 67 notice (CPA 1974) relating to new buildings liable to an abatement order and the noise was less than the specified level (sect. 80(9)).

Defaults

When an abatement notice has not been complied with the LA may, in addition to prosecuting for non-compliance, do whatever is necessary in the terms of the notice to abate or prevent the nuisance (sect. 81(3)). This may include the confiscation of sound amplification equipment (FC33e). The special procedures for dealing with noise from vehicles, machinery and plant in streets are also dealt with below.

The costs of abating or preventing the nuisance are recoverable from the person/s by whose act or default the nuisance was caused (sect. 80(4)). Where the owner of any premises is the person responsible for the nuisance, the LA may, with prior notification to that person and subject to appeal, recover costs in executing notices by a charge on the property (sect. 81A). A LA may also recover its costs by instalments (sect. 81B).

Noise complaints

CIEH has issued two guides to assist LAs in developing their policies to deal with noise complaints:

(a) *Noise Management Guidance for LAs*; and
(b) *Good Practice Guidance for Police and LA Cooperation.*

Both are available from CIEH Publications (tel. 020 7827 5882).

Control over 'noisy parties'*

Whilst certain types of these activities may be controlled through the public entertainment licensing procedure (FC99a and b) or through the Private Places of Entertainment (Licensing) Act 1967, many noisy parties are private, domestic, and held in the home and for no private gain. In such cases the use of the statutory noise nuisance legislation is one possible means of controlling noise.

It is unlikely that these provisions were drafted with this particular use in mind, nevertheless they can form an effective remedy and the DoE/Home Office has issued guidance to LAs: 'Control of Noisy Parties – A Joint Guide Note – 1992'. The way in which the statutory nuisance provisions can be used in this way is shown in FC33d.

However, specific powers are now provided by the adoptive provisions of the Noise Act 1996 set out in FC34 to deal with night-noise from dwellings.

Noise in streets (FC33c)

The Noise and Statutory Nuisance Act 1993 brought the definition of a new statutory nuisance from noise arising from a vehicle, machinery or equipment (definitions) in streets. This was done to allow LAs to deal with problems that could give rise to considerable nuisance but which were not covered by previous legislation, which was restricted to noise from premises. In particular this was the case with DIY car repairs, cooling engine noise from refrigerated lorries, misfiring of car alarms, buskers, etc.

The provisions also introduced a special procedure for dealing with noise which amounts to a statutory nuisance from vehicles, machinery or equipment in a street which is unattended and where the person responsible cannot be readily found.

The special procedure allows for the fixing of the abatement notice to the unattended vehicle but the AO must spend up to 1 hour attempting to trace the person responsible (for appeals against these notices see page 133).

---

* Also see police powers for dealing with 'raves' in sect. 66(1) of the Criminal Justice and Public Order Act 1994 and the Police (Disposal of Sound Equipment) Regulations 1995 which provide for the disposal of sound equipment by the police under the 1994 Act. The Police (Retention and Disposal of Vehicles) Regulations 1995 deal with vehicles seized under the 1994 Act.

Where he is successful in so doing, that person must then comply with the notice, there being subsequent penalties for non-compliance and the LA having default powers to take any steps necessary to abate the nuisance. To allow sufficient time for the person responsible, having been contacted, to abate the nuisance the LA may indicate on the notice attached to the unattended vehicle etc., that the time limit for compliance will be extended if it is possible for the person responsible to be served with a copy of the notice. A new notice can then be served with a different period allowed.

Where the person responsible cannot be traced within the 1 hour the LA may then enforce the notice. Before any action is taken the police must be informed but the AO has power to enter or open the vehicle etc., if necessary by force or remove it to a safe place. Having secured the abatement of the nuisance the AO must secure the vehicle etc., e.g. reset the vehicle alarm. The AO must not cause any more damage than is necessary to execute the notice (sect. 80A).

Guidance on these provisions is given in DETR Circular 8/97.

High Court proceedings

If a LA considers that the taking of summary proceedings (service of abatement notice etc.) would afford an inadequate remedy in relation to the abatement, prohibition or restriction of any statutory nuisance, it may take proceedings in the High Court (sect. 81(5)).

In such cases respecting noise nuisance, it will be a defence to show that the noise was authorized by either a notice under sect. 60 or a consent under sect. 61 CPA 1974 dealing with construction sites. Otherwise the defences identified above relating to summary proceedings are not applicable to the High Court procedure (sect. 81(6)).

Summary proceedings by persons aggrieved (FC33b)

It is possible for an aggrieved person or persons to proceed directly to abate or prevent the recurrence of a statutory nuisance by making a complaint to the magistrates' court. For example, this would include the tenant of a house, local authority or privately owned, seeking a remedy to disrepair. At least 21 days' notice shall be given to the person responsible for the nuisance (3 days for noise nuisances including noise from vehicles, machinery or equipment in streets).

If the court is satisfied, it may make an order for either or both the abatement or prohibition of the nuisance by the carrying out of specified works within the time specified. The court may also fine the defendant up to a maximum of level 5. The defences available parallel those where the LA serves the abatement notice as identified above.

In the event of non-compliance with the order the court may impose further fines of up to level 5 and one-tenth of that level daily for its continuation. The court may also, after giving the LA an opportunity of being heard, order the LA to undertake whatever works or steps are necessary to abate or prohibit the nuisance.

At the discretion of the court and upon the making of an order, compensation may be payable by the defendant to the aggrieved persons making the complaint (sect. 82).

Powers of entry

Specific powers relating to power of entry to deal with statutory nuisances are provided in Schedule 3 to the EPA 1990.

Authorized officers of the LA upon producing if required their authority, may enter premises at any reasonable time to see whether or not a statutory nuisance exists or to execute works. Unless in an emergency, 24 hours notice of entry is required to the occupier of residential premises. Where admission is refused or apprehended, where premises are unoccupied, in an emergency or where application for entry would defeat the object, application may be made to the magistrates' court for a warrant.

The maximum penalty for obstruction is level 3 (Sch. 3, paras. 2 and 3).

Definitions

Chimney includes structures and openings of any kind from or through which smoke may be emitted.

Dust does not include dust emitted from a chimney as an ingredient of smoke.

Equipment includes a musical instrument.

Fumes means any airborne solid matter smaller than dust.

Gas includes vapour and moisture precipitated from vapour.

Industrial, trade or business premises means premises used for any industrial, trade or business purposes or premises not so used on which matter is burnt in connection with any industrial, trade or business process, and premises are used for industrial purposes where they are used for the purpose of any treatment or process as well as where they are used for the purposes of manufacturing.

Local authority means:

(a) in Greater London, a London borough council, the Common Council of the City of London and, as respects the Temples, the Sub-Treasurer of the Inner Temple and the Under-Treasurer of the Middle Temple respectively;
(b) in England, outside Greater London, a district council;
(c) in Wales, a county council or county borough council;
(d) in Scotland, a unitary authority; and
(e) the Council of the Isles of Scilly; and
(f) port health authorities (except for noise nuisance).

Noise includes vibration.

Person responsible

(a) in relation to a statutory nuisance, means the person to whose act, default or sufferance the nuisance is attributable;
(b) in relation to a vehicle, includes a person in whose name the vehicle is for the time being registered under the Vehicles (Excise) Act 1994 and any other person who is for the time being the driver of the vehicle;
(c) in relation to machinery or equipment, includes any person who is for the time being the operator of the machinery or equipment.

Prejudicial to health means injurious, or likely to cause injury, to health.

Premises includes land and any vessel (other than one powered by steam reciprocating machinery).

Private dwelling means any building, or part of a building, used or intended to be used, as a dwelling.

Smoke includes soot, ash, grit and gritty particles emitted in smoke.

Street means a highway and any other road, footway, square or court that is for the time being open to the public (sect. 79(8)).

Best practicable means is to be interpreted by reference to the following provisions:

(a) practicable means reasonably practicable having regard among other things to local conditions and circumstances, to the current state of technical knowledge and to the financial implications;
(b) the means to be employed include the design, installation, maintenance and manner and periods of operation of plant and machinery, and the design, construction and maintenance of buildings and structures;
(c) the test is to apply only so far as it is compatible with any duty imposed by law;
(d) the test is to apply only so far as it is compatible with safety and safe working conditions, and with the exigencies of any emergency or unforeseeable circumstances;

and, in circumstances where a code of practice under sect. 71 CPA 1974 (noise minimization) is applicable, regard shall also be had to guidance given in it.
(sect. 79(7), (8) and (9))

SEIZURE OF EQUIPMENT USED TO MAKE NOISE UNLAWFULLY

References

Noise Act 1996 (sect. 10 and Schedule).
Environmental Protection Act 1990 (sects. 79(1)(g) and 81(3)).

FC33e Seizure of equipment used to make unlawful noise

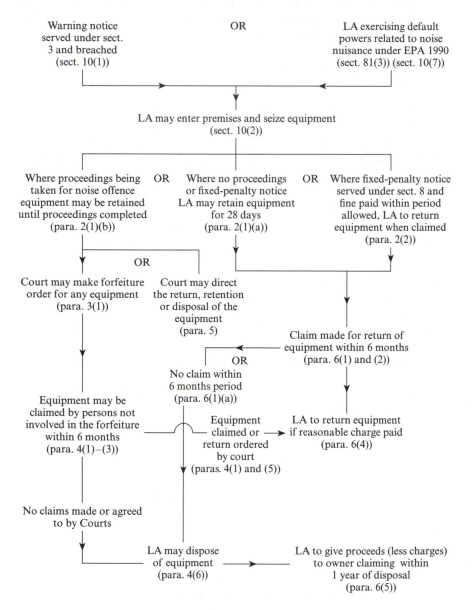

Note

1.   Paragraph numbers refer to the Schedule to the NA 1996.

DoETR circular 8/97. The Noise Act 1996.
CIEH Good Practice Guide on the Management of LA Noise Services 1997.

Extent

In addition to England and Wales the Act applies, with some minor amendments in Northern Ireland but not in Scotland (sect. 14(4)).

Scope

This procedure is available (without adoption) to AOs of LAs where:

(a) a warning notice under the NA 1996 has been served (FC34b) and noise emitted from the offending dwelling has exceeded the permitted level as measured from within the complainants' dwelling; or
(b) the LA is using its powers under sect. 81(3) of the EPA 1990 to itself abate or prevent a statutory noise nuisance under sect. 79(1)(g) following the service of an abatement notice (FC33a).

Seizure of equipment

An officer of the LA or someone authorized by the LA has the power to enter the dwelling from which the noise is being or has been emitted and may seize and remove any equipment which is being or has been used to make the noise. Any person exercising these powers must produce his or her authority, if required to do so (sects. 10(2) and (3)).

If entry is in connection with enforcement under the 1990 Act then 24 hours notice of intention to enter must be given.

Warrants

Where a JP is satisfied on sworn information that:

(a) a warning notice has been served;
(b) noise has been emitted in breach of the requirements of the notice; and
(c) entry to the dwelling has been refused, such refusal is apprehended or a request for entry would defeat the object of admission;

the JP may by warrant authorize the LA to enter, if necessary by force. Persons authorized by the LA to effect the warrant may take with them any other person or equipment as may be necessary but must ensure when leaving that any unoccupied premises are left as effectively secured against trespass as they found them.

Warrants continue in force until the purpose for entry has been satisfied (sects. 10(2)–(6)).

Offences

Any person wilfully obstructing someone exercising the powers of this procedure is liable to a fine not exceeding level 3 (sect. 10(8)).

Retention of equipment

Equipment seized under these powers may be retained by the LA for 28 days unless it was:

(a) the subject to proceedings for a noise offence – in which case it may be retained until the proceedings have been dealt with; or
(b) equipment used to emit noise which has been the subject of a fixed-penalty notice under sect. 8 – in which case it must be returned on request.

(Sch. para. 2)

Forfeiture

When a person is convicted by a magistrates' court of a noise offence (either the breach of a warning notice under the NA or an abatement notice for a statutory noise nuisance under the EPA) the court may order the forfeiture of any seized equipment. In these cases, third parties (e.g. hiring companies) may make application to the court within 6 months for its return. Persons subject to the forfeiture order have no further rights to the equipment named (para. 4).

The court may also give directions (where a forfeiture order is not made) as to the return, retention or disposal of the equipment (para 5).

Return of seized equipment

Equipment which is claimed within the required periods or is ordered to be returned to its owner by a magistrates court must be returned by the LA once a reasonable charge has been paid to it (para. 6(5)).

Notification of claim rights

LAs have a duty to take reasonable steps to bring to the attention of persons who may be entitled to do so their rights to make claims for the return of seized equipment (paras. 4(4) and 6(3)).

Definitions

Local Authority – see page 149.

NOISE FROM DWELLINGS AT NIGHT

References

Noise Act 1996.
DoETR circular 8/97 The Noise Act 1996.
CIEH Good Practice Guidance on the Management of LA Noise Services 1997.
NB. The general provisions of the EPA 1990 at page 92 do not apply to this procedure.
NB. Following a review of the Noise Act 1996 the Environment Minister announced on 20 December 2001 that it was intended to amend the provisions by making them less prescriptive and non-adoptive. No such legislation has yet been passed.

Extent

In addition to England and Wales, the Act applies with some minor modification in Northern Ireland, but does not apply in Scotland (sect. 14(4)).

Application

The application of the provisions of this Act is unusual in that while its powers are adoptive by LAs, nevertheless the Secretary of State has reserve powers to direct individual LAs to take them on. However, it should be noted that in either case the use of the powers themselves, e.g. the service of warning notices, is discretionary (sect. 1).

Adoption

The voluntary adoption of powers by a LA (definitions below) is by resolution of the Council followed by publicity in a local newspaper in two consecutive weeks, the last of which must be at least two months before the commencement date. The resolution must cover the whole of the LA area. The commencement date must be at least three months after the date of the resolution. The published notice must state:

(a) that the resolution has been passed;
(b) the commencement date; and
(c) the general effect of the adopted provisions of the Act (sect. 1(1)–(3)).

The Secretary of State may by Order direct that the powers become available in the area of a particular LA although the circumstances in which this might happen are unspecified and unclear. Such an order is subject to annulment by resolution of either House of Parliament, but there is no provision for formal consultation with the LA concerned. Such Orders become effective on the stated date which may not be less than three months from the making of the Order. There is no requirement for local publicity although this should obviously be undertaken (sects. 1(1) and (4)).

Scope

The procedure enables a LA to deal with complaints from any individual present in a dwelling (definitions below) during night hours that excessive noise is being emitted from another dwelling.

Night hours are defined as from 11 p.m. to 7 a.m. the following morning (sects. 2(2) and (6)).

Complaints

Once the powers have been adopted (or applied by Order) the LA must investigate complaints of night-noise which appear to be covered by the provisions. Such investigations must be undertaken by an officer of the LA (sect. 2(1)).

Warning notices

Where upon investigation the officer is satisfied that:

(a) noise is being emitted from the offending dwelling during night hours; and

(b) the noise if measured within the complainants' dwelling would or might exceed the permitted level;

he may serve a warning notice (sect. 2 (4)).

At this stage actual measurement of the noise level is not obligatory.

Noise above the permitted level does not necessarily create a statutory nuisance.

The permitted level of noise is

(a) where the background noise is 25dB or less, the permitted level is 35dB; and

(b) where the underlying noise is >25dB, 10dB in excess of the background figure measured in accordance with directions in circular 8/97 (sect. 6 and DoETR circular 8/97).

Complaints made by a resident of one LA area about noise arising from a dwelling in an adjoining LA area may be pursued by the first LA and warning notices served even if the provisions of the Act have not been adopted in the area of the offending dwelling (sect. 2(7)).

In judging whether or not the noise complained about is, would or might exceed the permitted level, the officer has discretion whether or not measurements are taken and whether the assessment of the noise is from within or outside the complainants' dwelling (sect. 2(5)).

FC34a Noise from dwellings at night – (a) adoption/application of powers of Noise Act 1996

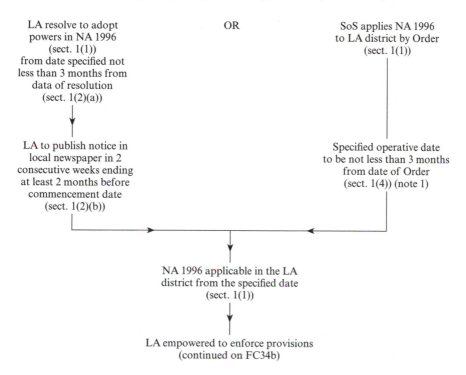

LA resolve to adopt
powers in NA 1996
(sect. 1(1))
from date specified not
less than 3 months from
data of resolution
(sect. 1(2)(a))

OR

SoS applies NA 1996
to LA district by Order
(sect. 1(1))

LA to publish notice in
local newspaper in 2
consecutive weeks ending
at least 2 months before
commencement date
(sect. 1(2)(b))

Specified operative date
to be not less than 3 months
from date of Order
(sect. 1(4)) (note 1)

NA 1996 applicable in the LA
district from the specified date
(sect. 1(1))

LA empowered to enforce provisions
(continued on FC34b)

Notes

1. There is no requirement for a SoS order to be published locally, but this should be arranged.
2. For noise which is a statutory nuisance, see FC33a which may be used as an alternative remedy.

FC34b Noise from dwellings at night – (b) enforcement by LAs

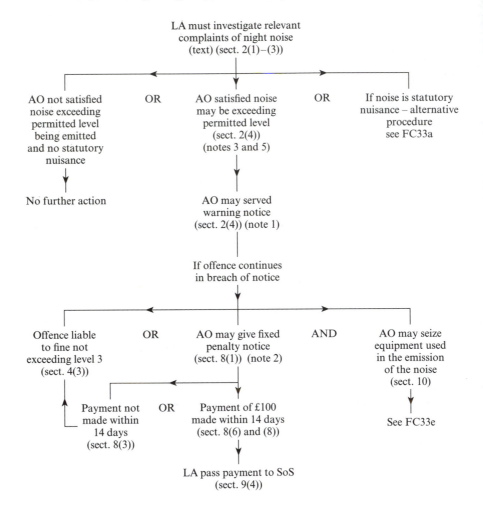

Notes
1. There is no prescribed form for a warning notice but see CIEH guidance.
2. No form yet prescribed but see CIEH guidance.
3. Action may be taken against a dwelling within an adjoining LA area whether or not that LA has adopted these powers (sect. 2(7)).
4. For noise which is a statutory nuisance see FC33a.
5. This may or may not be a statutory nuisance.

Content of warning notices

The notice must:

(a) state that the officer considers that there is night-noise from the offending dwelling which exceeds, or may exceed, the permitted level as measured within the complainants' dwelling; and

(b) give warning that any person responsible for noise emitted above the permitted level in the period specified in the notice may be guilty of an offence (sect. 3(1)).

The period specified must begin not earlier than 10 minutes from the notice being served and end at the following 7 a.m. Officers may use discretion as to how much 'grace' should be given before the notice comes into effect given the circumstances of the situation. The notice must also state the time at which it is served (sect. 3(2) and (4)).

Service of warning notices is effected either by delivering to any person present at or near the offending dwelling and appearing to be responsible for the noise or, where identification of such a person is not reasonably practicable, by leaving it at the offending dwelling (sect. 3(3)).

A person is deemed as responsible for the noise if he is a person by whose act, default or sufferance the emission of the noise is wholly or partly attributable (sect. 3(5)).

Offences

Once a warning notice has been served, if noise is emitted from the offending dwelling in excess of the permitted level as measured from within the complainants' dwelling any person responsible (see above) is guilty of an offence and a fine not exceeding level 3 (sect. 4).

It should be noted that at this stage the noise must be measured and the measurement taken from within the dwelling (as distinct from the situation at the investigative stage).

Evidence of measurements taken will not be admissible in court proceedings unless they were taken using a device approved by the Secretary of State and used in accordance with conditions prescribed by him (sect. 6).

There are detailed provisions about the way in which evidence must be gathered and presented (sect. 7).

It is a defence to show that there was a reasonable excuse for the act, default or sufferance in question (sect. 4(2)).

Fixed-penalty notices

Where the AO believes that an offence has been committed by the infringement of a warning notice he may serve a fixed-penalty notice as an alternative to prosecution.

The notice offers the person the opportunity of discharging liability to conviction by the payment of £100 within 14 days. The notice is served either by giving it to the person or by leaving it, addressed to him at the offending dwelling. The notice must state:

(a) details of the offence;
(b) the period during which proceedings will not be taken, i.e. 14 days;
(c) the amount of the fixed penalty (£100); and
(d) the ways in which the payment can be made (these are not 'on the spot' fines).

Where a payment is not received during the period specified (if by post the date of payment is deemed to be the time at which the letter would be delivered in the ordinary course of post), the LA may proceed by way of prosecution for the offence (sect. 8).

Only one fixed-penalty notice may be given to the same person for a particular dwelling in the same night (sect. 9(2)).

Seizure of equipment/powers of entry

Where noise exceeding the permitted level has been emitted from a dwelling during the currency of a warning notice, officers may enter and seize or remove any equipment which is being or has been used in the emission of the noise (also see FC33e).

In exercising this power of entry the AO must, if requested, produce his authority. There are provisions for application to a JP for a warrant authorising entry, if need be by force, where:

(a) a Warning Notice has been served;
(b) noise has exceeded the permitted level in contravention of the Notice; and
(c) entry has been refused, refusal is apprehended or a request for admission would defeat the object of entry.

Warrants are effective until their purpose has been satisfied. Persons executing the entry may take other persons or equipment as necessary and must leave unoccupied premises secured. The penalty for obstructing an AO is a fine not exceeding level 3.

This power is not adoptive and also applies to offences under the statutory noise nuisance procedure under the EPA 1990 (see FC33a) (sect. 10).

Definitions

'Dwelling' means any building, or part of a building, used or intended for use as a dwelling and references to noise emitted from a dwelling include noise emitted from any garden, yard, outhouse or other appurtenance belonging to or enjoyed within the dwelling (sect. 11(2)).

'Local Authority' means:

(a) in Greater London, a London borough council, the Common Council of the City of London and, as respects the Temples, the Sub-Treasurer of the Inner Temple and the Under-Treasurer of the Middle Temple; and
(b) outside Greater London:
    (i) any district council;
    (ii) the Council of any county so far as they are the Council for any area for which there are no district councils;
    (iii) in Wales, the Council of a county or county borough; and
(c) the Council of the Isles of Scilly (sect. 11(1)).

LITTER

POWERS OF ENTRY

By virtue of the repeal of sect. 69 of the EPA 1990 there do not appear to be any statutory powers of entry available to authorized agencies of a LA in enforcing the procedures in this section.

DEFINITIONS

Principal litter authority (also page 155)

In England and Wales the following are PLAs:

(a) a county council;
(b) a county borough council;
(c) a district council;
(d) a London borough council;
(e) the Common Council of the City of London; and
(f) the Council of the Isles of Scilly.

This definition includes the unitary authorities in England and Wales. The SoS may, by order, designate other descriptions of local authorities as litter authorities for the purposes of this Part; and any such authority shall also be a PLA (sect. 86(2)).

In Scotland the PLAs are councils constituted under the Local Government etc. (Scotland) Act 1994, ie unitary councils, and joint boards.

Litter/Refuse

While these words are not defined, the Litter (Animal Droppings) Order 1991 includes as refuse for the whole of Part 4 of the EPA 1990 dog faeces* on land of the following description which is not heath or woodland used for the grazing of animals:

(a) any public walk or pleasure ground;

(b) any land, whether enclosed or not, on which there are no buildings or of which no more than one-twentieth part is covered with buildings, and the whole or the remainder of which is laid out as a garden or is used for the purposes of recreation;

(c) any part of the seashore (that is to say every cliff, bank, barrier, dune, beach, flat or other land adjacent to and above the place to which the tide flows at mean high water springs) which is:

(i) frequently used by large numbers of people; and

(ii) managed by the person having direct control of it as a tourist resort or recreational facility;

(d) any esplanade or promenade which is above the place to which the tide flows at mean high water springs;

(e) any land not forming part of the highway, or, in Scotland, a public road, which is open to the air, which the public are permitted to use on foot only, and which provides access to retail premises;

(f) a trunk road picnic area provided by the minister under sect. 112 of the Highways Act 1980 or, in Scotland, by the SoS under sect. 55 of the Roads (Scotland) Act 1984;

(g) a picnic site provided by a local planning authority under sect. 10(2) of the Countryside Act 1968 or, in Scotland, a picnic place provided by an islands or district council or a general or district planning authority under sect. 2(2)(a)(i) of the Local Government (Development and Finance) (Scotland) Act 1964;

(h) land (whether above or below ground and whether or not consisting of or including buildings) forming or used in connection with off-street parking places provided in accordance with sect. 32 of the Road Traffic Regulation Act 1984.

Relevant land

The duties placed on various public bodies, statutory undertakers and others to keep their land clear, as far as practicable, from litter and refuse relate to the 'relevant land' of those organizations and individuals. Relevant land is defined as:

'(4) Subject to subsection (8) below, land is "relevant land" of a Principal Litter Authority if, not being relevant land falling within subsection (7) below, it is open to the air and is land (but not a highway or in Scotland a public road) which is under the direct control of such an authority to which the public are entitled or permitted to have access with or without payment (see note at (8) below).

---

* For other controls over the fouling of land by dogs see FC42.

(5) Land is Crown Land if it is land:
  (a) occupied by the Crown Estate Commissioners as part of the Crown Estate;
  (b) occupied by or for the purposes of a government department or for naval, military or air force purposes; or
  (c) occupied or managed by anybody acting on behalf of the Crown; and is "relevant Crown land" if it is Crown land which is open to the air and is land (but not a highway or in Scotland a public road) to which the public are entitled or permitted to have access with or without payment; and "the appropriate Crown authority" for any Crown land is the Crown Estate Commissioners, the Minister in charge of the government department or the body which occupies or manages the land on the Crown's behalf, as the case may be (see note at (8) below).

(6) Subject to subsection (8) below, land is "relevant land" of a designated statutory undertaker if it is the land which is under the direct control of any statutory undertaker of any description which may be designated by the Secretary of State, by order, for the purposes of this Part, being land to which the public are entitled or permitted to have access with or without payment or, in such cases as may be prescribed in the designation order, land in relation to which the public have no such right or permission. (See the Litter (Statutory Undertakers) (Designation and Relevant Land) Order 1991 (as amended 1992).)

(7) Subject to subsection (8) below, land is "relevant land" of a designated educational institution if it is open to the air and is land which is under the direct control of the governing body of or, in Scotland, of such body or of the educational authority responsible for the management of, any educational institution or educational institution of any description which may be designated by the Secretary of State, by order, for the purposes of this Part. (The Litter (Designated Educational Institutions) Order 1991.)

(8) The Secretary of State may, by order, designate descriptions of land which are not to be treated as relevant Crown land or as relevant land of Principal Litter Authorities, of designated statutory undertakers or of designated educational institutions or of any description of any of them.
NB. The Litter (Relevant Land of Principal Litter Authorities and Relevant Crown Land) Order 1991 excludes from the interpretation of relevant land as respects both PLAs and the Crown, land below the high water spring tide level.

(9) Every highway maintainable at the public expense other than a trunk road which is a special road is a "relevant highway" and the local authority which is, for the purposes of this Part, "responsible" for so much of it as lies within its area is, subject to any order under subsection (11) below (The Highway Litter Clearance and Cleaning (Transfer of Duties) Order 1991):

(a) in Greater London, the Council of the London borough or the Common Council of the City of London;

(b) in England, outside Greater London, the Council of the district;

(c) in Wales, the Council of the County or County borough; and

(d) the Council of the Isles of Scilly.

(10) In Scotland, every public road other than a trunk road which is a special road is a "relevant road" and the local authority which is, for the purposes of this Part, "responsible" for so much of it as lies within its area is, subject to any order under subsection (11) below, the council constituted under sect. 2 of the Local Government etc. (Scotland) Act 1994.

(11) The Secretary of State may, by order, as respects relevant highways or relevant roads, relevant highways or relevant roads of any class or any part of a relevant highway or relevant road specified in the order, transfer the responsibility for the discharge of the duties imposed by sect. 89 below from the local authority to the highway or roads authority; but he shall not make an order under this subsection unless:

(a) (except where he is the highway or roads authority) he is requested to do so by the highway or roads authority;

(b) he consults the local authority; and

(c) it appears to him to be necessary or expedient to do so in order to prevent or minimize interference with the passage or with the safety of traffic along the highway or, in Scotland, road in question; and where, by an order under this subsection, responsibility for the discharge of those duties is transferred, the authority to which the transfer is made is, for the purposes of this Part, "responsible" for highway, road or part specified in the order. (See the Highway Litter Clearance and Cleaning (Transfer of Duties) Orders 1991 and 1997.)

(12) Land is "relevant land within a litter control area of a local authority" if it is land included in an area designated by the local authority under sect. 90 of the EPA 1990 to which the public are entitled or permitted to have access with or without payment (FC35).

(13) A place on land shall be treated as "open to the air" notwithstanding that it is covered if it is open to the air on at least one side.

(14) The Secretary of State may, by order, apply the provisions of this Part which apply to refuse to any description of animal droppings in all or any prescribed circumstances subject to such modifications as appear to him to be necessary. (See the Litter (Animal Droppings) Order 1991.)

(15) Any power under this section may be exercised differently as respects different areas, different descriptions of land or for different circumstances.'

(sect. 86)

Statutory undertaker

Statutory undertaker means:

(a) any person authorized by any enactment to carry on any railway, light railway, tramway or road transport undertaking;
(b) any person authorized by any enactment to carry on any canal, inland navigation, dock, harbour or pier undertaking; or
(c) any relevant airport operator (within the meaning of Part V of the Airports Act 1986) (sect. 98(6)).

The position of Railtrack in relation to this definition is unclear and may be determined by the courts.

LITTER OFFENCES

References

Environmental Protection Act 1990 sects. 87 and 88.
The Litter (Fixed Penalty Notices) Order 1991.
The Litter (Fixed Penalty) Order 1996.
The Litter (Fixed Penalty) Order 2002.

Litter offence

Any person who throws down, drops or otherwise deposits litter and leaves anything which causes the defacement of a place by litter is guilty of an offence (sect. 87(1)).

No offence is committed if the deposit etc., of the litter was either:

(a) authorized by law, or
(b) done with the consent of the owner, occupier, etc., having control of the place where the deposit occurred (sect. 87(2)).

The places in which the litter offence may be committed are:

(i) any public open space;
(ii) any relevant highway or relevant road and any trunk road which is a special road;
(iii) any place on relevant land of a principal litter authority;
(iv) any place on relevant Crown land;
(v) any place on relevant land of any designated statutory undertaker;
(vi) any place on relevant land of any designated educational institution;
(vii) any place on relevant land within a litter control area of a local authority.

Public open space means a place in the open air to which the public are entitled or permitted to have access without payment; and any covered place

FC35 Litter offences

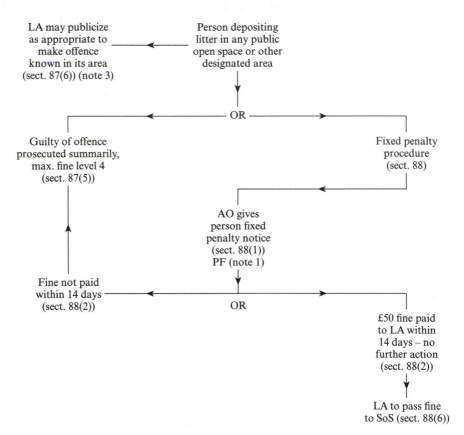

Notes
1. The notice to be used is prescribed in the Schedule to the Litter (Fixed Penalty Notices) Order 1991.
2. In this procedure, LA is used as an abbreviation for litter authority which is defined in the text.
3. Publicity is not a prerequisite to action under sects. 87 or 88.
4. For controls over the fouling of land by dogs, see FC42.

open to the air on at least one side and available for public use shall be treated as a public open place (sect. 87(3) and (4)).

Fixed penalty schemes

As an alternative to proceeding summarily for the littering offence, an AO of a litter authority may serve a fixed penalty notice on the offender (sect. 88(1)). The form of this notice is prescribed in the Litter (Fixed Penalty Notices) Order 1991 and requires the payment of the fine (£50 from 1 April 2002) to the LA within 14 days (sect. 88(2)).

If paid the LA passes the fine to the SoS and no further action ensues (sect. 88(2) and (6)). If not paid the LA may prosecute summarily for the offence (sects. 87(1) and 88(2)).

Definitions

Litter authority

for the purpose of this procedure is:

(a) any principal litter authority, other than an English county council or a joint board;
(b) any English county council or joint board designated by the SoS, by order (no such orders have been made), in relation to such area as is specified in the order (not being an area in a National Park);
(c) the Broads Authority (sect. 88(9)).

Authorized officer means an officer of a litter authority who is authorized in writing by the authority for the purpose of issuing notices under this section (sect. 88(10)).

LITTER CONTROL AREAS

References

Environmental Protection Act 1990 sect. 90.
Litter Control Areas Order 1991 (as amended 1997).

Scope

A PLA other than an English county council, or joint board may designate any land within its area, which is of a description set out by the SoS, as a litter control area (sect. 90(3)).

Such a designation does not in itself impose additional duties on land-owners or occupiers but does give the PLA the power to serve written abatement notices (FC38).

FC36 Litter control areas

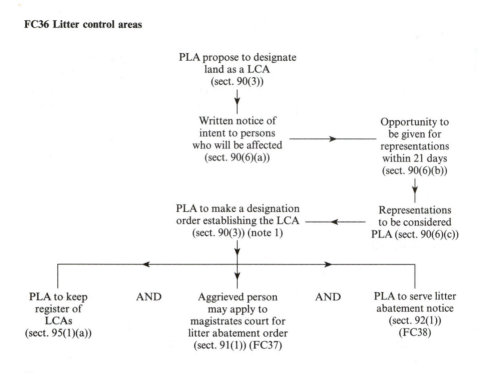

PLA propose to designate
land as a LCA
(sect. 90(3))

Written notice of
intent to persons
who will be affected
(sect. 90(6)(a))

Opportunity to
be given for
representations
within 21 days
(sect. 90(6)(b))

PLA to make a designation
order establishing the LCA
(sect. 90(3)) (note 1)

Representations
to be considered
PLA (sect. 90(6)(c))

PLA to keep
register of
LCAs
(sect. 95(1)(a))

AND

Aggrieved person
may apply to
magistrates court for
litter abatement order
(sect. 91(1)) (FC37)

AND

PLA to serve litter
abatement notice
(sect. 92(1))
(FC38)

Notes

1. The form of the designation is specified in the Schedule to the Litter Control Areas Order 1991.
2. For controls over the fouling of land by dogs, see FC42.

Type of land to be covered by a litter control area

The Litter Control Areas Order 1991 details the following land which may be designated as a LCA:

(a) car parks to which the public are entitled or permitted to have access;

(b) land forming a retail shopping development other than the land within that development which is retail floorspace or ancillary space used directly with retail floorspace;

(c) land to which the public are entitled or permitted to have access, which is open to the air, and which forms part of a business or office park or an industrial or trading estate;

(d) land used as a cinema, theatre, concert hall, bingo hall, casino, dance hall, swimming bath, skating rink, gymnasium or area for other indoor or outdoor sports or recreations, or as an amusement arcade or centre. (NB. The DETR advised that this includes fairgrounds, show-grounds and a circus.)

(e) any part of an inland beach or the seashore (that is to say every cliff, bank, barrier, dune, beach, flat or other land adjacent to and above the place to which the tide flows at mean high water springs) which is:

　(i) frequently used by large numbers of people, and

　(ii) managed by the person having direct control of it as a tourist resort or recreational facility;

(f) any esplanade or promenade which is above the place to which the tide flows at mean high water springs;

(g) land which is, or is part of, an aerodrome licensed under Part IX of the Air Navigation Order 1989, other than an aerodrome operated by a relevant airport operator within the meaning of Part V of the Airports Act 1986;

(h) land which is, or is part of, a marina, or other similar recreational boating facility and is above the place to which the tide flows at mean high water springs, other than an area used solely for repairing boats;

(i) land which is, or is part of, a motorway service station;

(j) land to which the public are entitled or permitted to have access, which is open to the air, and which is under the direct control of:

　(i) a parish or community council or parish trustees,

　(ii) an urban development corporation established under Part XVI of the Local Government, Planning and Land Act 1980,

　(iii) a new town development corporation established under section 3 of the New Towns Act 1981 or section 2 of the New Towns (Scotland) Act 1968,

　(iv) the Development Board for Rural Wales,

　(v) the Commission for the New Towns,

　(vi) an authority established under section 10 of the Local Government Act 1985 (waste disposal authorities),

(vii) a joint authority established by Part IV of the Local Government Act 1985 (police, fire services, civil defence and transport),

(viii) a residuary body established under section 57(1) of the Local Government Act 1985 or any body established pursuant to an order under section 67 of that Act (successors to residuary bodies),

(ix) a housing action trust established under section 62 of the Housing Act 1988,

(x) the Broads Authority,

(xi) a National Park authority, or

(xii) a health service body as defined in section 60(7) of the National Health Service and Community Care Act 1990 or an NHS trust established under section 5 of that Act or under section 12A of the National Health Service (Scotland) Act 1978;

(k) land on which a market is held, other than land forming part of a highway or, in Scotland, a public road;

(l) land forming, or forming part of, a camping or caravan site (including a mobile home site) which is used for more than 28 days in one year;

(m) a trunk road picnic area provided by the Minister under section 112 of the Highways Act 1980 or, in Scotland, by the Secretary of State under section 55 of the Roads (Scotland) Act 1984, or a picnic site provided by a local planning authority under section 10(2) of the Countryside Act 1968 or, in Scotland, a picnic place provided by an islands or district council or a general or district planning authority under section 2(2)(a)(i) of the Local Government (Development and Finance) (Scotland) Act 1964.

Land as respects which section 89(1)(a) to (f) of the Environmental Protection Act 1990 imposes a duty on public bodies and statutory undertakers may not be designated as, or as part of, a litter control area.

Designation

The PLA must be satisfied that the designation of an area as a LCA is necessary because the condition of the land is, and without the LCA will continue to be, detrimental to the amenities of the locality by reason of the presence of litter or refuse (sect. 90(4)). Consideration may include fouling by dog faeces (page 149), but see FC42

Before making the designation order the PLA must:

(a) notify in writing persons who will be affected;

(b) allow them to make representations within 21 days of the notice being served; and

(c) take account of any representations (sect. 90(6)).

The designation must be in the form set out in the Schedule to the Litter Control Areas Order 1991.

PLA powers within a LCA

Within the designated area the PLA (other than English county councils and joint boards) may serve litter abatement notices (FC38) (sect. 92(1)(d)).

Aggrieved persons (the public) may also complain to a magistrates' court in relation to conditions within a LCA under the procedure in FC37 (sect. 91(1)(g)), with a view to a litter abatement order being made.

Public register

The PLA must keep a register showing all orders designating LCAs (sect. 95(1)(a)).

The register must be available to the public free of charge at all reasonable times, and the PLA must provide copies of entries on payment of reasonable charge (sect. 95(4)).

The register can be in documentary or computerized form (sect. 95(5)).

LITTER ABATEMENT ORDERS

References

Environmental Protection Act 1990 sects. 89 and 91.
Code of Practice on Litter and Refuse issued June 1999 by the DETR under sect. 89 of Environmental Protection Act.

Duty to keep land clear of litter

The following are placed under a duty to keep their relevant land (definition on page 150) clear of litter and refuse, as far as is practicable:

(a) local authorities as respects highways;
(b) the SoS as respects trunk roads or other highways for which he is responsible;
(c) PLAs as respects their relevant land;
(d) the Crown as respects its relevant land;
(e) designated statutory undertakers as respects their relevant land;
(f) governing bodies as designated educational institutions as respects their relevant land; and
(g) occupiers of land within litter control areas (FC36 (sect. 89(1)).

In respect of (a) and (b), each LA and the SoS is also required to keep highways and roads clean, as far as is practicable (sect. 89(2)).

There is no definition of 'litter' or 'refuse' but, through the Litter (Animal Droppings) Order 1991, the word 'refuse' includes dog faeces on certain land (page 149) but also see FC42 for the controls over the fouling of land by dogs.

FC37 Litter abatement orders

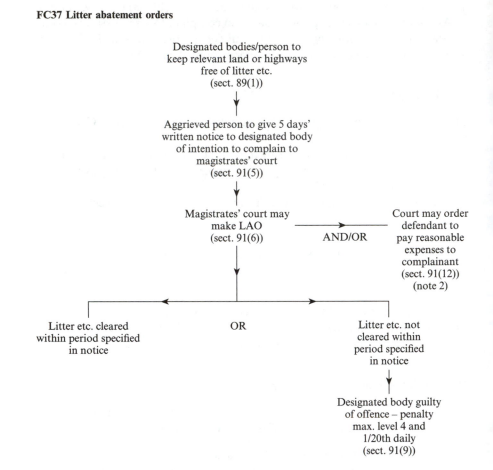

Designated bodies/person to
keep relevant land or highways
free of litter etc.
(sect. 89(1))

↓

Aggrieved person to give 5 days'
written notice to designated body
of intention to complain to
magistrates' court
(sect. 91(5))

↓

Magistrates' court may
make LAO
(sect. 91(6))   AND/OR   Court may order
defendant to
pay reasonable
expenses to
complainant
(sect. 91(12))
(note 2)

Litter etc. cleared   OR   Litter etc. not
within period specified   cleared within
in notice   period specified
                                         in notice

↓

Designated body guilty
of offence – penalty
max. level 4 and
1/20th daily
(sect. 91(9))

Notes
1. This procedure allows the public to bring to the attention of a court the fact that a responsible body is not complying with its responsibilities under sect. 89. For powers of a PLA in relation to other bodies, see FC38.
2. Even if a LAO is not made, the court may require the complainant to be compensated if, although the litter etc. had been cleared away, it had been present at the time the complaint was made and there were reasonable grounds for bringing the complaint (sect. 91(12)).

Litter standards

Regard must be had to the character and use of the land in question as well as to the measures which are practicable in the circumstances (sect. 89(3)).

The SoS has prescribed standards to which the responsible bodies must have regard in discharging their duties in a code of practice made under sect. 89(7). These establish what are reasonable and generally acceptable levels of cleanliness to be met and are admissible in court in any proceedings for a LAO (sect. 91(11)).

Litter abatement order

Any person aggrieved (other than a PLA) by the defacement by litter or refuse of any of the land set out under 'Duty to keep land clear of litter' (above), or where appropriate is aggrieved by a highway not being clear, may complain to a magistrates' court (sect. 91(1) and (2)).

The person must give the responsible person or body 5 days' written notice of his intention to make the complaint.

If the Court is satisfied that the land or highway is defaced by litter and refuse or the highway is not clear and that the person has not complied with his statutory duty, taking due account of the code of practice, it may make a LAO (sect. 91(6)).

Effect of a LAO

The LAO requires the litter or refuse to be cleared away or cleaning to be undertaken within a specified period failing which the person or body to whom the order is addressed is guilty of an offence (sect. 91(9)).

Payment of expenses

Whether or not the court makes a LAO, it may order the defendant authority to pay expenses to the complainant where it feels that there were reasonable grounds for bringing the complaint and that the defacement or uncleanliness existed at the time the complaint was made (sect. 91(12)).

LITTER ABATEMENT NOTICES

References

Environmental Protection Act 1990 sects. 89 and 92.
Code of Practice on Litter and Refuse issued in June 1999 by the DETR under sect. 89 EPA.

FC38 Litter abatement notices

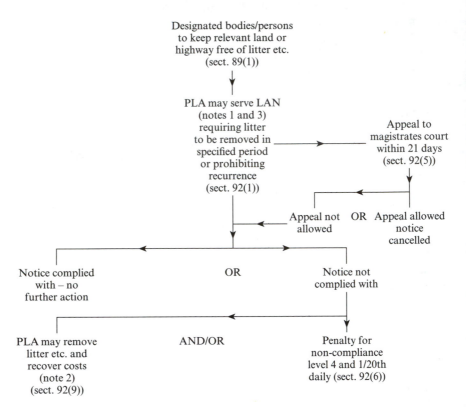

Notes
1. The power to serve LANs is given to PLAs other than English county councils and joint boards (sect. 92(1)).
2. The PLAs' powers of default may not be exercised in respect of relevant Crown land or relevant land of statutory undertakers (sect. 92(10)).
3. For the service of notices under the EPA see page 92.

Duty to keep land clear of litter

Designated bodies/persons are placed under a duty to keep their land or highways free of litter or certain highways clean, as far as is reasonably practicable and subject to guidance in the code of practice (sect. 89).

These duties are outlined more fully under litter abatement orders on page 159.

Relationship of litter abatement orders and litter abatement notices

LAOs are made by a magistrates' court against the designated body/person (including PLAs) upon complaint by an aggrieved person other than a PLA. LANs are served by PLAs (other than an English county or joint boards) against any other designated bodies/persons.

Designated bodies/persons other than PLAs are therefore subject to action for non-compliance with their duty to keep land etc. free of litter and clean both from certain PLAs and the public.

Scope

Where a PLA (other than an English county or joint board) is satisfied that any of the following land is defaced by litter or refuse, or that such deface-ment is likely to recur to the extent that the statutory duty under sect. 89(1) is breached, it may serve a LAN.

The land is any relevant land (definition on page 150) of:

(a) the Crown;
(b) a designated statutory undertaker;
(c) a designated educational institution; or
(d) within a litter control area (sect. 92(1)).

The words 'litter' and 'refuse' are not defined but through the Litter (Animal Droppings) Order 1991 'refuse' includes dog faeces on certain land (page 149) but FC 42 deals with controls over the fouling of land by dogs.

Litter abatement notices

The notice may either require the litter etc. to be cleared within a specified period or prohibit the land from becoming defaced again (sect. 92(2)). Fail-ure to comply brings a maximum penalty of level 4 and one-twentieth daily, and the PLA has powers of default and cost recovery (sect. 92(6) and (9)).

Notices are served on:

(a) the Crown authority;
(b) the designated statutory undertaker;
(c) the education authority and in all other cases; on
(d) the occupier of the land, or where the land is unoccupied, the owner (sect. 92(3)).

Appeals

Appeal is to the magistrates' court within 21 days of the date of service and compliance with the duty under sect. 89(1) constitutes a statutory defence against the LAN (sect. 92(4) and (5)).

STREET LITTER CONTROL NOTICES

References

Environmental Protection Act 1990 sects. 93, 94 and 95.
The Street Litter Control Notices Order 1991 (amended 1997).

Scope

PLAs (other than an English county or joint board) may, in order to prevent accumulations of litter or refuse in and around any street or land in the open air adjacent to any street, serve SLNs.

The circumstances which must appertain before a notice can be issued are:

(a) the premises are one of a description which has been prescribed by the minister (see below); and
(b) the premises must have a street frontage; and either
(c) there is recurrent defacement by litter or refuse of the street or open land in the vicinity; or
(d) the condition of open land in the vicinity of the premises is detrimental to the amenities of the neighbourhood; or
(e) the conditions at the premises produce quantities of litter or refuse which causes defacement of the street or open land in the vicinity of the premises (sect. 93(1) and (2)).

Through the Street Litter Control Notices Order 1991 made under sect. 94(1) of the EPA the Minister has designated the following descriptions of land and premises which may be the subject of SLNs.

Prescribed commercial and retail premises

A street litter control notice may be issued in respect of commercial and retail premises of the following descriptions:

(a) premises used wholly or partly for the sale of food or drink for consumption off the premises;
(b) premises used wholly or partly for the sale of food or drink for consumption on a part of the premises forming open land adjacent to the street;
(c) service stations and other premises on which fuel for motor vehicles is sold to the public;

FC39 Street litter control notices

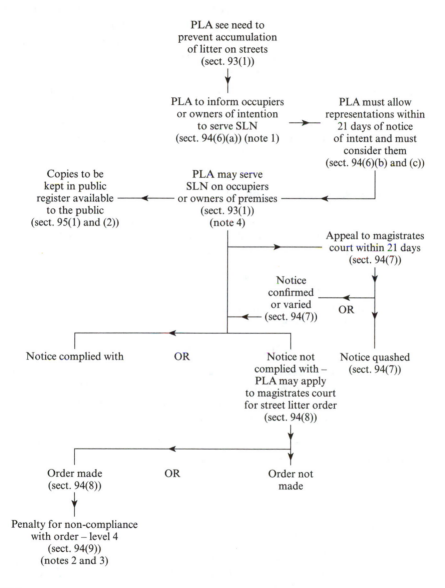

PLA see need to
prevent accumulation
of litter on streets
(sect. 93(1))

PLA to inform occupiers
or owners of intention
to serve SLN
(sect. 94(6)(a)) (note 1)

PLA must allow
representations within
21 days of notice
of intent and must
consider them
(sect. 94(6)(b) and (c))

Copies to be
kept in public
register available
to the public
(sect. 95(1) and (2))

PLA may serve
SLN on occupiers
or owners of premises
(sect. 93(1))
(note 4)

Appeal to magistrates
court within 21 days
(sect. 94(7))

Notice
confirmed
or varied
(sect. 94(7))

OR

Notice complied with

OR

Notice not
complied with –
PLA may apply
to magistrates court
for street litter order
(sect. 94(8))

Notice quashed
(sect. 94(7))

Order made
(sect. 94(8))

OR

Order not
made

Penalty for non-compliance
with order – level 4
(sect. 94(9))
(notes 2 and 3)

Notes
1. This notification does not need to be in writing but it is suggested that this should be the case.
2. No daily penalty is specified.
3. There is no provision for action in default and recharging.
4. For the service of notices under the EPA see page 92.

(d) premises used wholly or partly as a cinema, theatre, concert hall, bingo hall, casino, dance hall, swimming bath, skating rink, gymnasium or area for other indoor or outdoor sports or recreations, or as an amusement arcade or centre;

(e) banks, building society offices or other premises with automated teller machines located on an outside wall of the premises;

(f) premises in respect of which there is for the time being in force a betting office licence granted under Schedule 1 to the Betting, Gaming and Lotteries Act 1963(c);

(g) premises used wholly or partly for the sale of tickets or changes in any lottery; or

(h) premises used wholly or partly for the sale of goods of any description which are displayed on open land adjacent to the street, or on the street.

Prescribed descriptions of land

Land which may be included in an area of open land specified in a street litter control notice is land which is part of the premises in respect of which the notice is issued and, subject to below, land of the following descriptions:

(a) land which is part of a street, other than a carriageway when it is open to vehicular traffic;

(b) relevant land of a principal litter authority; and

(c) land under the direct control of any other local authority.

The land described above may be specified:

(a) in a street litter control notice issued in respect of premises described in category (e) above if the land is within 10 metres of those premises;

(b) in a street litter control notice issued in respect of any other premises, if the land is within 100 metres of the premises.

The words 'litter' and 'refuse' have not been defined but through the Litter (Animal Droppings) Order 1991, the word 'refuse' includes dog faeces on certain land (page 149) but see FC 42 for controls over the fouling of land by dogs.

Consultation

Before serving the SLN, the PLA must notify those on whom notices are to be served and consider any representation made by them within 21 days of that notification (sect. 94(6)).

Responsible person

Notices are to be served on the occupier of the premises or, where the premises are unoccupied, on the owner (sect. 93(2)).

Content of notices

The notice **must**:

(a) specify the premises;
(b) state the grounds upon which the notice is issued;
(c) specify all areas of open land adjoining or in the vicinity of the frontage;
(d) specify in relation to that area the PLA's reasonable requirements (sect. 93(3)).

The requirements in (d) must relate to the clearing of litter and refuse from the specified area of open land and may include:

(a) the provision and emptying of litter receptacles;
(b) the carrying out of actions within a specified period;
(c) regular actions at such times and intervals as specified (sect. 94(4)).

In forming its requirements, the PLA must have regard to its own duties under Part 4 of the EPA including those under sect. 85 in respect of its own land and highways (sect. 94(5)).

Appeals

Recipients of an SLN may appeal to the magistrates' court who may confirm, vary or quash the notice (sect. 94(7)).

Enforcement

If any requirement of the SLA is not complied with, the PLA may apply to the magistrates' court for an order which, if made, will require the person to comply with the requirement(s) within a specified time (sect. 94(8)).

Failure to comply with the order is an offence punishable by a maximum penalty of level 4 (sect. 94(9)).

Public register

The PLA must keep a register including copies of all SLNs (and associated court orders). The register must be available to the public free of charge at all reasonable times and the PLA must provide copies of entries on payment of a reasonable charge (sect. 95(1) and (2)).

Definition

Street means a relevant highway, a relevant road or any other highway or road over which there is a right of way on foot (sect. 93(4)).

ABANDONED TROLLEYS

Reference

Environmental Protection Act 1990 sect. 99 and Sch. 4.

Adoption of scheme

This procedure is discretionary and the powers may only be used after the scheme has been adopted by resolution of the LA (sect. 99(1)).

Consultation

Before passing a resolution adopting the scheme the LA must consult with persons likely to be affected or their representatives, e.g. supermarket operators (sect. 99(3)).

There is also a duty on the LA once a scheme has been adopted to consult the same people from time to time on the operation of the scheme (sect. 99(4)).

Publicity

Following the resolution of adoption the scheme must be publicized in at least one local newspaper (sect. 99(2)).

Voluntary schemes

The LA is able to reach separate agreements with each owner of trolleys whereby the operators agree to arrange collection themselves outside of the statutory scheme. In such cases, if abandoned trolleys are collected by the LA, for any trolleys removed by them within the times specified in the statutory scheme no charge is payable (Sch. 4, para. 4(2)).

Trolleys

The scheme applies to both shopping and luggage trolleys defined as:

Luggage trolley means a trolley provided by a person carrying on an undertaking of any railway, light railway, tramway, road transport undertaker or airport operator to travellers for use by them for carrying their luggage to, from or within the premises used for the purpose of the undertaking, not being a trolley which is power-assisted.

Shopping trolley means a trolley provided by the owner of a shop to customers for use by them for carrying goods purchased at the shop, not being a trolley which is power-assisted (Sch. 4, para. 5).

FC40 Abandoned trolleys

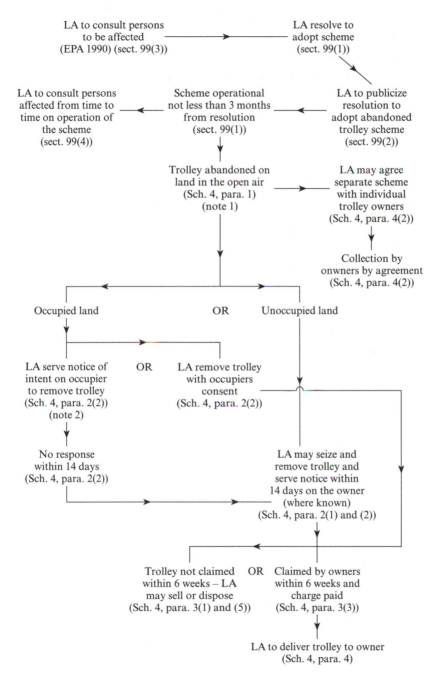

Notes
1. For exceptions, see text.
2. For service of notices under EPA 1990, see page 92.

Abandoned trolleys

Trolleys which may be dealt with by the LA under a scheme are those found to be abandoned in the open air except:

(a) on land in which the trolley owner has a legal estate;
(b) in a trolley-park in an off-street parking place;
(c) in any other trolley park designated by the LA; or
(d) on land used for the purposes of the undertaking by the owner of the luggage trolley (Sch. 4, para. 1).

Removal

On unoccupied land the LA may remove the trolley immediately. However, on occupied land a notice of intention must be served on the occupier who has 14 days to respond. In the absence of a response within that period the LA may remove the trolley (Sch. 4, para. 2(2)).

Retention, return and disposal

Once seized and removed, the LA must serve notice within 14 days on the owner notifying him of the removal of the trolley, the place where it is being kept and notification that, if not claimed within 6 weeks, it will dispose of it (Sch. 4, para. 3(2)).

If claimed within the 6 weeks, the trolley is delivered back to its owner by the LA and a charge made (Sch. 4, para. 3(3) and (4)).

If not claimed within the 6-week period and the LA has made reasonable enquiries to ascertain ownership, the LA may sell or otherwise dispose of the trolley (Sch. 4, para. 3(1) (b) and (5)).

Charges

Charges to be levied by the LA should be such as to recover the costs of removing, storing and disposing of trolleys under the scheme (Sch. 4, para. 4(2)).

Local authority

The LAs with the ability to adopt and operate these schemes are:

(a) district councils;
(b) London borough councils;
(c) the Common Council of the City of London;
(d) the Council of the Isles of Scilly;
(e) in Scotland a council constituted under sect. 2 of the Local Government etc. (Scotland) Act 1994, i.e. unitary authorities;
(f) in Wales, the council of a county or county borough, i.e. unitary authorities.

DOGS

SEIZURE OF STRAY DOGS

References

Environmental Protection Act 1990 sects. 149 and 150.
Dogs Act 1906 (as amended).
The Environmental Protection (Stray Dogs) Regulations 1992.
The Control of Dogs Order 1992.

Extent

This procedure applies in England, Wales and Scotland but not in Northern Ireland where there is separate legislation (sect. 164(4)).

Duty on LAs

LAs must appoint an officer to deal with stray dogs (sect. 149(1)).
There are no particular powers of entry provided for any authorized LA officers.

Scope

Where the AO finds a stray dog in a public place or on any other land or premises he must seize it and detain it provided, in the case of other than a public place, the owner or occupier agrees (sect. 149(3)). Dogs not wearing collars with the name and address of the owner may be seized and treated as a stray dog under this procedure (Control of Dogs Order 1992).

A member of the public may take possession of a stray dog and must then take it either to the LA or to the police. Unless the finder wishes to keep the dog, the LA is required to deal with it by the procedure in the chart. The police are required to use a similar procedure under the Dogs Act 1906.

Using a procedure set out in sect. 27 of the Road Traffic Act 1988 and the Control of Dogs on Roads Order (Procedure) (England and Wales) Regulations 1995, LAs may require dogs on designated roads to be held on leads – see FC43.

Notification to owner

Where the AO knows the dog's owner or this is shown on its collar, the AO must give that person notice in writing that he will dispose of the dog unless it is claimed within 7 days and the LAs expenses are paid (sect. 149(4)).

FC41a Seizure of stray dogs by LAs

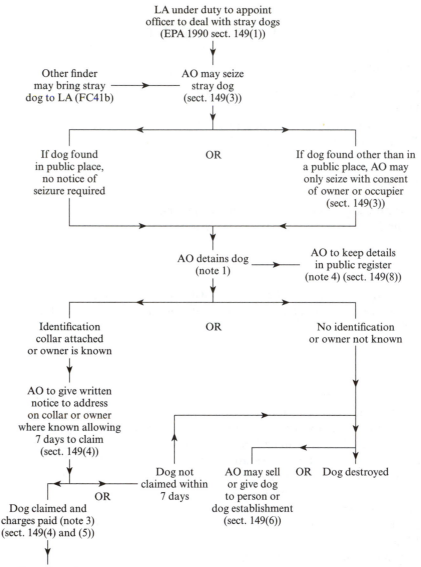

Notes

1. The AO may arrange for the dog to be destroyed immediately if this is necessary to avoid suffering (sect. 149(10)).
2. The charges to be made are controlled by the Environmental Protection (Stray Dogs) Regulations 1992.
3. The details to be kept in the register are prescribed in the Environmental Protection (Stray Dogs) Regulations 1992.
4. For the control of dogs on roads (leash orders), see FC43.
5. For the control over the fouling of land by dogs, see FC42.

FC41b Seizure of stray dogs by the public

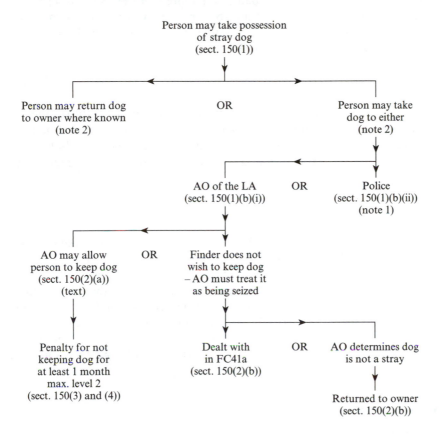

Notes

1. Dogs taken to the police are to be dealt with by them in accordance with the provisions of the Dogs Act 1906 (as amended) which contains a procedure similar to that exercised by LAs (FC41a).
2. Having taken possession of a stray dog, the public must either bring it to the owner, the LA or the police – maximum penalty for not doing so level 2 (sect. 150(1) and (4)).
3. For the control of dogs on roads (leash orders) see FC43.

Identification of dog owners

The Control of Dogs Order 1992, which is enforced by LAs, requires dogs, with certain exceptions, e.g. pack hounds, whilst on a highway or in a public place to wear a collar with the name and address of the owner inscribed on it or on a place attached to it. If not so worn, the owner commits an offence under the Animal Health Act 1981 under which LAs under the EPA have been given power to prosecute (EPA 1990 sect. 151).

Detention of dogs

The LA must provide or arrange for the detention of dogs and for them to be properly fed and confined. Costs involved in the kennelling of dogs are recoverable from owners who require return of the animal (sect. 149(9)) and additionally the owner is required to pay a sum prescribed by regulations, currently £25 (Reg. 2).

Dog to be kept by finder

The finder may indicate to the AO of the LA a wish to keep the dog. The AO must record specific details and contact the dog's owner where this proves to be possible. The owner may elect to collect it. However, where the owner is not known or does not wish to collect the dog, the AO can allow the finder to keep the dog if he is satisfied that the person is fit and proper and is able to feed and care for it. The AO must inform the finder both verbally and in writing that he must keep the dog for at least 1 month and that failure to do so is a criminal offence (sect. 150(2) and Environmental Protection (Stray Dogs) Regulations 1992 Reg. 4).

Disposal of dogs

The AO may dispose of the dog by either:

    (a) giving it or selling it to:
        (i) any person who will care for it;
        (ii) an establishment for the reception of stray dogs; or
    (b) destroying it so as to cause as little pain as possible.

No dog may be sold or given away for vivisection (sect. 149(6)).

Definitions

Local authority means in England:

    (a) a district council;
    (b) a London borough council;
    (c) the Common Council of the City of London;
    (d) the Council of the Isles of Scilly;

in Wales, a county or county borough council; and in Scotland a council constituted under sect. 2 of the Local Government etc. (Scotland) Act 1994, i.e. unitary authorities.

Public place means:

(a) as respects England and Wales, any highway and any other place to which the public are entitled or permitted to have access;
(b) as respects Scotland, any road (within the meaning of the Roads (Scotland) Act 1984) and any other place to which the public are entitled or permitted to have access (sect. 149(11)).

CONTROLS OVER THE FOULING OF LAND BY DOGS

References

The Dogs (Fouling of Land) Act 1996.
The Environmental Protection Act 1990 sect. 88.
The Dogs (Fouling of Land) Regulations 1996.
The Dog Fouling (Fixed Penalties) Order 1996.
The Dog Fouling (Fixed Penalty) (England) Order 2002.
DoE Circular 18/96 2 December 1996.

NB. The general provisions of the EPA at page 92 do not apply to this procedure.

Extent

The Act applies in England and Wales but not in Scotland or Northern Ireland (sect. 8).

Scope

By the use of designation orders LAs are able, at their discretion, to create an offence by a person in charge of a dog of not cleaning the faeces forthwith after the dog has defecated on any land designated in the order (sects. 2 and 3).

The LAs empowered to make designation orders are unitary and district councils in England and Wales (sect. 7).

Land to which orders may be applied

The Act applies to any land in the open air, including covered land which is open on at least one side, and to which the public have access (with or without payment) (sect. 1(1) and (5)).

There are, however, certain types of land which are exempt from this general definition:

FC42 Controls over fouling of land by dogs

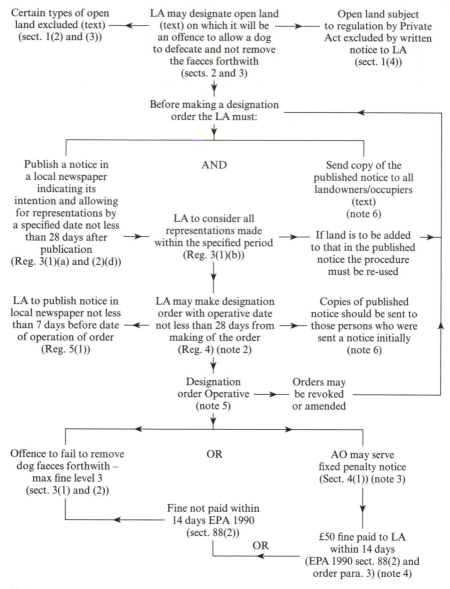

Certain types of open land excluded (text) (sect. 1(2) and (3)) ← LA may designate open land (text) on which it will be an offence to allow a dog to defecate and not remove the faeces forthwith (sects. 2 and 3) → Open land subject to regulation by Private Act excluded by written notice to LA (sect. 1(4))

↓

Before making a designation order the LA must:

Publish a notice in a local newspaper indicating its intention and allowing for representations by a specified date not less than 28 days after publication (Reg. 3(1)(a) and (2)(d)) → **AND** LA to consider all representations made within the specified period (Reg. 3(1)(b)) → Send copy of the published notice to all landowners/occupiers (text) (note 6) / If land is to be added to that in the published notice the procedure must be re-used →

↓

LA to publish notice in local newspaper not less than 7 days before date of operation of order (Reg. 5(1)) ← LA may make designation order with operative date not less than 28 days from making of the order (Reg. 4) (note 2) → Copies of published notice should be sent to those persons who were sent a notice initially (note 6)

↓

Designation order Operative (note 5) → Orders may be revoked or amended

↓

Offence to fail to remove dog faeces forthwith – max fine level 3 (sect. 3(1) and (2)) **OR** AO may serve fixed penalty notice (Sect. 4(1)) (note 3)

Fine not paid within 14 days EPA 1990 (sect. 88(2)) ← / **OR** ← £50 fine paid to LA within 14 days (EPA 1990 sect. 88(2) and order para. 3) (note 4)

Notes
1. Regulation numbers refer to Dogs (Fouling of Land) Regulations 1996.
2. The form of designation orders is set out in the schedule of the Dog (Fouling of Land) Regulations 1996.
3. The form of fixed penalty notice is prescribed in the schedule of the Dog Fouling (Fixed Penalties) Order 1996.
4. The fine is to be passed to the SoS (EPA 1990 sect. 88(6)).
5. Orders may subsequently be amended or revoked, presumably (it is not made clear) by the operation of the full procedure (sect. 2(1)).
6. This action is recommended (not required) where the number of persons is small.
7. For the seizure of stray dogs see FC41a and b.
8. For the control of dogs on roads see FC43.

(a) carriageways with a speed limit of more than 40 mph and land which runs alongside;
(b) land used for agriculture (definitions) or woodlands used as such;
(c) land which is predominantly marshland, moor or heath;
(d) rural common land; and
(e) land where powers of regulation are conferred by a Private Act after written notice has been given to the LA to exclude the application of this Act (sects. 1(2)–(4)).

Where the Act does not apply it may be possible for the LA to seek local dog fouling byelaws (see below).

Byelaws

Prior to this Act, LAs dealt with the problem of requiring dog faeces to be removed by the use of 'poop scoop' byelaws. With the availability of this procedure existing byelaws cease to have effect when any land is designated and, in any event where no designation orders have been made, on 17 August 2006. For land to which the Act does not apply, byelaws will continue beyond the 10-year period unless they are specifically revoked (sect. 6).

Defining the land to be designated

LAs are able to designate land either specifically or by description. This means that every area to be covered does not have to be separately listed or defined. For example the areas may be clearly marked on a map (sects. 2(2) and Reg. 3(2)).

Procedure before making a designation order

The LA must first publish a notice (at least once) in a local newspaper circulating in the area where the land is to be designated. The notice must:

(a) identify the land;
(b) state the effect of the proposed designation;
(c) where a map is used to describe the land, say where the map may be inspected; and
(d) state that written representation must be made within a defined period (not less than 28 days) (Reg. 3).

The LA are recommended to send a copy of the notice to all landowners and occupiers, where the numbers are small.

Designation orders

All valid representations must be considered by the LA before proceeding. The LA in finalizing its proposal may delete land previously identified in the notice but if land is to be added the publicity procedure must be undertaken again (Regs. 3(1)(b) and 4).

The form of designation order is prescribed in the schedule to the regulations and orders must be sealed in accordance with the LAs standing orders.

After the making of a designation order the LA is required not less than 7 days before it becomes operative to:

(a) publish a notice in a local newspaper circulating in the area of the land to be covered stating that the order has been made;
(b) make available copies of the order at a place identified in the notice (Reg. 5).

LAs are recommended to send a copy of the published notice to any affected landowners or occupiers to whom a copy of the initial public notice was sent.

Once made designation orders may subsequently be revoked or amended (sect. 2(2)). LAs should follow the same procedure.

Offences

An offence is committed if a person in charge of a dog does not clear up forthwith after the dog has defecated on any land which is the subject of a designation order. A person who habitually has a dog in his possession is taken to be in charge of the dog at any time unless at the time of the offence some other person is in charge.

However, an offence is not committed if:

(a) the person has reasonable excuse for failing to remove the faeces; or
(b) the owner, occupier or a person having control of the land gave consent to the failure to remove the faeces.

Placing the faeces in a receptacle on the land provided for the purpose is acceptable but there is no requirement on the LA or anyone else to provide such bins. No offences are committed by blind persons.

The maximum penalty on conviction is level 3 on the Standard Scale (sect. 3(2)).

Fixed penalty notices

As an alternative to prosecuting for the offence an AO of the LA may serve a fixed penalty notice on the offender using the procedure set out in sect. 88 of the EPA 1990 (which deals with fixed penalties for litter offences). The form of notice is prescribed in the Schedule to the Dog Fouling (Fixed Penalties) Order 1996. Using this procedure requires the payment of a fine of £50 (From 1 April 2002) to be made to the LA within 14 days. If the fine is not paid within that period the LA may prosecute for the offence. Fines paid must be passed by the LA to the SoS.

Authorized officers are defined (see below) in such a way that they may be either employees of the LA or contractors working for them (sect. 4).

Powers of entry

There are no powers of entry for authorized officers provided in respect of this procedure.

Definitions

Agriculture means horticulture, fruit growing, seed growing, dairy farming or livestock breeding and keeping, and the use of land as grazing land, meadow land, osier land, market gardens and nursery ground (sect. 1(6)).

Authorized officer in relation to a LA means any employee of the authority who is authorized in writing by the authority for the purpose of serving notices under this section (fixed penalty notices). Reference to any employee of the authority includes reference to:

(a) any person by whom, in pursuance of arrangements made with the authority, any functions relating to the enforcement of this Act fall to be discharged; and
(b) any employee of any such person (sects. 4(4) and (5)).

CONTROL OF DOGS ON ROADS (LEASH ORDERS)

References

Road Traffic Act 1988, sect. 27.
The Control of Dogs on Roads Orders (Procedure) (England and Wales) Regulations 1995.
NB. The general provisions of the EPA 1990 at page 92 do not apply to this procedure.

Extent

These provisions apply in England, Wales and Scotland.

Scope

LAs (see definitions) may make Orders which create an offence where a person causes or permits a dog to be on a designated road (i.e. a road specified in the Order) without being held on a lead (sect. 27(1)–(3)).
There are no powers of entry provided in relation to this procedure.

FC43 Control of dogs on roads (leash orders)

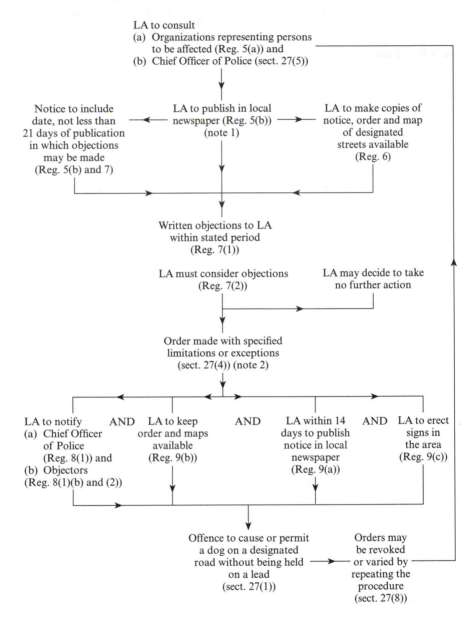

LA to consult
(a) Organizations representing persons
　　　to be affected (Reg. 5(a)) and
(b) Chief Officer of Police (sect. 27(5))

Notice to include date, not less than 21 days of publication in which objections may be made (Reg. 5(b) and 7)

LA to publish in local newspaper (Reg. 5(b)) (note 1)

LA to make copies of notice, order and map of designated streets available (Reg. 6)

Written objections to LA within stated period (Reg. 7(1))

LA must consider objections (Reg. 7(2))

LA may decide to take no further action

Order made with specified limitations or exceptions (sect. 27(4)) (note 2)

LA to notify
(a) Chief Officer of Police (Reg. 8(1)) and
(b) Objectors (Reg. 8(1)(b) and (2))

AND

LA to keep order and maps available (Reg. 9(b))

AND

LA within 14 days to publish notice in local newspaper (Reg. 9(a))

AND

LA to erect signs in the area (Reg. 9(c))

Offence to cause or permit a dog on a designated road without being held on a lead (sect. 27(1))

Orders may be revoked or varied by repeating the procedure (sect. 27(8))

Notes
1. The form of notice is prescribed in part 1 of the Schedule to the Regulations.
2. The form of notice is prescribed in part 2 of the Schedule to the Regulations.
3. For the seizure of stray dogs, see FC33a and b.
4. For the control of fouling of land by dogs, see FC 42.

Consultation

Before making an Order the LA must consult:

(a) the Chief Officer of Police; and
(b) any organizations representing persons likely to be affected;

and must also publish at least once a notice in a local newspaper giving details of the proposed Order as set out in part 1 to the Schedule of the Regulations (sect. 27(5) and Reg. 5).

Inspection of documents

During the period within which objections may be made (see below) the LA must make available, at least at the principal offices during normal office hours, the following documents:

(a) the published notice;
(b) a copy of the Order;
(c) a map clearly showing all lengths of road to which the Order relates (Reg. 6).

Objections

Written objections to the proposed Order may be made within the period to be specified by the LA in the published notice (not less than 21 days from the date of its first publication). Objections must state the grounds on which they are made. The LA must consider all objections duly made before reaching its decision (Reg. 7 and Schedule Part 1).

Orders

If the LA decides to proceed, Orders may be subject to such limitations and exceptions as specified in the Order and will indicate by a map the designated roads (these may relate to the whole or parts of the LA area). Dogs which are:

(a) kept for driving or tending sheep in the course of a trade or business; or
(b) under proper control for sporting purposes;

are excepted from the operation of any Order (sect. 27(4)).

Orders may subsequently be varied or revoked using the same procedure (sect. 27(8)).

Notification of the making of an order

The LA decision must be notified in writing to:

(a) the Chief Officer of Police; and
(b) any person who made a valid objection which has not been withdrawn. In this case reasons for the decision must be given (Reg. 8).

In addition the LA must:

(a) within 10 days publish at least once in a local newspaper a notice containing the information specified in part 2 to the Schedule;
(b) keep available at the principal offices a copy of the Order and the map showing the designated roads; and
(c) erect signs on or near the designated roads to ensure that adequate information is given to persons likely to be affected (Reg. 9).

Local authorities

The authorities with the power to make these Orders are:

(a) in England, county councils, unitary councils (but not other districts), London boroughs and the Common Council;
(b) in Wales, county and county borough councils; and
(c) in Scotland, the unitary authorities (sect. 27(7)).

CROP RESIDUES

BURNING OF CROP RESIDUES

References

Environmental Protection Act 1990 sect. 152.
The Crop Residues (Burning) Regulations 1993.

Extent

The procedure applies in England, Wales and Scotland but not in Northern Ireland (sect. 164).

Scope

There is a prohibition on the burning of the following crop residues:

(a) cereal straw;
(b) cereal stubble;
(c) residues of:
  (i) oil-seed rape;
  (ii) field beans harvested dry; and
  (iii) peas harvested dry.

(Reg. 4 and Sch. 1)

Exemptions

The burning of the crop residues specified above is allowed however for the purposes of:

FC44 Burning of crop residues

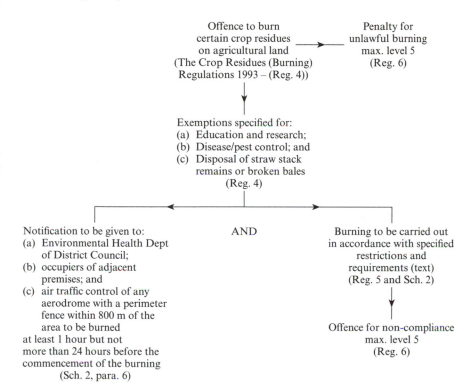

Offence to burn
certain crop residues
on agricultural land
(The Crop Residues (Burning)
Regulations 1993 – (Reg. 4))

Penalty for
unlawful burning
max. level 5
(Reg. 6)

Exemptions specified for:
(a) Education and research;
(b) Disease/pest control; and
(c) Disposal of straw stack
remains or broken bales
(Reg. 4)

AND

Notification to be given to:
(a) Environmental Health Dept
of District Council;
(b) occupiers of adjacent
premises; and
(c) air traffic control of any
aerodrome with a perimeter
fence within 800 m of the
area to be burned
at least 1 hour but not
more than 24 hours before the
commencement of the burning
(Sch. 2, para. 6)

Burning to be carried out
in accordance with specified
restrictions and
requirements (text)
(Reg. 5 and Sch. 2)

Offence for non-compliance
max. level 5
(Reg. 6)

(a) education and research;
(b) disease control or the elimination of plant pests where a notice has been served under article 22 of the Plant Health (Great Britain) Order 1993;
(c) the disposal of straw stack remains or broken bales (Reg. 4).

Following a review of the exemptions from these regulations in 1996, MAFF announced that the non-inclusion of linseed residues would continue. Farmers are, however, urged to follow the safety rules on allowable burning.

Restriction on allowable burning

Where any of the specified crop residues are to be burned this must be carried out in strict compliance with the following restrictions and requirements:

1. No crop residue may be burned:
   (a) during the period between one hour before sunset and the following sunrise; or
   (b) on any Saturday, Sunday or bank holiday.
2. No crop residue may be burned if the area to be burned extends, in the case of cereal straw or cereal stubble, to more than 10 hectares, and in any other case to more than 20 hectares.
3. No crop residue may be burned unless:
   (a) the area to be burned is surrounded by a fire-break, which borders on that area and which, in the case of cereal straw or cereal stubble, shall be at least 10 metres wide and in any other case at least 5 metres wide;
   (b) any building, structure or other thing mentioned in paragraph 4(b) or (c) below which lies within the area to be burned is surrounded by a fire-break of the relevant width referred to in sub-paragraph (a) above which borders on any crop residues which are to be burned; and
   (c) in the case of any land ('intervening land') between a fire-break to be established in accordance with sub-paragraph (a) or (b) above and any other land or any building, structure or other thing mentioned in paragraph 4(b), (c) or (d) below which lies within the relevant distance there mentioned of the area to be burned or, as the case may be, of any crop residues within that area:
      (i) the intervening land is cleared of all crop residues; or
      (ii) all crop residues on the intervening land are incorporated into the soil before burning takes place.
4. No crop residue may be burned:
   (a) if the area to be burned is less than 150 metres from any other area in which crop residues are being burned;

(b) in the case of cereal straw or cereal stubble, less than 15 metres, and in any other case less than 5 metres from:
  (i) the trunk of any tree (including any tree in coppice or scrubland);
  (ii) any hedgerow;
  (iii) any fence not the property of the occupier of the land upon which the burning is carried out;
  (iv) any pole which is or may be used to carry telegraph or telephone wires;
  (v) any electricity pole, pylon or substation;
(c) in the case of cereal straw or cereal stubble, less than 50 metres, and in any other case less than 15 metres from:
  (i) any residential building;
  (ii) any structure having a thatched roof;
  (iii) any building, structure, fixed plant or machinery which could be set alight or damaged by heat from the fire;
  (iv) any scheduled monument which could be set alight by the fire;
  (v) any stack of hay or straw;
  (vi) any accumulation of combustible material other than crop residues removed in the making of a fire-break;
  (vii) any mature standing crop;
  (viii) any woodland or land managed as a nature reserve;
  (ix) any building or structure containing livestock;
  (x) any oil or gas installation on or above the surface of the ground; or
(d) less than 100 metres from:
  (i) any motorway;
  (ii) any dual carriageway;
  (iii) any A-road;
  (iv) any railway line.
5. No crop residue may be burned unless all persons concerned in the burning operation are familiar with the provisions of the Regulations and, except where an emergency arising during the operation renders it impracticable, each area to be burned is supervised by at least two responsible adults, one of whom having experience of burning crop residues shall be in general control of the operation.
6. No crop residue may be burned unless there is available at the area being burned:
  (a) not less than 1000 litres of water in one or more mobile containers together with means of dispensing the water for fire-fighting purposes in a spray or jet at a rate of 100 litres per minute; and
  (b) not fewer than five implements suitable for use in fire-beating.
7. No crop residue may be burned unless every vehicle used in connection with the burning is equipped with a suitable and serviceable fire extinguisher.
8. No crop residue may be burned unless reasonable precautions have been taken to ensure that the fire will not cross a fire-break.

9. Ashes of burnt cereal straw or cereal stubble shall not, without reasonable excuse, be allowed to remain on the soil for longer than 24 hours after the time of commencement of the burning, but shall be incorporated into the soil:

(a) within that period; or

(b) in a case where to do so would be likely, having regard to wind conditions, to cause nuisance, as soon as conditions allow.

(Sch. 2)

Notification of intended burning

Before any allowable burning takes place and so far as is reasonably practicable, notice must be given to:

(a) the Environmental Health Department of the District Council where the burning is to take place;

(b) occupiers of any adjacent premises; and

(c) the air traffic control of any aerodrome with a perimeter fence within 800 metres of the area to be burned.

Such notice is to be given at least 1 hour or not more than 24 hours before the commencement of the burning (Sch. 2, para. 6).

Power of entry

No such powers are provided in respect of this procedure.

Definition

Crop residue means straw or stubble or any other crop residue remaining on the land after harvesting of the crop grown thereon (Reg. 2(1)).

CONTAMINATED LAND

THE IDENTIFICATION AND REMEDIATION OF CONTAMINATED LAND

References

Environmental Protection Act 1990 Part 2A (inserted by sect. 57 of the Environment Act 1995).

The Contaminated Land (England) Regulations 2000.

The Environmental Information Regulations 1992.

Statutory Guidance DETR Circular 02/2000: Environmental Protection Act 1990: Part 2A Contaminated Land.

FC45 Identification of contaminated land

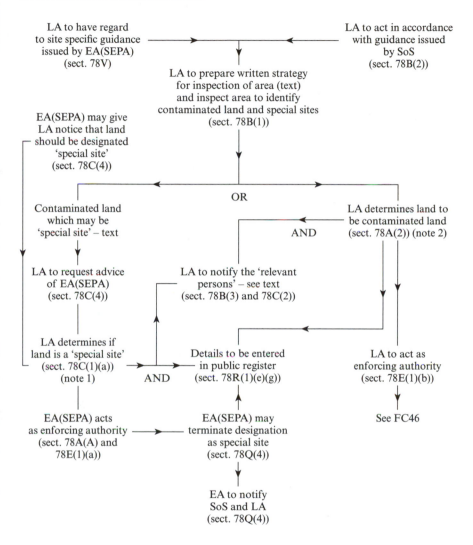

LA to have regard to site specific guidance issued by EA(SEPA) (sect. 78V)

LA to act in accordance with guidance issued by SoS (sect. 78B(2))

LA to prepare written strategy for inspection of area (text) and inspect area to identify contaminated land and special sites (sect. 78B(1))

EA(SEPA) may give LA notice that land should be designated 'special site' (sect. 78C(4))

OR

Contaminated land which may be 'special site' – text

AND

LA determines land to be contaminated land (sect. 78A(2)) (note 2)

LA to request advice of EA(SEPA) (sect. 78C(4))

LA to notify the 'relevant persons' – see text (sect. 78B(3) and 78C(2))

LA determines if land is a 'special site' (sect. 78C(1)(a)) (note 1)

AND

Details to be entered in public register (sect. 78R(1)(e)(g))

LA to act as enforcing authority (sect. 78E(1)(b))

EA(SEPA) acts as enforcing authority (sect. 78A(A) and 78E(1)(a))

EA(SEPA) may terminate designation as special site (sect. 78Q(4))

See FC46

EA to notify SoS and LA (sect. 78Q(4))

Notes

1. Disputes between the LA and EA(SEPA) about the designation of special sites is referred to the SoS whose decision must be notified to the relevant persons as well as to the parties involved (sect. 78D).
2. This information is subject to the Environmental Information Regulations 1992 subject to commercial confidentiality.

FC46 Remediation of contaminated land

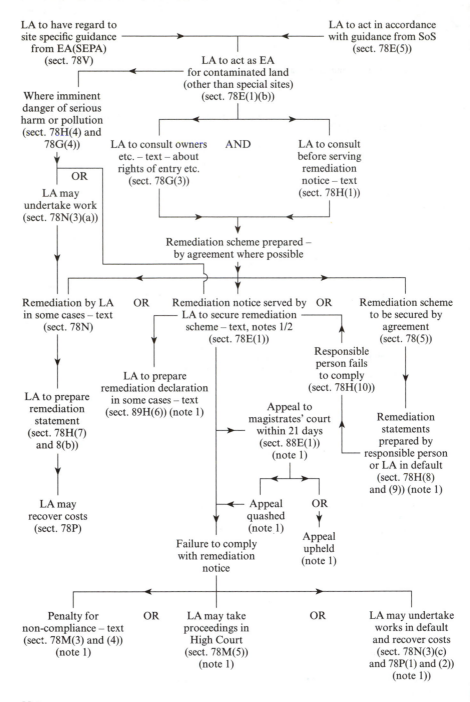

LA to have regard to
site specific guidance
from EA(SEPA)
(sect. 78V)

LA to act in accordance
with guidance from SoS
(sect. 78E(5))

LA to act as EA
for contaminated land
(other than special sites)
(sect. 78E(1)(b))

Where imminent
danger of serious
harm or pollution
(sect. 78H(4) and
78G(4))

LA to consult owners
etc. – text – about
rights of entry etc.
(sect. 78G(3))

AND

LA to consult
before serving
remediation
notice – text
(sect. 78H(1))

OR

LA may
undertake work
(sect. 78N(3)(a))

Remediation scheme prepared –
by agreement where possible

Remediation by LA
in some cases – text
(sect. 78N)

OR

Remediation notice served by
LA to secure remediation
scheme – text, notes 1/2
(sect. 78E(1))

OR

Remediation scheme
to be secured by
agreement
(sect. 78(5))

Responsible
person fails
to comply
(sect. 78H(10))

LA to prepare
remediation
statement
(sect. 78H(7)
and 8(b))

LA to prepare
remediation declaration
in some cases – text
(sect. 89H(6)) (note 1)

Appeal to
magistrates' court
within 21 days
(sect. 88E(1))
(note 1)

Remediation
statements
prepared by
responsible person
or LA in default
(sect. 78H(8)
and (9)) (note 1)

LA may
recover costs
(sect. 78P)

Appeal
quashed
(note 1)

OR

Appeal
upheld
(note 1)

Failure to comply
with remediation
notice

Penalty for
non-compliance – text
(sect. 78M(3) and (4))
(note 1)

OR

LA may take
proceedings in
High Court
(sect. 78M(5))
(note 1)

OR

LA may undertake
works in default
and recover costs
(sect. 78N(3)(c)
and 78P(1) and (2))
(note 1))

Notes
1. Details to be included in remediation register.
2. Remediation notices may be used several times on the same site if necessary to produce a staged result or to react to changing assessments and circumstances.

BS 10175:2001 Investigation of potentially contaminated sites – code of practice.

Scope

These procedures provide for the identification of contaminated land and formal determination by LAs. They also provide for what is to be done by way of remediation, which may be required for any contaminated land where the LA is the enforcing authority i.e. all except special sites that are the responsibility of the EA. The guidance identifies three categories of remediation action i.e. assessment, remedial treatment and monitoring in paras. C64–C71.

Local authorities

The LAs with the duty to carry out inspections and enforce the provisions relating to remediation (other than for special sites) are:

(a) in Greater London, the London Borough Councils, the City of London and the Temples;
(b) in the rest of England, the Borough or District Councils or, where there is none (unitary areas) the County Council;
(c) in Scotland, the Unitary Councils; and
(d) in Wales, the Unitary Councils (sect. 78A(9)).

The role of the LAs is to:

(a) Prepare a written strategy regarding inspection of their area (sect. 78B(1) and para. B12 of statutory guidance)
(b) Identify any potentially contaminated land (sect. 78B(1)) and any special sites (sect. 78C(1) and regs. 2 and 3)
(c) Determine whether land is contaminated land (sect. 78A(2)–(5) and paras. B37–51 of statutory guidance)
(d) Establish who is the owner of the land (sect. 78A(9)), who appears to be in occupation, who is the appropriate person to bear responsibility for any necessary remediation action (sects. 78E(1) and 78F)
(e) Notify such persons and the EA that land is determined as contaminated and establish special sites (sect. 78B(3) and (4))
(f) Require appropriate remediation defined in sect. 78A(7), implement under sects. 78E, 78H and 78N. Effect urgent remediation and establish where remediation is the responsibility of the EA for processes covered by the PPC Regulations 2000
(g) Maintain a register (sect. 78R) in accordance with reg. 15 and schedule 3 of the 2000 regulations.

In undertaking these roles the LAs are required to act in accordance with guidance issued by the SoS (sect. 78B(2)) and to have regard to any site specific advice given to it by the EA (SEPA in Scotland) (sect. 78V).

Contaminated land

This is defined as:

'any land which appears to the LA in whose area it is situated to be in such a condition by reason of substances in, on or under the land that:

(a) significant harm is being caused or there is a significant possibility of such harm being caused or:
(b) pollution of controlled waters is being, or is likely to be, caused' (sect. 78A(2)).

In this context 'harm' means harm to the health of living organisms or other interference with the ecological systems of which they form part and, in the case of man, includes harm to his property (sect. 78A(4)).

Sect. 78(5) provides that the following are to be determined by the LA in accordance with the statutory guidance (Chapter A parts 3 and 4):

(a) What harm is to be regarded as significant (part 3 paras. A23–A26 and table A)
(b) Whether the possibility of significant harm is significant (part 3 paras. A27–A34 and table B)
(c) Whether pollution of controlled waters is being, or is likely to be caused (part 4 paras. A35–A39).

Controlled waters are defined in sect. 78A(9) by reference to part 3 sect. 104 of the Water Resources Act 1991 and this embraces territorial and coastal waters, inland fresh waters and groundwaters.

Special sites

Wherever the LA has determined land to be contaminated land, it must also decide whether it meets the description of special sites prescribed for the purposes of sect. 78C(8). Regs. 2 and 3 and schedule 1 of the 2000 Regs. and identify special sites for which the EA is the enforcing authority.

If the LA considers that the land might be designated a special site, it should seek the advice of the EA (sect. 78C(8)). The EA also needs to consider whether any contaminated land should be designated as a special site. This might be based on information from its other pollution control functions. There is a duty on the EA to notify the LA (sect. 78C(4)). Where the contaminated land meets one or more of the prescribed descriptions, the LA must designate it a special site. Disputes between a LA and the EA are to be settled by the SoS.

Having made such a designation, the LA must give written notice to:

(a) the EA
(b) the owner of the land
(c) any person who appears to be the occupier of all or part of the land and
(d) each person who appears to be an appropriate person (see below).

Following this process the EA and not the LA is responsible for securing any necessary remediation (sects. 78C, 78D and 78E(1)).

The identification of contaminated land

Part 3 of chapter B of the statutory guidance sets out the inspection duty. The LA has sole responsibility for determining whether land is contaminated and it cannot delegate this except in accordance with sect. 101 of the LGA 1972. The duty to inspect under sect. 78B(1) is extended by the statutory guidance to include a strategic approach. Every LA must therefore have set out its approach to this duty in a written strategy and published it by July 2001 (chapter B paras. B12 and B15). This approach should enable the LA to identify, in a rational, ordered and efficient manner, the land which merits detailed individual inspection, identifying the most pressing and serious problems first and concentrating resources on the areas where contaminated land is most likely to be found (chapter B para. B9).

Paras. B18–25 of the statutory guidance cover detailed inspections and use of powers of entry. If land has been determined as contaminated land and is likely to be designated a special site, the LA should always make arrangements for the EA to carry out the inspection on its behalf. Sect 108 of the EA 1995 (see page 220) can also be used to authorize EA staff.

The National House Building Council has also published guidance in NHBC Standards chapter 4.1 'Land Quality: managing ground conditions'. This is to be used by all NHBC registered builders to investigate sites for development and involves consultation with the LAs.

BS 10175:2001 provides guidance for LAs on investigation techniques, sampling and on-site testing and laboratory analysis. It supersedes DD 175:1988 which has been withdrawn.

To enable a LA to make a judgement regarding significant harm (sect. 78A(2)) scientific guidance was published by DEFRA in March 2002. This consists of the Contaminated Land Exposure Assessment methodology (CLEA) together with accompanying reports C7–10. The key element of CLEA is the soil guidance values (SGVs) which represent 'intervention' levels above which unacceptable risks may exist and requires further investigation or remediation.

Appropriate persons

These are the persons responsible for any remediation of contaminated land that the LA may require. Part 2A of the Act defines two categories of 'appropriate persons' and sets out the circumstances in which they might be liable.

The first category is created by sect. 78F(2) – 'any person or any persons who caused or knowingly permitted the substances, or any of the substances, by which the contaminated land in question is such land, to be in, or under that land, is an appropriate person'. These are referred to as Class A persons by the statutory guidance. Such a person will be the appropriate person only in respect of any remediation, which is referable to particular substances, which he caused or knowingly permitted to be in, on or under the land (sect. 78F(3)).

The second category arises where it is not possible to find a Class A person. In these circumstances the owner (definitions) or occupier for the time being of the land is an appropriate person, referred to as a Class B person. Occupier is not defined but DETR circular 02/2000 states that this would normally be the person in occupation e.g. tenant or licensee.

The statutory guidance in part 3 of chapter D guides LAs in the circumstances where two or more appropriate persons are liable for remediation. It allows a LA to determine who should bear the liability and identifies five distinct phases in this procedure.

Remediation (see definition page 195)

Having identified land as contaminated land the LA has a statutory duty to ensure that appropriate remediation is carried out (sect. 78E(1)(b)).

It is the Government's intention that as far as possible remediation should be carried out by agreement rather than through the use of the formal notice procedures. In any event, the LA is required to consult with 'appropriate persons' (see above), owners etc. about what is to be done by way of remediation before serving any remediation notices (sect. 78H(1)) and to produce a remediation scheme.

From a practical point of view, remediation is likely to be phased with different remediation actions being required at different times, e.g. assessment actions, remedial treatment actions and monitoring actions.

The LA must have regard to the standard of remediation i.e. it should result in land being 'suitable for use'. Paras. C17–C28 of the statutory guidance give general advice on the best practicable techniques, multiple pollution linkages, volunteered remediation and to the circumstances where land is no longer contaminated.

The practicability, effectiveness and durability of remediation are governed by sect. 78E(5)(b) and paras. C44–C63 of the guidance set out the general criteria to meet these objectives.

Sect. 78E(5)(c) requires that regard should be had to the reasonableness of remediation. Paras. C29–C33 set out the criteria for this. Paras. C34–C37 set out the matters which the LA should take into account when considering the costs involved and compensation that is to be paid in accordance with sect. 78G(5) involving compliance with the remediation notice.

Urgent remediation action

Where there is imminent danger of serious harm or serious pollution of controlled water being caused the LA can either:

(a) serve an urgent remediation notice without going through the normal consultation and other procedural requirements (sect. 78H(4)); or

(b) where an urgent remediation notice would not result in remediation happening soon enough, carry out the urgent remediation itself and recover its costs where possible (sect. 78H(7)).

Depending upon the circumstances the LA may take such action either before any remediation work has commenced by normal procedures or during the course of such work.

If the LA carries out the work itself it must produce a remediation statement describing the actions it has carried out (sect. 78H(7)).

'Imminent' and 'serious' are not defined in part 2A of the Act. The statutory guidance states that a LA needs to judge each case on the normal meaning of the words and the facts of the case. Paras. C39–C43 guide LAs with regard to the seriousness of any significant harm. A22 A34 and Tables A and B assist in the interpretation of significant harm.

Remediation notices

Where it has not been possible to secure a remediation scheme by agreement, the LA is required to serve a remediation notice on each 'appropriate person' (see above) specifying what needs to be done by way of remediation and the time periods within which each action must be taken (sect. 78E(1) and reg. 4).

Reg. 5 requires that copies of the notice are sent at the same time to those persons required to be consulted under sects. 78G(3) and 78H(1) and to the EA.

Remediation notices may be served on different persons for action relating to different substances (sect. 78E(2)) and, where served on more than one person, the notice must state the proportion of the costs to be borne by each person (sect. 78E(3)).

In specifying the works to be undertaken the LA must consider any work to be reasonable having regard to:

(a) the costs; and

(b) the seriousness of the harm or of the pollution (sect. 78E(4)).

In determining what is to be done by way of remediation the LA must also have regard to the standard to which land is to be remediated and to what is regarded as reasonable (sect. 78E(5)). Remediation should not be required for the purpose of achieving any aims other than those set out in paras. C18–C24 of the guidance. In particular, not for dealing with matters which do not themselves form part of a significant pollution linkage or making the land suitable for any uses other than its current use.

Remediation declarations

Where the LA has identified works which could be carried out but is precluded from including them on a remediation notice because of the 'cost' and 'seriousness' tests above, it must produce and publish a remediation declaration which indicates:

  (a) the work in question;
  (b) why the LA would have otherwise specified that work; and
  (c) the grounds on which the LA feels justified in not specifying the work on the notice (sect. 78H(6)).

Such declarations must be included in the remediation register (sect. 78R(1)(c)).

Appeals

The grounds for appeal against a remediation notice are set out in reg. 7 and the procedures involving the magistrates' court in reg. 8. The person has 21 days to appeal and, where it is duly made, the notice is suspended pending the decision. In these circumstances the LA needs to consider whether it should carry out urgent remediation itself using its powers under sect. 78N. Appeals against any decision of the magistrates' court lie in the High Court (reg. 13).

Offences

Persons failing to comply with a remediation notice are liable, on summary conviction, to a fine not exceeding level 5 and to 1/5th of level 5 for each day on which the offence continues.

Where the land concerned is industrial, trade or business premises, the maximum fine is increased to £20 000 or 1/10th daily.

As an alternative to proceeding summarily, the LA may go to the High Court when it feels the former would not secure an effectual remedy (sect. 78M).

In addition the LA may itself undertake the works in default and recover costs (sect. 78N(3)(c)).

Remediation statements

Where, without the service of a remediation notice, remediation is to take place, the person who is to carry it out (including the LA where this is the case) is required to prepare and publish a remediation statement which records:

  (a) the things which have been, are being or are to be done;
  (b) the name and address of the person undertaking the works; and
  (c) the time which the work is expected to take (sect. 78H(7) and (8)).

These statements must be entered in the remediation register (sect. 78(1)(c)).

Where the requirement for a remediation statement is not met by the person concerned, the LA may produce it and recover its costs in so doing (sect. 78H(9)).

Remediation registers

Each LA must maintain a remediation register which will contain details of formal notices and other specified documents in relation to each area of contaminated land for which the LA is responsible (sect. 78R(1) and reg. 15).

Certain information is excluded from this requirement where this is in the interest of national security or it is commercially confidential (sects. 78S and 78T).

The register is available for inspection at the LA's principal offices at all reasonable times (sect. 78R(8)(a)) and there must be a facility for members of the public to obtain copies of entries at a reasonable charge (sect. 78R(8)(b)).

Schedule 3 of the regulations prescribes full particulars of the following matters to be included in the register by the LA:

(a) remediation notices
(b) appeals against remediation notices
(c) remediation declarations
(d) appeals against charging notices
(e) designation of special sites
(f) notification of claimed remediation
(g) convictions for offences under sect. 78M
(h) guidance issued under sect. 78V(1) and
(i) other environmental controls.

Definitions

Contaminated land see page 190.
Remediation means:
(a) the doing of anything for the purpose of assessing the condition of:
　(i) the contaminated land in question;
　(ii) any controlled waters affected by that land; or
　(iii) any land adjoining or adjacent to that land;
(b) the doing of any works, the carrying out of any operations or the taking of any steps in relation to any such land or waters for the purpose:
　(i) of preventing or minimizing, or remedying or mitigating the effects of, any significant harm, or any pollution of controlled waters, by reason of which the contaminated land is such land; or
　(ii) of restoring the land or waters to their former state; or

(c) the making of subsequent inspections from time to time for the purpose of keeping under review the condition of the land or waters;

(sect. 78A(7))

Controlled waters see page 190.

Local authority see page 189.

Notice means a notice in writing.

Owner, in relation to any land in England and Wales, means a person (other than a mortgagee not in possession) who, whether in his or her own right or as trustee for any other person, is entitled to receive the rackrent of the land, or, where the land is not let at a rackrent, would be so entitled if it were so let.

Owner, in relation to any land in Scotland, means a person (other than a creditor in a heritable security not in possession of the security subjects) for the time being entitled to receive or who would, if the land were let, be entitled to receive the rents of the land in connection which the work is used and includes a trustee, factor, guardian or curator and in the case of public or municipal land includes the persons to whom the management of the land is entrusted (sect. 78A(9)).

Chapter 7

POLLUTION PREVENTION AND CONTROL ACT 1999
with the Pollution Prevention and Control (England and Wales) Regulations 2000

INTEGRATED POLLUTION PREVENTION AND CONTROL (IPPC)

Objective

The intention of the IPPC regime is to achieve a higher level of protection for the environment as a whole. It is also to meet the requirements of the EC Directive 96/61/EC by applying an integrated environmental approach for discharges to air, water (including sewers) and land; this also involves regard for energy conservation and site restoration which is linked to the contaminated land regime.

General

The existing procedures under Part 1 of the Environmental Protection Act 1990 (pages 93 to 116) are to be repealed and replaced by those in the Pollution Prevention and Control (England and Wales) Regulations 2000 (the PPC Regs.) made under the PPCA 1999.

This will create a coherent new framework to prevent and control pollution with two parallel regimes. The essence of IPPC is that operators should use best available techniques (BAT – definitions page 202) to control the impact of their installations (definitions page 201) on the environment as a whole. The first part of the framework encompasses the former IPC system enforced by the EA but extends the issues that regulators must consider alongside emissions (definitions) into areas such as energy use and site restoration and involves LAs as regulators for some activities. The regime relates to activities (definitions) at installations (definitions page 200) rather than processes under original part 1 of the EPA 1990. Mobile plant (definitions page 201) is also covered.

The second part of the framework is Local Air Pollution Prevention and Control (LAPPC) which is a continuation of the old LAAPC procedure (page 97). It is similar to IPPC from a procedural perspective but it still focuses on controlling emissions to air only and will continue to be enforced by LAs.

Activities covered

The activities covered by the regime are listed in schedule 1 of the Regulations. Each group of activities is divided into 3 parts – part A(1), to be dealt with by the EA under the IPPC regime, Part A(2), to be dealt with by LAs also under IPPC and part B to be dealt with by LAs under LAAPC. (Table 7.1). The schedule includes all activities that were previously authorized as Part B processes under the EPA 1990 (with some modifications) but also includes some new activities controlled for the first time.

Timetable for introduction

Operators of existing installations included in part A of schedule 1 are required to obtain permits according to the transitional timetable included in schedule 3 unless there are proposals for a substantial change (definitions page 200). For activities controlled by LAs this is outlined in table 7.1. Operators must apply for a permit for any new installations to be put into operation after 31 October 1999 or for existing installations subject to substantial change.

Table 7.1 Main IPPC activities and the transition timetable for existing installations to be dealt with by LAs – A(2) installations

Part A Activity Status	Relevant Section of Schedule 1 of PPC Regulations 2000	Relevant period for Part A(2) installation applications
Gasification, liquefaction and refining	1.2	1 June to 31 August 2006
Ferrous metals	2.1	1 May to 31 July 2003
Non-ferrous metals	2.2	1 May to 31 July 2003
Production of cement and lime	3.1	1 April to 30 June 2003
Other mineral activities	3.5	1 April to 30 June 2003
Manufacturing glass and glass fibre	3.3	1 May to 31 July 2003
Ceramic production	3.6	1 January to 31 March 2004
Paper, pulp and board manufacturing activities	6.1	1 April to 30 June 2003
Coating activities, printing and textile treatments	6.4	1 May to 31 July 2003
Activities involving rubber	6.7	1 April to 30 June 2003
Treatment of animal and vegetable matter and food industries	6.8	1 June to 31 August 2004

Note: Existing installations are those put into operation before 31/10/99. New installations are those put in after that date.

Permitting of Part B Activities

Operators must apply for a permit according to the timetable in schedule 3 para. 10 of the regulations. The earliest date is 1 April 2002 for combustion activities described in para. 1.1 of schedule 1. As a result of later amendment the timetable for the following activities has been deferred to the dates shown below:

1 April 2002 deferred to 1 April 2003

Combustion, incineration and cremation
Bulk handling and blending cement
Industrial finishing of asbestos and firing of clay goods
Crushing and grinding of minerals and timber treatments

1 April 2003 deferred to 1 April 2004

Ferrous and non-ferrous metals
Solvents and coatings, manufacture of printing ink and rubber
Surface treatment of metals and plastics

1 April 2004 deferred to 1 April 2005

Service stations, terminals, natural gas odorising and animal vegetable processes
Glass manufacture inc. glass fibre reinforced plastic and use of isocynates
Bulk chemical storage, tar and bitumen activities

GENERAL PROCEDURAL PROVISIONS

Extent

The procedures in this chapter apply only in England and Wales (reg. 1(2)). Separate systems will be introduced for Scotland and Northern Ireland.

Enforcement

Part A(1) installations are regulated by the EA and part A(2) by the relevant LA (see below). For both groups one body is a statutory consultee of the other. Part B installations are controlled by the LA. (reg. 8(1)–(4))

Local authorities

The authorities with responsibility for enforcement are:

(a) in Greater London, London borough councils, the Common Council of the City of London the Sub-Treasurer of the Inner Temple and the Under Treasurer of the Middle Temple
(b) in England outside Greater London, borough and district councils or, where there is a county council but no district council, the county council and the council of the Isles of Scilly
(c) in Wales, county councils or county borough councils (reg. 8(15))

Additionally port health authorities may be empowered by Order. (reg. 8(16)). All of these authorities are called regulators for the purpose of these procedures.

Definitions

Activities means activities of any nature, whether:

(a) industrial or commercial or other activities or
(b) are carried on in particular premises or otherwise, and includes (with or without other activities) the depositing, keeping or disposal of any substance.

Environmental pollution means pollution of the air, water or land which may give rise to any harm; and for the purposes of this definition (. . . . .)

(a) pollution includes pollution caused by noise, heat or vibrations or any other kind of release of energy, and
(b) air includes air within buildings and air within other natural or man-made structures above or below ground.

In this definition '**harm**' means:

(a) harm to the health of human beings or other living organisms,
(b) harm to the quality of the environment, including
    (i) harm to the quality of the environment taken as a whole,
    (ii) harm to the quality of the air, water or land,
    (iii) other impairment of, or interference with, the ecological systems of which any living organisms form part.
(c) offence to the senses of human beings,
(d) damage to property,
(e) impairment of, or interference with, amenities or other legitimate uses of the environment. (Expressions used in this paragraph have the same meaning as in Council Directive 96/61/EC)
(PPCA 1999 sect. 1(1))

Change in operation means, in relation to an installation or mobile plant, a change in the nature or functioning or an extension of the installation or mobile plant which may have consequences for the environment and **substantial change in operation means**, in relation to an installation or mobile plant, a change of operation which, in the opinion of the regulator, may have significant negative effects on human beings or the environment.
Emission means:

(i) in relation to part A installations, the direct or indirect release of substances, vibrations, heat or noise from individual or diffuse sources in an installation into the air, water or land;

(ii) in relation to part B installations, the direct release of substances or heat from individual or diffuse sources in an installation into the air;

(iii) in relation to part A mobile plant, the direct or indirect release of substances, vibrations heat or noise from the mobile plant into the air, water or land;

(iv) in relation to part B mobile plant, the direct release of substances or heat from the mobile plant into the air.

Emission limit value means the mass, expressed in terms of specific parameters, concentration or level of an emission, which may not be exceeded during one or more periods of time.

Installation means:

(i) a stationary technical unit where one or more activities listed in part 1 of schedule 1 are carried out; and

(ii) any other location on the same site where any other directly associated activities are carried out which have a technical connection with the activities carried out in the stationary technical unit and which could have an effect on pollution, and, other than in schedule 3, references to an installation include references to part of an installation.

Mobile plant means plant which is designed to move or to be moved whether on roads or otherwise and which is used to carry out one or more activities listed in part 1 of schedule 1.

Operator subject to (a) and (b) below means, in relation to an installation or mobile plant, the person who has control over its operation:

(a) Where an installation or mobile plant has not been put into operation, the person who will have control over the operation of the installation or mobile plant when it is put into operation shall be treated as the operator of the installation or mobile plant.

(b) Where an installation or mobile plant has ceased to be in operation, the person who holds the permit which applies to the installation or mobile plant shall be treated as the operator of the installation or mobile plant.

Pollution means emissions as a result of human activity which may be harmful to human health or the quality of the environment, cause offence to any human senses, result in damage to material property, or impair or interfere with amenities and other legitimate uses of the environment; and 'pollutant' means any substance, vibration, heat or noise released as a result of such an emission which may have such an effect.

Substance includes any chemical element and its compounds and any biological entity or micro-organism, with the exception of radioactive substances within the meaning of Council Directive 80/836, Euratom, genetically modified micro-organisms within the meaning of Council Directive 90/219 and genetically modified organisms within the meaning of Council directive 90/220 (PPC Regs. 2000 reg. 2(1)).

Best available techniques (BAT)

Operators are required to use BAT to achieve a high level of protection of the environment taken as a whole. BAT is defined in Regulation 3(1) of the PPC Regulations as 'the most effective and advanced stage in the development of activities and their methods of operation which indicates the practical suitability of particular techniques for providing in principle the basis for emission limit values designed to prevent and, where that is not practicable, generally to reduce emissions and the impact on the environment as a whole.'

It should be noted that management systems are an integral part of BAT. It is not sufficient for an operator to have adequate technical control. Operating staff must be properly trained and regulators should recognize and encourage environmental management systems, e.g. Eco-Management Audit Scheme (EMAS) and ISO 14001 Environmental Management Systems.

Guidance notes

To meet the requirements of BAT it will be normal practice to set permit conditions in accordance with Sectoral Technical Guidance (STG), which is based on the EC Best Available Technique Reference Documents (BREF). These STGs lay down, at national level, indicative BAT requirements. If the operator complies with these no further assessment is required. STGs will contain indicative standards for both new and existing installations and upgrading timetables. These documents will replace the PG notes of the IPC regime where the activity is to be subject to IPPC. Where the STG does not exist reference should be directly to the relevant BREF note. If the relevant BREF document has not been prepared the General Technical Guidance Note Version 2 issued 12.6.01 is designed to be used in conjunction with existing technical guidance.

Notices

Any notices must be given in writing (reg. 6(1)).

Notices may be served on or given to a person by leaving it at his proper address or by sending it by post to him at that address. In the case of a body corporate, the notice is served on the secretary or clerk. In the case of a partnership, it is served or given to a partner or person having the control or management of the partnership.

The proper address is the last known address except:

(a) in the case of a body corporate, it is the registered or principal office and

(b) in the case of a partnership, the principal office of the partnership.

(reg. 6(2)–(4))

Powers of entry

These are provided for in the EA 1995 as amended by the schedule 10 para. 16 of the PPC Regs. 2000 and are detailed on page 220.

Application to the Crown

Whilst the Crown is bound by the PPC Regs. it is not criminally responsible and cannot be prosecuted for non-compliance with notices etc. However, LAs may apply to the High Court to have the Crown's actions, or lack of them, declared illegal (reg. 5).

Obtaining of information

By notice served on any person, LAs may require to be given such information as is specified in the notice and in such form and time/period as is specified (reg. 28(2)).

PERMITTING OF ACTIVITIES BY LAs

References

Environment Act 1995
Pollution Prevention and Control Act 1999
Pollution Prevention and Control (England and Wales) Regulations 2000
Pollution Prevention and Control (England and Wales) (Amendment) Regulations 2002
Integrated Pollution Prevention and Control: A Practical Guide DETR August 2000 (rev. March 2001)

Scope

These procedures apply to activities, installations and mobile plant, prescribed in parts A(2) and B of schedule 1 to the PPC Regs. 2000 – see table 7.1. These activities etc. are controlled and permitted by the LA (see page 198). In each case the operator (definitions page 201) is required to obtain a permit from the LA before operating the installation or plant after the date prescribed for that activity in Table 7.1 (reg. 9).

Applications (FC47)

Applications are to be made in writing to the LA, with the prescribed fee (see page 216), and must contain all of the detailed information specified in para. 1(1–3) of part 1 of schedule 4. For part B activities the information required is less detailed than for part A activities (reg. 10(1)). For novel or complex installations staged applications are possible by agreement between the operator and the LA.

FC47 Permitting of scheduled activities by LAs

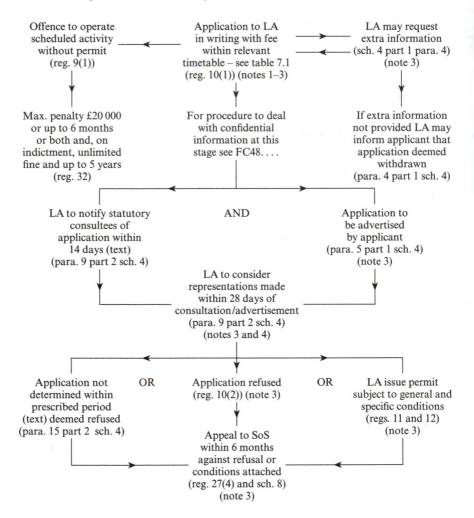

Notes

1. For details to be included see paras. 1–3 part 1 sch. 4.
2. For complex sites staged applications are possible, see Guide paras. 5.5–5.9.
3. Details of these events to be included in public register (reg. 29 and sch. 9).
4. LA to take account of Environmental Impact Assessment Directive 85/537/EEC as amended by 97/11/EEC.

LAs are empowered to request by notice additional information sufficient for them to determine the application. If the information requested has not been provided within the time specified in the notice the application is deemed to have been withdrawn (sched. 4 part 1 para. 4).

Publicity

In the case of part A installations and mobile plant, but subject to provisions concerning national security and confidentiality (FC48), and within a period of 28 days beginning 14 days after the application is made, the applicant must advertise the application in the London Gazette and in at least one newspaper circulating in the locality (para. 5 part 1 schedule 4). In addition the LA must, as soon as possible, place details of the application in the public register (see below) (sched. 9 para. 1(a)).

Consultations

Subject to matters of national security and confidentiality set out in part 3 of schedule 4 (see FC48), the LA must send copies of the application within 14 days of receipt to the statutory consultees listed in para. 9 of part 2 of schedule 4. Consultees are allowed 28 days to make representations and the LA must have regard to these before determining the application. The list of consultees is slightly different as between part A and part B activities (para. 12 part 2 schedule 4).

Determination of applications

The general principles against which all applications, parts A and B, are to be considered are:

(i) that all appropriate preventative measures must be taken against pollution and in particular the application of BAT (see page 202);
(ii) that no significant pollution will be caused (definitions page 201).
(reg. 11(2))

The additional general principles to be considered for part A activities are:

(a) that satisfactory methods will be in place to avoid waste production and, where it is produced, for its recovery and disposal;
(b) there is efficient use of energy;
(c) that measures are taken to prevent accidents and to limit their consequences;
(d) that, on the cessation of operation of the installation, satisfactory measures will be taken to avoid pollution and return the site to a satisfactory state.
(reg. 11(3))

The LA must determine applications within 4 months of submission unless a longer period is agreed. Failure to do so is deemed to be a refusal (para. 15 part 2 schedule 4).

HANDLING OF COMMERCIALLY CONFIDENTIAL INFORMATION BY LAs

FC48 Handling of commerically confidential information by LAs

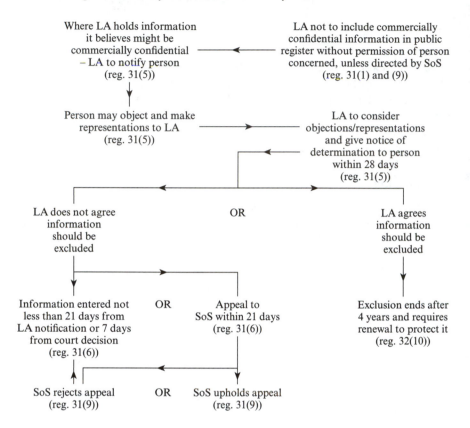

Permit with conditions

If the LA decides to issue a permit it must attach conditions appropriate to the following issues and in accordance with sectoral technical guidance. For all activities these are:

(a) Emission limit values (EMLs) (definitions) or equivalent parameters for pollutants, in particular in relation to those listed in schedule 5 and likely to be emitted in significant quantities. These will normally be based on BAT (see page 202) taking account of the particular characteristics and the local environment of the plant.
(b) There is an implied condition that the operator must use BAT for preventing or reducing emissions.
(reg. 12(10))

Additionally for part A activities conditions must cover:

(a) Long distance and transboundary pollution;
(b) The protection of soil and groundwater and the management of waste;
(c) Precautions to protect the environment when the installation is not operating normally e.g. during start up;
(d) Site monitoring and remediation;
(e) The ongoing monitoring of emissions and the submission of reports to the LA;
(f) Notification procedures to deal with incidents or accidents;
(g) Must take account of conditions for emissions to water specified by the EA; and
(h) Must avoid conflict with other legislation prescribing release levels e.g. Radioactive Substances Act 1993, Water Resources Act 1991 (discharges) etc.

For part B activities the additional conditions, if any, must relate to those considered necessary to ensure compliance with the implied condition above. (regs. 12(1)(c) and 13).

Operators are under a general duty to comply with the conditions of a permit (reg. 23).

General Binding Rules (GBRs)

The SoS is empowered to make GBRs for certain types of installation which can be used by a LA instead of site-specific conditions. GBRs will, by their nature, be suitable for industry sectors where installations share similar characteristics. The SoS has so far made none (reg. 14).

Refusals

A permit cannot be granted by the LA unless it considers that the applicant will be the person who will have control after the permit is issued and will ensure that the installation or mobile plant is operated so as to comply with the conditions which would be included in the permit (reg. 10(3)). This may be, for example, where there is reason to believe that the operator lacks the management systems or competence to run the installation according to the application or any permit conditions (page 27 of Guide).

There is an appeal procedure against refusal – see below.

Changes to permitted installations

Operators of permitted installations and mobile plant must notify the LA of their intention to make a 'change in operation' (definitions) and must do so in writing at least 14 days before the change is made. The LA must acknowledge receipt of the notification in writing (reg. 16).

Permit transfers (FC49)

Transfers of permits between operators must be the subject of a joint application to the LA who must agree to the transfer unless it considers that the conditions attached will not be complied with. Unless a longer period is agreed, the LA must determine the application within 2 months and failure to do so is deemed to be a refusal. There is an appeal against refusal (see below) (reg. 18).

Permit reviews

The LA must periodically review the conditions of permits and may do so at any time. Such reviews are mandatory where:

(a) the installation causes such significant pollution that the LA must change the ELVs;
(b) substantial changes in BAT make it possible to reduce emissions significantly without excessive costs and
(c) operators must change techniques for reasons of safety.
(reg. 15)

Variation of conditions (FC50)

The LA may vary the conditions at any time either on its own initiative or because of an application from the operator. The prescribed fee must accompany applications (see 'Charges' below). The LA must serve notice on the operator specifying the variations proposed having taken account of the issues listed under 'Conditions' above. There is an appeal against the change of conditions and against a refusal to grant an application for change (see below) (reg. 17).

Surrender of permits (FCs 51 and 52)

Where the operator of a part A activity intends to cease operation he must apply to the LA with the prescribed fee and information. The LA may

TRANSFER OF PERMITS

FC49 Transfer of permits

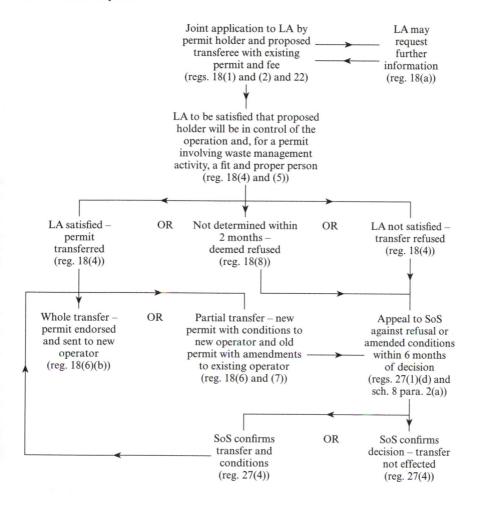

Joint application to LA by permit holder and proposed transferee with existing permit and fee (regs. 18(1) and (2) and 22) → LA may request further information (reg. 18(a))

↓

LA to be satisfied that proposed holder will be in control of the operation and, for a permit involving waste management activity, a fit and proper person (reg. 18(4) and (5))

LA satisfied – permit transferred (reg. 18(4)) **OR** Not determined within 2 months – deemed refused (reg. 18(8)) **OR** LA not satisfied – transfer refused (reg. 18(4))

Whole transfer – permit endorsed and sent to new operator (reg. 18(6)(b)) **OR** Partial transfer – new permit with conditions to new operator and old permit with amendments to existing operator (reg. 18(6) and (7)) → Appeal to SoS against refusal or amended conditions within 6 months of decision (regs. 27(1)(d) and sch. 8 para. 2(a))

SoS confirms transfer and conditions (reg. 27(4)) **OR** SoS confirms decision – transfer not effected (reg. 27(4))

VARIATIONS TO PERMIT CONDITIONS

FC50 Variations to permit conditions

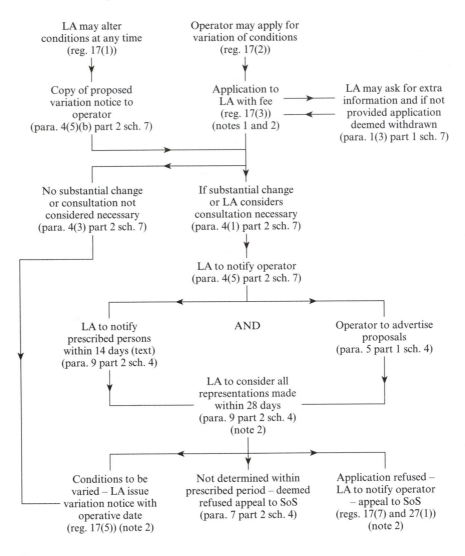

LA may alter conditions at any time (reg. 17(1))

Operator may apply for variation of conditions (reg. 17(2))

Copy of proposed variation notice to operator (para. 4(5)(b) part 2 sch. 7)

Application to LA with fee (reg. 17(3)) (notes 1 and 2)

LA may ask for extra information and if not provided application deemed withdrawn (para. 1(3) part 1 sch. 7)

No substantial change or consultation not considered necessary (para. 4(3) part 2 sch. 7)

If substantial change or LA considers consultation necessary (para. 4(1) part 2 sch. 7)

LA to notify operator (para. 4(5) part 2 sch. 7)

LA to notify prescribed persons within 14 days (text) (para. 9 part 2 sch. 4)

AND

Operator to advertise proposals (para. 5 part 1 sch. 4)

LA to consider all representations made within 28 days (para. 9 part 2 sch. 4) (note 2)

Conditions to be varied – LA issue variation notice with operative date (reg. 17(5)) (note 2)

Not determined within prescribed period – deemed refused appeal to SoS (para. 7 part 2 sch. 4)

Application refused – LA to notify operator – appeal to SoS (regs. 17(7) and 27(1)) (note 2)

Notes
1. The details to be included in the application are set out in para. 1 part 1 sch. 7.
2. Information to be included in public register (reg. 29 and sch. 4).

SURRENDER OF PERMITS

FC51 Surrender of permits for part A installations and mobile plant

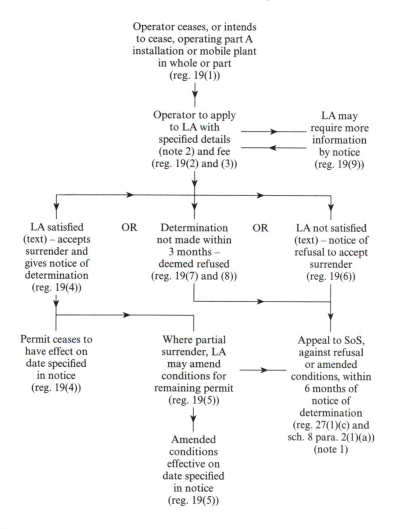

Operator ceases, or intends
to cease, operating part A
installation or mobile plant
in whole or part
(reg. 19(1))

Operator to apply
to LA with
specified details
(note 2) and fee
(reg. 19(2) and (3))

LA may
require more
information
by notice
(reg. 19(9))

LA satisfied
(text) – accepts
surrender and
gives notice of
determination
(reg. 19(4))

OR

Determination
not made within
3 months –
deemed refused
(reg. 19(7) and (8))

OR

LA not satisfied
(text) – notice of
refusal to accept
surrender
(reg. 19(6))

Permit ceases to
have effect on
date specified
in notice
(reg. 19(4))

Where partial
surrender, LA
may amend
conditions for
remaining permit
(reg. 19(5))

Appeal to SoS,
against refusal
or amended
conditions, within
6 months of
notice of
determination
(reg. 27(1)(c) and
sch. 8 para. 2(1)(a))
(note 1)

Amended
conditions
effective on
date specified
in notice
(reg. 19(5))

Notes
1. SoS may affirm or quash LA determination or alter amended conditions (reg. 27(4)).
2. To include site report and must identify any changes in condition of site declared at application stage and details of steps taken to avoid any pollution risk (regs. 11(3) and 19).

FC52 Surrender of permits for part B installations and mobile plant

Operator ceases or
intends to cease operating
whole or part of a part B
installation or mobile plant
(reg. 20(1))

↓

Operator to notify LA with
specified details (text)
and a date of surrender
not less than 28 days
from notification
(reg. 20(4))

↓

Permit ceases to be
effective on date
specified in notification
(reg. 20(4))

↓

For partial surrenders
LA may vary
conditions for
remaining permit
by serving a
variation notice
(see FC50)
(reg. 20(5))

accept the surrender of the permit if it is satisfied that the steps to be taken to implement the closure are appropriate to avoid risk of pollution and will return the site to a satisfactory state. Applications must be determined within 3 months and failure to do so is deemed to be a refusal. There is an appeal against refusal to accept surrender (reg. 19).

In relation to part B activities, the operator must notify the LA of his intention with specified information but, in this case, the LA has no right to refuse to accept the surrender.

Enforcement notices (ENs) (FC53)

If the LA is of the opinion that a condition has been or is likely to be contravened they may serve an enforcement notice on the operator. The notice must specify:

(a) the contravention;
(b) the steps necessary to remedy it and
(c) the time period allowed.

There is an appeal against the notice and non-compliance is an offence – see below.
(reg. 24)

Suspension notices (SNs) (FC53)

Where the LA believes that the continued operation of the activities will involve an imminent risk of serious pollution it must serve a suspension notice on the operator. This applies whether or not the activities are permitted and may deal with a failure to comply with conditions of a permit or a failure to comply with an EN. Alternatively the situation may be dealt with by use of its powers under reg. 26 – see below.

The suspension notice must:

(a) specify what the imminent risk is;
(b) specify the steps to be taken;
(c) state that any permit shall cease to have effect in relation to the installation or mobile plant as a whole or specified activities in it;
(d) where the activities will continue in part, state any additional measures to (b) above that must be taken.

The LA may withdraw the notice at any time and must do so when the imminent risk has been removed.
(reg. 25)

There is an appeal against these actions by the LA – see below.

Prevention or remedying of pollution (FC54)

Where the LA believes that the operation of an activity involves an imminent risk of serious pollution or there has been the commission of an offence by the operator that causes pollution, it may arrange itself for steps to be taken to remove that risk or pollution.

ENFORCEMENT AND SUSPENSION NOTICES

FC53 Enforcement and suspension notices

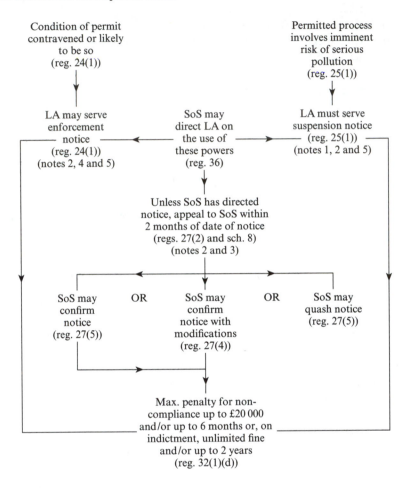

Notes

1. Where these circumstances exist the LA must serve a suspension notice unless taking action itself under reg. 26 – see FC54.
2. Details to be entered in public register (reg. 29 and sch. 9).
3. The notices are not suspended during the appeal process (reg. 27(8)).
4. Instead of serving an enforcement notice the LA may prosecute for the contravention (reg. 32(1)(b)).
5. An enforcement or suspension notice may be withdrawn at any time (regs. 24(4) and 25(6)).

LA POWERS TO PREVENT OR REMEDY POLLUTION

FC54 LA powers to prevent or remedy pollution

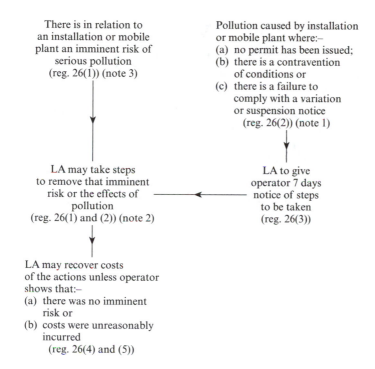

There is in relation to
an installation or mobile
plant an imminent risk of
serious pollution
(reg. 26(1)) (note 3)

Pollution caused by installation
or mobile plant where:–
(a) no permit has been issued;
(b) there is a contravention
    of conditions or
(c) there is a failure to
    comply with a variation
    or suspension notice
    (reg. 26(2)) (note 1)

LA may take steps
to remove that imminent
risk or the effects of
pollution
(reg. 26(1) and (2)) (note 2)

LA to give
operator 7 days
notice of steps
to be taken
(reg. 26(3))

LA may recover costs
of the actions unless operator
shows that:–
(a) there was no imminent
    risk or
(b) costs were unreasonably
    incurred
    (reg. 26(4) and (5))

Notes
1. This power may be implemented in addition to the committed offences.
2. There is no provision for appeal against these actions, only a challenge to the recovery of costs.
3. Where these circumstances exist the LA must implement his procedure or serve a suspension notice – see FC53.

Before taking any steps the LA must give the operator 7 days notice. The costs of the necessary works may be recovered from the operator unless the operator can show that there was no imminent risk or the costs, or part of them, were incurred unnecessarily (reg. 26).

Revocation notices (RNs) (FC55)

The LA may revoke a permit at any time, in whole or in part, by the service of a RN. This is a wide power that can be used whenever the LA considers it to be appropriate. The circumstances in which this may be done include:

(i) in relation to permitted waste management activities, the operator has ceased to be a fit and proper person to hold a permit, and
(ii) the holder of the permit has ceased to be the operator of the installation or mobile plant.

The notice must state the extent to which the permit is being revoked and the date on which this will take effect, being at least 28 days from service of the notice.

Where in relation to part A activities the LA considers that, when the activities cease, steps will be required to avoid a risk of pollution or return the site to a satisfactory condition, the notice may specify the steps to be taken.

There is an appeal procedure – see below.

Public registers

Local authorities must maintain public registers containing information on all Part A(2) and Part B permits issued by them for installations in their areas and also details of Part A(1) installations in their area regulated by the EA (reg. 29). The content of the register is prescribed by Schedule 9 para. 1 (a)–(w). The Register may be in the form of a computer file/record.

Where the operator of an installation feels that the application for a permit would contain information regarded as commercially confidential, the operator may apply to the LA to have it withheld from the register (FC56). Information is commercially confidential 'if its being contained in the register would prejudice to an unreasonable degree the commercial interests of that individual or other person'. If the LA refuses the request, the operator has a right of appeal – see below.
(reg. 31)

Charges

The SoS under the powers of reg. 22 determines the type and level of charges that may be made by LAs in relation to their activities for Part A installations and mobile plant. This scheme is amended from time to time but currently is set out in The Local Authority Permits for Part A Installations

REVOCATION OF PERMITS BY LA

FC55 Revocation of permits by LA

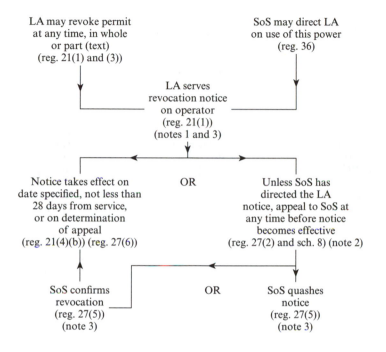

LA may revoke permit
at any time, in whole
or part (text)
(reg. 21(1) and (3))

SoS may direct LA
on use of this power
(reg. 36)

LA serves
revocation notice
on operator
(reg. 21(1))
(notes 1 and 3)

Notice takes effect on
date specified, not less than
28 days from service,
or on determination
of appeal
(reg. 21(4)(b)) (reg. 27(6))

OR

Unless SoS has
directed the LA
notice, appeal to SoS at
any time before notice
becomes effective
(reg. 27(2) and sch. 8) (note 2)

SoS confirms
revocation
(reg. 27(5))
(note 3)

OR

SoS quashes
notice
(reg. 27(5))
(note 3)

Notes
1. The LA may withdraw the revocation notice at any time before it becomes effective (reg. 21(9)).
2. The revocation is suspended pending determination of the appeal or withdrawal of the notice (reg. 21(6)).
3. Details to be entered in public register (reg. 29 and sch. 8).

APPLICATIONS TO LA TO EXCLUDE COMMERCIALLY CONFIDENTIAL INFORMATION FROM PUBLIC REGISTER

FC56 Applications to LA to exclude commercially confidential information from public register

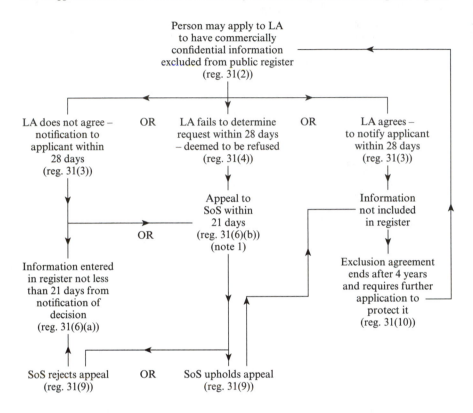

Note

1. The information concerned may not be entered in the register until 7 days after the appeal is finally determined (reg. 31(6)).

and Mobile Plant (England and Wales) Fees and Charges Scheme 2001. The charges may be raised relating to:

(a) Permit applications under reg. 10(1)
(b) Variations under reg. 17(2)
(c) Transfers under reg. 18(1)
(d) Surrenders under reg. 19(2)
(e) Subsistence charges on an annual basis.

In relation to part B activities the charging scheme is made by the SoS using the powers in sect. 8(2) of the EPA 1990 and is set out currently in the Local Enforcing Authorities Air Pollution Fees and Charges Scheme (England) 2001 and covers:

(a) fees for applications and variations, and
(b) annual subsistence charges.

Appeals

Unusually for environmental health enforcement work, rights of appeal are to the SoS and not to a court of law. The operator may appeal against LA decisions relating to:

(a) refusal of an application for a permit (reg. 10)
(b) refusal of an application for the variation of a permit (reg. 17(2))
(c) the service of a revocation, enforcement or suspension notice (regs. 21, 24 and 25)
(d) determination that information is not commercially confidential (reg. 31)
(e) refusal of an application to transfer or surrender a permit (regs. 18 and 19)
(f) the service of a variation notice on the LA's initiative (reg. 17(1)).

The full procedure with timescales is set out in schedule 8 to the PPC Regs. The SoS has the power to affirm or quash the LA decisions and to alter the terms of any conditions (reg. 27 and schedule 8).

Offences

Offences committed against any of the requirements of the PPC Regs. are punishable, on summary conviction, of a fine up to £20 000 and/or up to 6 months imprisonment. Conviction in the Crown Court may lead to an unlimited fine and/or to imprisonment for up to 5 years. (reg. 32)

Chapter 8

ENVIRONMENT ACT 1995

GENERAL PROCEDURAL PROVISIONS

Notices

(a) **Form** Notices must be in writing (sect. 124(1)).

(b) **Service** (i) Notices to be served on any person may be served by delivering it to him or by leaving it at his proper address or by sending it by post to him at that address.

(ii) Notices in the case of a body corporate may be served on the Secretary or Clerk.

(iii) Notices in the case of a partnership may be served on a partner or a person having the control or management of the partnership business.

(iv) The proper address is the last known address except for a body corporate where it is the registered or principal office or a partnership where it is the principal office of the partnership.

(v) If after reasonable enquiry the name and address of the person in occupation cannot be ascertained or the premises appear to be unoccupied, the notice may be served by leaving it with a person who is or appears to be resident or employed on the premises or by leaving it conspicuously affixed to some building or object on the premises (sect. 123).

Powers of entry

An enforcing authority may authorize suitable persons to exercise any of the following powers:

(a) to enter at reasonable times (or in emergency at any time) any premises which he has reason to believe it is necessary for him to enter (7 days' notice required for residential premises);

(b) to take with him other persons and equipment (7 days' notice to take heavy equipment);

(c) to make examination and investigations;

(d) to direct premises to be left undisturbed for the purpose of examination or investigation;

(e) to take measurements, photographs and take recordings for the purpose of examination and investigation;

(f) to take samples of any substances or of the air, water or land in and around premises;

(g) to subject to process or test any article or substance which appears to have caused or be likely to cause pollution of the environment or harm to human health;

(h) to take possession of substances etc., as in (g) above to examine it, ensure that it is not tampered with or to use as evidence;

(i) to require persons to give information;

(j) to require the production of records;

(k) to require persons to afford facilities and assistance;

(l) any other power necessary to carry out the purposes of entry.

The purpose for which the AO may demand entry (and exercise the above powers) are to determine whether any provision of the pollution control enactment is being, or has been complied with, to exercise a pollution control function of the authority or to determine whether and how such a function should be exercised.

In relation to local enforcing authorities, the pollution control powers to which these provisions relate are:

(a) Part 1 (Air Pollution Control by LAs) and Part 2A (Contaminated Land) of the EPA 1990;

(b) Regulations made under Part 4 (Air Quality) of the EA 1995 e.g. the Air Quality (England) Regulations 2000; or

(c) Regulations made under the European Communities Act 1972 so far as they relate to pollution e.g. the Air Quality Limit Values Regulations 2001.

(d) Regulations made under sect. 2 of the PPCA 1999 i.e. the PPC Regs. 2000.

Where entry has been refused or is apprehended the AO may apply to a JP for a warrant. In exercising a warrant the AO must secure any unoccupied premises and there are also provisions for the payment of compensation in certain cases (sects. 108 and 110 and Schedule 18).

Extent

Part 4 of the Act applies in England, Scotland and Wales but not in Northern Ireland (sect. 125(7)).

LA REVIEWS OF AIR QUALITY

References

Environment Act 1995 Sect 80 and Schedule 11.
The Air Quality (England) Regulations 2000.
The Air Quality (Scotland) Regulations 2000.
The Air Quality (Wales) Regulations 2000.
DETR Guidance Notes:
  (a) Framework for review and assessment of air quality LAQM. G1(97);
  (b) Developing air quality strategies and action plans LAQM. G2(97);
  (c) Air quality and traffic management LAQM. G3(97);
  (d) Air quality and land and use planning LAQM. G4(97).
DETR Technical Guidance Notes (Table 6.2).

Scope

This procedure deals with the statutory requirement on LAs (see below) to
review the present quality, and the likely future quality up to the year 2005,
of air within the authority's area (sect. 82(1)).

National Air Quality Strategy (NAQS)

The Secretary of State is required to prepare and publish a strategy containing policies with respect to the assessment and management of the quality
of air.

Such a strategy takes account of UK obligations under the EC treaties
and of international agreements and includes statements of:

  (a) air quality standards – concentrations of pollutants in the atmosphere
      which can broadly be taken to achieve a certain level of environ-
      mental quality;
  (b) air quality objectives – which provide a framework for determining
      the extent to which policies should aim to improve air quality, these
      indicate the progress which can be made towards air quality standards
      by 2005; and
  (c) measures to be taken by LAs and others to achieve the objectives
      (sect. 80(1)–(5)).

The strategy was first published in March 1997 and a revised version in
January 2000 (CM 4548). It sets out certain standards and specific objectives
to be achieved by 2005.

London Air Quality Strategy

Working within the NAQS, the Greater London Authority is required to
produce an Air Quality Strategy for London which will explain how, working

FC57 LA reviews of air quality

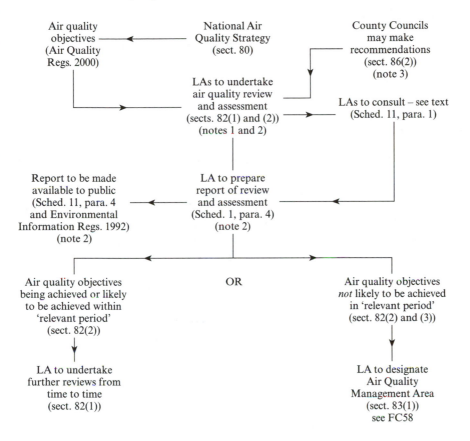

Notes

1. There is no statutory timescale but LAs were expected to have completed reviews by 31/12/00.
2. The SoS or SEPA have reserve powers in the event of LA default (sect. 85).
3. The County Council (where one exists) should be involved throughout the air quality management process.

with London borough councils and other bodies, the national objectives will be achieved. A draft strategy for consultation was issued by the GLA in November 2001.

LA reviews and assessments

Each of the following LAs (as enforcing authorities) is required to review periodically air quality in its area:

(a) Unitary authorities (in England, Wales and Scotland (including London borough councils);
(b) District Councils which are not unitary authorities (but see 'functions' of County Councils below) (sects. 82(1) and 91).

The reviews have to consider both the air quality for the time being and the likely future air quality during the 'relevant period'. They must also include an assessment of whether prescribed air quality standards or objectives are being achieved within that 'relevant period'. The three sets of Air Quality Regulations 2000 prescribe the relevant periods for each of the polluting substances, which range from 31 December 2003 to 31 December 2005, together with the air quality objective levels for each. The objectives are the same as those set out in CM 4548.

There is no requirement for imposing any statutory deadlines upon LAs to complete their reviews and assessments but LAs were initially expected to have completed them by 31/12/00 but only 88% had done so. The SoS (SEPA in Scotland) does have reserve powers in default of an LA (sect. 85)

Guidance on the principal considerations which should underpin the development of a local air quality strategy is given in DETR guidance Developing Air Quality Strategies and Action Plans LAQM. G2(97).

Phasing of assessments

The guidance indicates that there should be a phased approach to review and assessment.

(a) Stage 1 – an initial screening of industrial, transport and other significant sources of pollutants in the locality. If no significant sources exist, the likelihood of failure to achieve the air quality objectives is negligible.

If such sources do exist and there is potential for exposure over the 'averaging periods' within the objectives, the LA should proceed to the second-stage assessment.
(b) Stage 2 – where there is a potential risk of elevated levels of a pollutant the LA will estimate ground level concentrations at the roadside and at industrial and urban background locations in the area. If it is likely that by 2005 objectives will not be achieved, the third stage should be undertaken.

(c) Stage 3 – the LA, using more sophisticated techniques, will predict whether an air quality objective is unlikely to be achieved by 2005.
(d) Stage 4 – pre-declaration of AQMAs.

Action following a review and assessment

If there are parts of the area where the prescribed objectives are not likely to be achieved within the 'relevant period', the LA is required to designate the area as an Air Quality Management Area – see FC58 (sect. 83(1)).

Consultation

In undertaking an air quality review and assessment the LA is required to consult:

(a) the SoS;
(b) the Environment Agency (SEPA in Scotland);
(c) in England and Wales the Highways Authority;
(d) all neighbouring LAs;
(e) the County Council (where applicable);
(f) the National Park Authority (where applicable);
(g) any other public authority which the LA considers appropriate; and
(h) bodies representing local business interests and such other bodies as the LA considers appropriate (Schedule 11, para. 1).

The DETR also recommended that the LA should involve community and environmental groups and that, in relation to the review, consultees should be informed of:

(a) proposed start date of the review;
(b) timescales involved;
(c) broad plan of approach;
(d) the arrangements for working with the other LAs.

In relation to their assessment of air quality the LA should invite comments on the report of the data collected, the interpretation of that data and the conclusions it has drawn from it (LAQM. G2(97)).

Public access to information

A report of the results of the air quality review and a report of the results of any assessment must be made available to the public free of charge at all reasonable times (Schedule 11, para. 4).

In any event, information which the LA holds in relation to the quality of air needs to be made available (subject to exceptions) by the Environmental Information Regulations 1992.

Functions of County Councils

The DETR expected that, while District Councils will be the lead LAs in the remaining two tier areas of England, they will fully involve counties in the review and assessment process.

The County Council may make recommendations to the District in relation to any review and assessment and the District must take account of them (sect. 86(2)).

In addition, Schedule 11, para. 2 makes provision for the exchange of information between District and County Councils throughout the air quality management process.

SoS powers

The Act provides for the SoS (or SEPA in Scotland) to undertake the air quality review and assessment in default of the LA and to go on to indicate these areas within the LA area where prescribed air quality standards/objectives are unlikely to be achieved by 2005 (sect. 86(2)).

AIR QUALITY MANAGEMENT AREAS (AQMAs)

References

Environment Act 1995 sect. 84 and Schedule 11.
The Air Quality Regulations (England) (Scotland) and (Wales) 2000.
National Air Quality Strategy for England, Scotland, Wales and Northern Ireland CM 4548 (January 2000).
DETR Guidance notes:
  (a) Framework for review and assessment of air quality LAQM. G1(97);
  (b) Developing air quality strategies and action plans LAQM. G2(97);
  (c) Air quality and traffic management LAQM. G3(97); and
  (d) Air quality and land-use planning LAQM. G4(97).
DETR Technical guidance notes. See table 6.2.
The National Society for Clean Air and Environmental Protection. Air Quality Action Plans Part 1 2000.

Scope

Where, following a review and assessment of air quality in its area, a LA determines that prescribed air quality objectives are not likely to be achieved within the 'relevant period' (page 224), it must identify the areas where this is likely and, by order, designate them as an AQMA (sect. 83(1)).

Such Orders may subsequently be varied or revoked by further Orders following further assessment and, in the case of revocation, where the prescribed standards/objections are likely to be met throughout the 'relevant period' (sect. 83(2)).

FC58 Air Quality Management Areas (AQMAs)

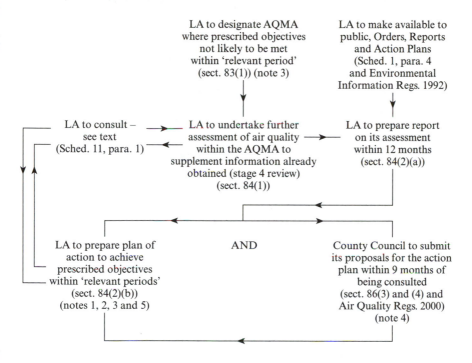

LA to designate AQMA
where prescribed objectives
not likely to be met
within 'relevant period'
(sect. 83(1)) (note 3)

LA to make available to
public, Orders, Reports
and Action Plans
(Sched. 1, para. 4
and Environmental
Information Regs. 1992)

LA to consult –
see text
(Sched. 11, para. 1)

LA to undertake further
assessment of air quality
within the AQMA to
supplement information already
obtained (stage 4 review)
(sect. 84(1))

LA to prepare report
on its assessment
within 12 months
(sect. 84(2)(a))

LA to prepare plan of
action to achieve
prescribed objectives
within 'relevant periods'
(sect. 84(2)(b))
(notes 1, 2, 3 and 5)

AND

County Council to submit
its proposals for the action
plan within 9 months of
being consulted
(sect. 86(3) and (4) and
Air Quality Regs. 2000)
(note 4)

Notes

1. The LA may revise the action plan periodically (sect. 84(4)) following periodic reviews and assessments of air quality.
2. There is no statutory period but LAQM. G1(00) suggests the action plan should be prepared within 12–18 months from the declaration of an AQMA.
3. The SoS or SEPA have powers of direction (text).
4. In the event of disagreement between district and county the matter is determined by the SoS (sect. 84(5)).
5. There are proposals to give powers to LAs with AQMAs to enforce fixed penalty procedures for exceeding prescribed vehicle emission limits.

The AQMA

While the Order needs to designate the part/s of the LA's area to which the AQMA status is to be attached, the DETR expected that delineation should make appropriate use of relevant physical and geographical boundaries.

Action by LA following designation of AQMA

The LA is first to undertake a further assessment of air quality (stage 4) within the AQMA itself to supplement information already obtained, identify likely sources and inform the preparation of the action plan (sect. 84(1)). These are intended to refine the outcomes of the earlier stages of review and assessments and provide a final check or clarification of the results from stage 3.

The LA is required to produce a report on this assessment within 12 months of the operation of the Order (sect. 84(2)(a)).

The LA is required to produce a plan of action (sect. 84(2)(b)) normally within 12 months of the completion of the further assessment.

The draft plan of action at the consultation stage (see page 225) should contain:

(a) details of the pollutants to be addressed and an indication of their source;
(b) the boundaries of the proposed action areas;
(c) the involvement of other LAs in the plan;
(d) proposals from the County Council (if applicable);
(e) the timescales over which each of the proposed measures are to be implemented;
(f) indication of where actions by other than the LA is required.

In October 2001 DEFRA issued consultative proposals which would allow LAs with AQMAs to issue fixed penalties of £60 to drivers whose vehicle emissions exceed prescribed limits and all LAs to issue £20 fixed penalties for leaving vehicle engines running unnecessarily when parked.

Consultation

In undertaking the further air quality assessment and in preparing its action plan the LA is required to consult widely. The bodies concerned are the same as those for the initial review and assessment on page 225 (Sched. 11, para. 1).

Public access to information

The LA is required to make available to the public free of charge copies of the Order designating the AQMA, the report of the further assessment, the action plan and the proposals of the County Council, where applicable (Sched. 11, para. 4).

In any event, information which the LA holds in relation to air quality must (with certain exceptions) be made available to the public under the Environmental Information Regulations 1992.

County Councils

The DETR expect that, whilst the lead LAs include district councils in the remaining two tier areas of England, county councils will be fully involved by the district in the air quality management process generally and Schedule 11, para. 2 makes provision for the exchange of information between the two LAs.

In relation to AQMAs, there is a specific requirement upon county councils that it should, within 9 months from the date it is first consulted on the action plan, submit to the district details of its proposals for the action plan (sects. 86(3) and (4) and Reg. 3(1) Air Quality Regulations 2000). Disputes between districts and county councils about the content of action plans are to be settled by the SoS or SEPA (sect. 84(5)).

SoS powers

The Act does provide for the SoS (or SEPA in Scotland) to direct LAs on the action it should take (including the designation of AQMAs) where prescribed air quality standards/objectives are unlikely to be achieved within the 'relevant periods', and the LA has either failed to discharge a duty imposed on it or has taken inappropriate action to deal with the situation (sects. 86(3) and (4)).

Chapter 9

CLEAN AIR ACT 1993

GENERAL PROCEDURAL PROVISIONS

In relation to England and Wales, Part 12 (General) of the PHA 1936 is incorporated in the CAA 1993 through sect. 62 of that Act with a parallel provision for Scotland.

Those notes on pages 12–15 under the following headings relating to the PHAs are therefore applicable to the procedures in this chapter:
Notices
Notices requiring execution of works
Recovery of costs and
Appeals

Extent

These procedures are applicable in England, Wales and Scotland, with the substitution of authorities and bodies appropriate to the structure of government in that country, but not to Northern Ireland (sect. 68).

Local authorities

The LAs charged with the enforcement of the Act are:

(a) in England the council of the district or a London borough, the Common Council of the City of London, the Sub-Treasurer of the Inner Temple and the Under-Treasurer of the Middle Temple;
(b) in Wales, the Council of a County or County borough; and
(c) in Scotland, a council constituted under sect. 2 of the Local Government (Scotland) Act 1994, i.e. unitary councils (sect. 64).

These LAs are under a duty to enforce Part 1 (dark smoke), Part 2 (smoke, grit, dust and fumes), Part 3 (smoke control areas), sect. 33 (cable burning) and Part 6 which deals with colliery spoil banks, railway engines and vessels (sect. 55).

Power of entry

An authorized officer of the LA may at any reasonable time enter any land or vessel in order to:

(a) perform functions conferred by the Act;
(b) determine if these functions should be performed; or
(c) carry out inspections, measurements and tests and to take away samples.

This power does not apply to entry into private dwellings except in relation to works of adaptation of fireplaces in a SCA where a notice had been served under sect. 24(1) and 7 days' notice of entry has been given. Where:

(a) admission has been refused; or
(b) refusal is apprehended; or
(c) the land or vessel is unoccupied; or
(d) the occupier is temporarily absent; or
(e) the case is one of emergency; or
(f) application for admission would defeat the object of entry

the LA, after giving (except in emergency) 7 days' notice to any occupier, may make application to a JP by way of sworn information in writing. The JP may by warrant authorize entry, if necessary by force (sects. 56 and 57).

The penalty for obstruction of an authorized officer is a fine not exceeding level 3 (sect. 57(6)).

Crown premises

Although the procedures in this chapter cannot be applied to Crown premises, the LA is required to bring the following situations to the attention of the responsible Minister where the premises or vessel concerned is Crown property:

(a) dark smoke or grit and dust outside a SCA;
(b) any smoke within a SCA;
(c) any smoke nuisance;
(d) any dark smoke from a vessel.

The Minister must inquire into the circumstances and take all practicable means of action (sect. 46).

Relationship to the EPA 1990 and the PP and CA 1999

The procedures in this chapter cannot be applied to any processes authorized under the EPA 1990 (see FC27) or to activities regulated under the PP and CA 1999 (see FC47) (sect. 41 and 41A CAA 1993) – NB. sect. 41A was inserted by para. 13 schedule 10 of the Pollution Prevention and Control (England and Wales) Regulations 2000).

DEFINITIONS

Black smoke means smoke which, if compared in the appropriate manner with a chart of the type known . . . as the Ringelmann chart, would appear to be as dark as or darker than shade 4 on the chart (Dark Smoke (Permitted Periods) Regulations 1958 Reg. 2(2)).

    Chimney includes structures and openings of any kind from or through which smoke, grit or fumes may be emitted and, in particular, includes flues, and references to the chimney of a building includes references to a chimney which serves the whole or part of a building but is structurally separate therefrom (sect. 64(1)).

    Dark smoke means smoke which, if compared in the appropriate manner with a chart of the type known . . . as the Ringelmann chart, would appear to be as dark as or darker than shade 2 on the chart (sect. 3(1)).

    Fireplace includes any furnace, grate or stove, whether open or closed (sect. 64(1)).

    Fumes means any airborne solid matter smaller than dust (sect. 64(1)).

    Practicable means reasonably practicable having regard, among other things, to local conditions and circumstances, to the financial implications and to the current state of technical knowledge and 'practicable means' includes the provision and maintenance of plant and its proper use (sect. 64(1)).

    Prejudicial to health means injurious, or likely to cause injury, to health (PHA 1936 sect. 343(1)).

    Private dwelling except as far as the context otherwise requires, means any building or part of a building used or intended to be used as such, and a building or part of a building shall not be deemed for the purposes of this Act to be used or intended to be used otherwise than as a private dwelling by reason that a person who resides or is to reside in it is, or is to be required or permitted to reside in it, in consequence of his employment or holding of an office (sect. 64(4)).

    Smoke includes soot, ash, grit and gritty particles emitted in smoke (sect. 64(1)).

SMOKE, GRIT, ETC.

PROHIBITION OF DARK SMOKE ETC. FROM CHIMNEYS

References

Clean Air Act 1993 sect. 1.
Dark Smoke (Permitted Periods) Regulations 1958.
Dark Smoke (Permitted Periods) (Vessels) Regulations 1958.

FC59 Prohibition of dark smoke etc. from chimneys

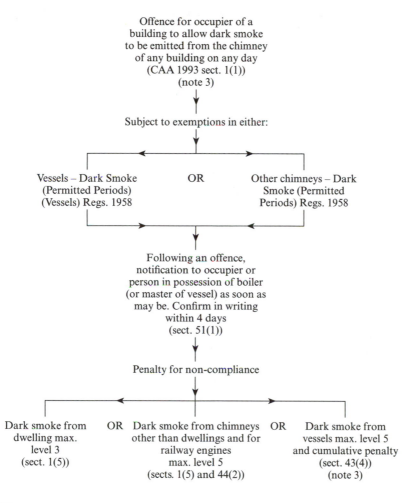

Offence for occupier of a
building to allow dark smoke
to be emitted from the chimney
of any building on any day
(CAA 1993 sect. 1(1))
(note 3)

Subject to exemptions in either:

Vessels – Dark Smoke
(Permitted Periods)
(Vessels) Regs. 1958

OR

Other chimneys – Dark
Smoke (Permitted
Periods) Regs. 1958

Following an offence,
notification to occupier or
person in possession of boiler
(or master of vessel) as soon as
may be. Confirm in writing
within 4 days
(sect. 51(1))

Penalty for non-compliance

Dark smoke from
dwelling max.
level 3
(sect. 1(5))

OR

Dark smoke from chimneys
other than dwellings and for
railway engines
max. level 5
(sects. 1(5) and 44(2))

OR

Dark smoke from
vessels max. level 5
and cumulative penalty
(sect. 43(4))
(note 3)

Notes

1. For dark smoke from industrial or trade premises (other than from chimneys), see sect. 1 CAA 1993 (FC60).
2. This procedure also applies to vessels and railway engines (CAA 1993 sects. 43 and 44).
3. In relation to offences which are a repetition or continuation of an earlier offence, there is a penalty of max. £50 daily (sect. 50).

Scope

It is an offence to emit dark smoke from the chimney of any building (including houses) on any one day and this includes chimneys serving the furnaces of boilers or industrial plant which are either attached to buildings or for the time being fixed to or installed on any land. Emissions lasting not longer than the following are, however, to be left out of account:

(a) Dark Smoke (Permitted Periods) Regulations 1958 – applies to emissions from chimneys other than on vessels:

No. of furnaces	Aggregate in any 8-hour period	
	without soot blowing	with soot blowing
1	10 min	14 min
2	18 min	25 min
3	24 min	34 min
4 or more	29 min	41 min

but in any event emissions must not exceed 4 minutes continuous dark smoke (other than by soot blowing) or 2 minutes black smoke in aggregate in any 30-minute period.

(b) Dark Smoke (Permitted Periods) (Vessels) Regulations 1958.

These regulations apply only to vessels and details of the exemptions for the emission of dark smoke are set out in the Schedule and vary depending upon the class of boiler. The exemptions do not permit the continuous emission of dark smoke for in excess of 4 minutes (10 minutes for certain natural-draught, oil-fired boilers) or the emission of black smoke for more than 3 minutes in any 30-minute period.

Defences

In any proceedings under sect. 1 it will be a defence to prove that the contravention was solely due to:

(a) lighting up from cold and all practicable steps were taken to prevent or minimize the emission; or

(b) failure etc. of the furnace that could not reasonably have been foreseen or prevented; or

(c) suitable fuel being unavailable and all practicable steps taken to prevent or minimize the emission; or

(d) any combination of (a)–(c) above; or

(e) there was a failure by the LA to comply with the notification provisions (below) (sects. 1(4) and 51(3)).

Person responsible

This is:

(a) for chimneys of buildings – the occupier of the building (sect. 1(1));
(b) for chimneys not attached to buildings – the person having possession of the boiler or plant (sect. 1(2));
(c) for vessels – the owner and master or other person in charge of the vessel (sect. 44(2));
(d) for railway engines – the owner (sect. 43(2)).

Notification of offences

Upon becoming aware that an offence under sect. 1 is being or has been committed an authorized officer of the LA must:

(a) notify the person responsible as soon as may be; and
(b) confirm the offence in writing before the end of the 4 days following the day on which the offence was committed (sect. 51).

PROHIBITION OF DARK SMOKE ETC. FROM INDUSTRIAL AND TRADE PREMISES (OTHER THAN FROM CHIMNEYS (E.G. TRADE BONFIRES))

References

Clean Air Act 1993 sect. 2.
Clean Air (Emission of Dark Smoke) (Exemption) Regulations 1969.

Scope

These provisions apply only to industrial and trade premises where the smoke is emitted other than from a chimney, e.g. bonfires etc. (sects. 2(1) and (2)).

Their main use is to control industrial bonfires.

Industrial or trade premises are defined as meaning premises used for any industrial or trade purposes or premises not so used on which matter is burnt in connection with any industrial or trade process (sect. 2(6)).

Where any material is burned on industrial and trade premises and this is likely to produce dark smoke, there is an assumption that dark smoke has been emitted unless the occupier or person who caused or permitted the emission can show that no dark smoke was emitted (sect. 2(3)).

Exemptions

These are specified in the Clean Air (Emission of Dark Smoke) (Exemption) Regulations 1969 as follows:

FC60 Prohibition of dark smoke etc. from industrial and trade premises (other than from chimneys)

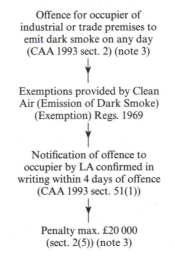

Offence for occupier of
industrial or trade premises to
emit dark smoke on any day
(CAA 1993 sect. 2) (note 3)

Exemptions provided by Clean
Air (Emission of Dark Smoke)
(Exemption) Regs. 1969

Notification of offence to
occupier by LA confirmed in
writing within 4 days of offence
(CAA 1993 sect. 51(1))

Penalty max. £20 000
(sect. 2(5)) (note 3)

Notes

1. For procedure relating to dark smoke emitted from chimneys, see FC59.
2. For special offences relating to cable burning, see sect. 33.
3. The maximum penalty was increased to this amount by Sch. 22, para. 195 of the Environment Act 1995 and came into effect on 1/4/96.

SCHEDULE 1

Exempted matter

Matter	Conditions
(1) Timber and any other waste matter (other than natural or synthetic rubber or flock or feathers) which results from the demolition of a building or clearance of a site in connection with any building operation or work of engineering construction (within the meaning of section 176 of the Factories Act 1961).	Conditions A, B and C.
(2) Explosive (within the meaning of the Explosives Act 1875) which has become waste: and matter which has been contaminated by such explosive.	Conditions A and C.

(3) Matter which is burnt in connection with:
    (a) research into the cause or control of fire
or
    (b) training in fire fighting.
	Condition C.

(4) Tar, pitch, asphalt and other matter which is burnt in connection with the preparation and laying of any surface, or which is burnt off any surface in connection with resurfacing, together with any fuel used for any such purpose.	Condition C.
(5) Carcases of animals or poultry which: (a) have died, or are reasonably believed to and have died, because of disease;	Conditions A and C, unless the burning is carried out by or on behalf of an inspector (within the meaning of the Animal Health Act 1981).

(b) have been slaughtered because of disease;
    or
    (c) have been required to be slaughtered pursuant to the Animal Health Act 1981.	

(6) Containers which are contaminated by any pesticide or by any toxic substance used for veterinary or agricultural purposes; and in this paragraph 'container' includes any sack, box, package or receptacle of any kind.	Conditions A, B and C.

SCHEDULE 2

Conditions

Condition A.   That there is no other reasonably safe and practicable method of disposing of the matter.

Condition B.   That the burning is carried out in such a manner as to minimize the emission of dark smoke.

Condition C.   That the burning is carried out under the direct and continuous supervision of the occupier of the premises concerned or a person authorized to act on his behalf.

Defences

It is a defence in proceedings to prove that:

(a) the contravention was inadvertent and that all practicable steps had been taken to prevent or minimize the emission of dark smoke; or

(b) the notification procedure had not been followed (sect. 2(4)).

Notification of offences

The notes on page 235 are also applicable here.

CONTROL OF GRIT AND DUST FROM FURNACES

References

Clean Air Act 1993 sects. 5–9.
Clean Air (Arrestment Plant) (Exemption) Regulations 1969.
Clean Air (Emission of Grit and Dust from Furnaces) Regulations 1971.
Clean Air (Units of Measurement) Regulations 1992

Requirement to fit arrestors

Furnaces, except those designed solely or mainly for domestic purposes and with a maximum heating capacity of less than 16.12 kilowatts, used in buildings to burn:

(a) pulverized fuel; or

(b) any other solid matter at a rate of 45.4 kilograms/h or more; or

(c) any liquid or gaseous matter at a rate of 366.4 kilowatts/h or more,

are to be provided with plant for arresting grit and dust which has been approved by the LA and be properly maintained and used (sect. 6(1)).

Domestic furnaces burning pulverized fuel or solid fuel at a rate of 1.02 tonnes/hour or more are also required to be provided with grit arrestment plant approved by the LA (sect. 8(1)).

FC61 Control of grit and dust from furnaces

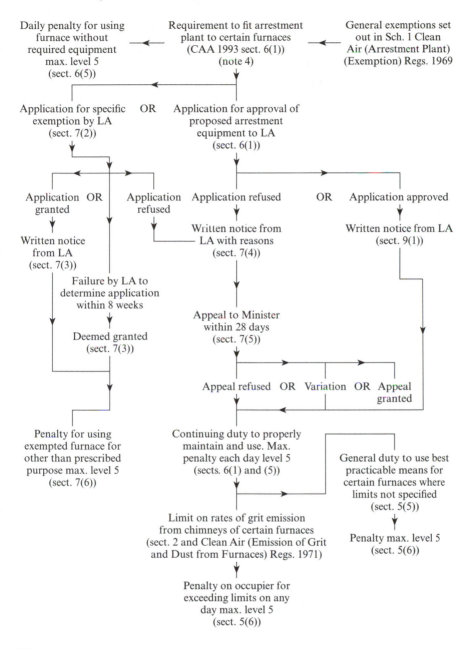

Notes

1. For grit, dust etc. emitted in smoke constituting statutory nuisance, see FC33.
2. For grit, dust etc. not emitted in smoke causing statutory nuisance, see FC33.
3. For measurement etc. of grit and dust, see FC62.
4. LA's powers to obtain information about furnaces and fuel under sect. 12 apply to this procedure.

Exemptions

Exemption from the requirement to fit arrestment plant may be by either:

(a) falling into the categories of exempted furnace set out in Sch. 1 Clean Air (Arrestment Plant) (Exemption) Regulations 1969, which also set out the information to be provided in the application for exemption; or

(b) specific application for exemption to the LA for furnaces falling outside the regulations in (a) above, in which case the LA will need to be satisfied that the furnace can be operated without being prejudicial to health or a nuisance without arrestment plant (sect. 7).

Approval by LA to proposed equipment

In situations where arrestment plant is required for either domestic or non-domestic furnaces the plant must be approved by the LA or installed in accordance with plans and specifications submitted to and approved by the LA (sects. 6(1) and 8(1)).

Emission of grit and dust

The occupier of any building containing a furnace, other than one designed solely or mainly for domestic purposes, with a maximum heating capacity less than 16.12 kilowatts/h which burns solid, liquid or gaseous matter must take all practicable means for minimizing the emission of grit and dust from the chimney (sect. 5(5)).

This general requirement does not, however, apply to furnaces where maximum emission limits have been specified by the Clean Air (Emission of Grit and Dust from Furnaces) Regulations 1971.

Although there is no similar requirement for those domestic furnaces which are required to have arrestment plant, that plant must be properly maintained and used (sect. 8(1)).

Defences

In proceedings for an offence under the regulations, it is a defence to prove that the best practicable means have been used to minimize the emission (sect. 5(4)).

Furnaces outside buildings

These provisions also apply to furnaces attached to a building or fixed to or installed on any land and in these cases the person responsible for compliance is the person having possession of the boiler or plant (sect. 13).

MEASUREMENT OF GRIT AND DUST FROM FURNACES

References

Clean Air Act 1993 sects. 10–12.
Clean Air (Measurement of Grit and Dust from Furnaces) Regulations 1971.
Clean Air (Units of Measurement) Regulations 1992.

Information

In connection with the provisions relating to grit arrestment plant and the measurement of grit and dust, LAs are empowered to require any occupier to give them reasonable information about any furnaces used and the fuels burned following written notice giving not less than 14 days to do so (12(l)).

Measurement by occupiers

These provisions apply only to furnaces used to burn:

(a) pulverized fuel; or
(b) any other solid matter at a rate of 45.4 kilograms/h or more; or
(c) any liquid or gaseous matter at a rate of 366.4 kilowatts/h or more.

The regulations detailing measurement etc. are applied to an individual furnace by written notice from the LA to the occupier (or, in relation to outside furnaces, the person having possession) (sects. 10(1) and 13(2)).

Requirement for LA to undertake measurement

This applies only to furnaces burning:

(a) solid matter, other than pulverized fuel, at a rate less than 1.02 tonnes/h; or
(b) liquid or gaseous matter at a rate less than 8.21 megawatts/h.

In these cases the occupier by written notice may request the LA to undertake the making and recording of measurements (sect. 11(2)).

The occupier remains responsible, however, for any necessary adaptations to the chimney to allow measurements to be taken.

Measurements etc.

The Clean Air (Measurement of Grit and Dust from Furnaces) Regulations 1971 lay down the detailed requirements. Although power is given to the SoS in the Act to include the measurement of fumes, the regulations apply only to grit and dust.

FC62 Measurement of grit and dust from furnaces

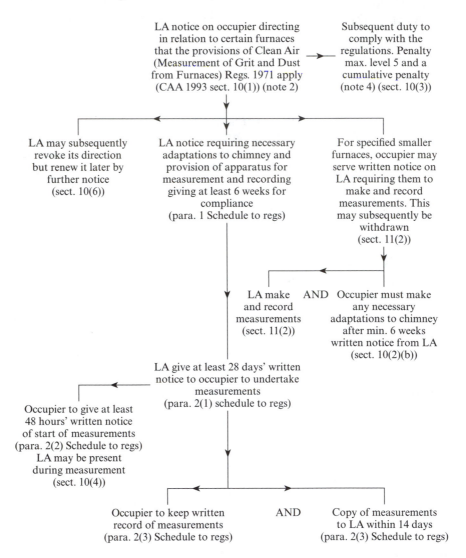

LA notice on occupier directing in relation to certain furnaces that the provisions of Clean Air (Measurement of Grit and Dust from Furnaces) Regs. 1971 apply (CAA 1993 sect. 10(1)) (note 2)

→ Subsequent duty to comply with the regulations. Penalty max. level 5 and a cumulative penalty (note 4) (sect. 10(3))

LA may subsequently revoke its direction but renew it later by further notice (sect. 10(6))

LA notice requiring necessary adaptations to chimney and provision of apparatus for measurement and recording giving at least 6 weeks for compliance (para. 1 Schedule to regs)

For specified smaller furnaces, occupier may serve written notice on LA requiring them to make and record measurements. This may subsequently be withdrawn (sect. 11(2))

LA make and record measurements (sect. 11(2))

AND Occupier must make any necessary adaptations to chimney after min. 6 weeks written notice from LA (sect. 10(2)(b))

LA give at least 28 days' written notice to occupier to undertake measurements (para. 2(1) schedule to regs)

Occupier to give at least 48 hours' written notice of start of measurements (para. 2(2) Schedule to regs) LA may be present during measurement (sect. 10(4))

Occupier to keep written record of measurements (para. 2(3) Schedule to regs)

AND Copy of measurements to LA within 14 days (para. 2(3) Schedule to regs)

Notes
1. For general powers of LA regarding information about atmosphere pollution, see FC66.
2. There is no appeal against the LA notice applying the provisions of the regulations.
3. These provisions also apply to outdoor furnaces, sect. 13.
4. In relation to offences which are a repetition or continuation of an earlier offence there is a penalty of £50 max. daily (sect. 50).

The notice from the LA requiring measurements may require them to be taken from time to time or at staged intervals, but the LA cannot normally require measurements from a chimney in excess of once each 3 months unless the LA thinks that this is necessary to obtain the true level of emission (para. 2(4) Schedule to regulations).

The records to be kept of measurements must show:

(a) the date;
(b) the number of furnaces discharging into the chimney;
(c) results of measurements in lb/h grit and dust emitted and the percentage of grit in the solids emitted (para. 2(3) Schedule to regulations).

CHIMNEY HEIGHTS

HEIGHT OF CHIMNEYS SERVING FURNACES

References

Clean Air Act 1993 sects. 14 and 15.
Clean Air (Height of Chimneys) (Exemption) Regulations 1969.
Clean Air (Height of Chimneys) (Prescribed Forms) Regulations 1969.
Clean Air (Measurement of Units) Regulations 1992.
AQ9(96) Air Quality Guidance Note DETR 1996.

Scope

This procedure applies to chimneys attached to furnaces used to burn:

(a) pulverized fuel; or
(b) any other solid matter at a rate of 45.4 kilograms/h or more; or
(c) any liquid or gaseous matter at a rate of 366.4 kilowatts/h or more.

Approval of chimney height

In relation to chimneys from these furnaces, the occupier of a furnace of a building or the person having possession of any fixed boiler or industrial plant outside of a building, must not use that furnace unless the height of its chimney has been approved by the LA and any conditions of approval are complied with (sects. 14(2)–(5)).

In relation to furnaces served by a chimney which was constructed before 1 April 1969, this requirement only applies where the combustion space of the furnace has been increased since that date or the furnace served by that chimney has been replaced by a larger one (Schedule 5, para. 7).

FC63 Height of chimneys serving furnaces

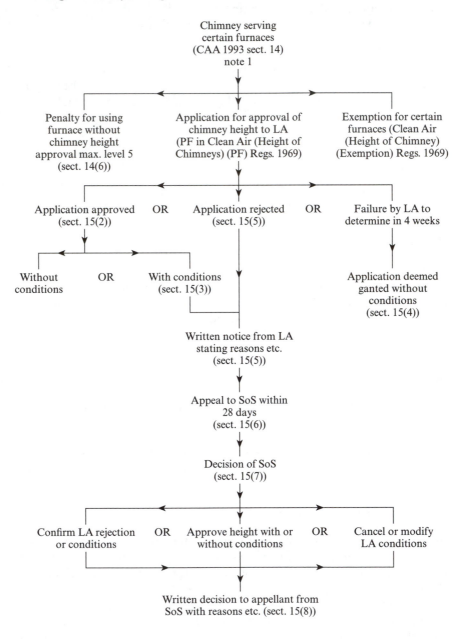

Note

1. For height of chimneys not serving furnaces, see FC64.

Exemptions

The Clean Air (Height of Chimneys) (Exemption) Regulations 1969 specify the following uses of boilers or industrial plant where chimney height approval is not required:

(a) temporarily replacing any other boiler or plant which is under inspection, maintenance or repair, or is being rebuilt or replaced;
(b) providing a temporary source of heat or power during building operations or for investigation and research;
(c) providing products of combustion to heat other plant to an operating temperature;
(d) providing heat or power for mobile agricultural plant.

Criteria for decision

The LA is required to reject the application unless it is satisfied that the height of the chimney will be sufficient to prevent as far as practicable, the smoke, grit, dust, gases or fumes from becoming prejudicial to health or a nuisance having regard to:

(a) the purpose of the chimney;
(b) the position and description of nearby buildings;
(c) the levels of neighbouring ground;
(d) any other matters requiring consideration in the circumstances (sect. 15(2)).

Refusals and conditions

If the application is not approved or conditions are attached to approval, the LA is required to give written notification of its reasons to the applicant. Conditions may relate to the rate and/or quality of the emissions from the chimney. Where the application is rejected, the LA is also required to state the lowest height which it is prepared to accept conditionally, unconditionally or both (sect. 15(4) and (5)).

Similar requirements are placed upon the SoS in relation to decisions following appeals (sect. 15(7)).

HEIGHT OF CHIMNEYS NOT SERVING FURNACES

Reference

Clean Air Act 1993 sect. 16.
Air Quality Guidance Note AQ9(96) DETR 1996.

Scope

No account can be taken of the radioactive state of any substance, article or premises (RSA 1993 sect. 40).

FC64 Height of chimneys not serving furnaces

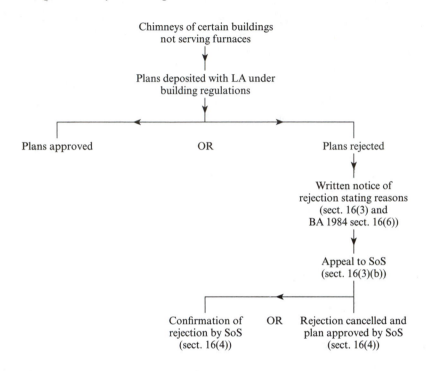

Chimneys of certain buildings
not serving furnaces

↓

Plans deposited with LA under
building regulations

↓

Plans approved   OR   Plans rejected

↓

Written notice of
rejection stating reasons
(sect. 16(3) and
BA 1984 sect. 16(6))

↓

Appeal to SoS
(sect. 16(3)(b))

Confirmation of   OR   Rejection cancelled and
rejection by SoS   plan approved by SoS
(sect. 16(4))   (sect. 16(4))

Note

1. For height of chimneys serving furnaces, see FC63.

This procedure applies where plans deposited under building regulations for buildings other than the following:

(a) residences; or
(b) shops; or
(c) offices;

show the proposed construction of a chimney, other than one serving a furnace for carrying smoke, grit dust or gases (sect. 16(1)).

Where plans are not deposited because the 'building notice' procedure is being used, the application of this chimney height approval procedure may be dealt with as a linked power.

Criteria for decision

The LA is required to reject the plans unless it is satisfied that the height of the chimney will be sufficient to prevent, as far as practicable, the smoke, grit, dust or gases from becoming prejudicial to health or a nuisance having regard to:

(a) the purpose of the chimney;
(b) the position and description of buildings nearby;
(c) the levels of neighbouring ground;
(d) any other matters requiring consideration in the circumstances (sect. 16(2)).

SMOKE CONTROL

SMOKE CONTROL AREAS

References

Clean Air Act 1993 sects. 18–29.
Smoke Control Areas (Exempted Fireplaces) Order 1996.
Smoke Control Areas (Authorized Fuels) Regulations (Various).

Directions by SoS

Where the SoS (or in Scotland the Scottish Environmental Protection Agency) considers it expedient to abate pollution by smoke and the LA concerned has not exercised its powers to make SCAs sufficiently or at all, he may, after consulting the LA, direct it to submit a programme to him. Following approval of its proposals by the SoS or, in its default, following the determination of a programme by the SoS, the LA is under a statutory obligation to comply (CAA 1993 sect. 19).

FC65a Smoke control areas: directions by the SoS

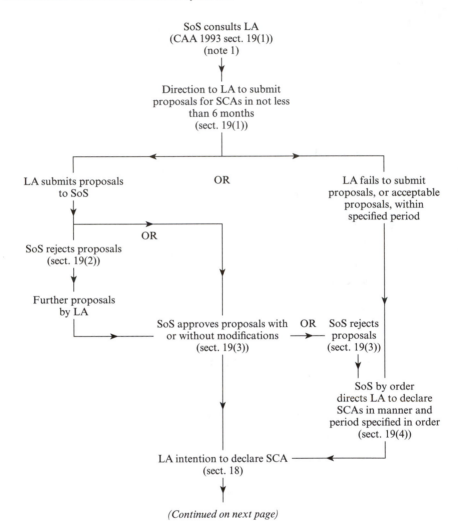

(Continued on next page)

Note

1. In Scotland, the powers of the SoS are exercised by the Scottish Environmental Protection Agency and the procedure chart needs to be read accordingly.

FC65b Smoke control areas: declaration by LA

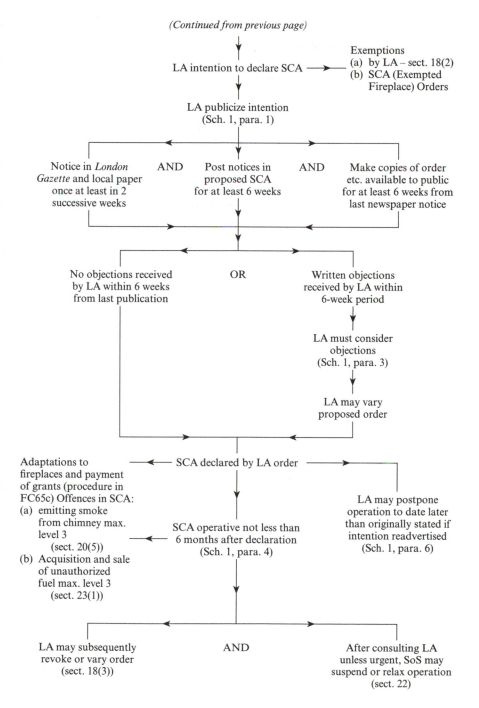

(Continued from previous page)

LA intention to declare SCA → Exemptions
(a) by LA – sect. 18(2)
(b) SCA (Exempted Fireplace) Orders

LA publicize intention
(Sch. 1, para. 1)

Notice in *London Gazette* and local paper once at least in 2 successive weeks
AND
Post notices in proposed SCA for at least 6 weeks
AND
Make copies of order etc. available to public for at least 6 weeks from last newspaper notice

No objections received by LA within 6 weeks from last publication
OR
Written objections received by LA within 6-week period

LA must consider objections
(Sch. 1, para. 3)

LA may vary proposed order

Adaptations to fireplaces and payment of grants (procedure in FC65c) Offences in SCA:
(a) emitting smoke from chimney max. level 3 (sect. 20(5))
(b) Acquisition and sale of unauthorized fuel max. level 3 (sect. 23(1))

SCA declared by LA order

LA may postpone operation to date later than originally stated if intention readvertised
(Sch. 1, para. 6)

SCA operative not less than 6 months after declaration
(Sch. 1, para. 4)

LA may subsequently revoke or vary order
(sect. 18(3))
AND
After consulting LA unless urgent, SoS may suspend or relax operation
(sect. 22)

FC65c Smoke control areas: adaptations to fireplaces etc.

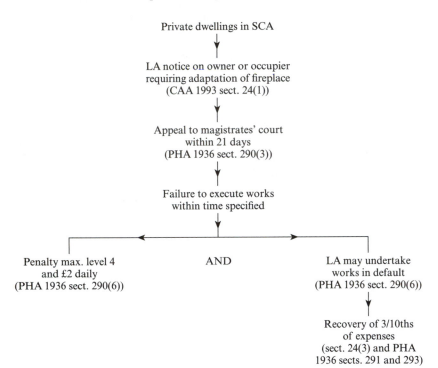

Private dwellings in SCA

↓

LA notice on owner or occupier
requiring adaptation of fireplace
(CAA 1993 sect. 24(1))

↓

Appeal to magistrates' court
within 21 days
(PHA 1936 sect. 290(3))

↓

Failure to execute works
within time specified

↓

Penalty max. level 4
and £2 daily
(PHA 1936 sect. 290(6))

AND

LA may undertake
works in default
(PHA 1936 sect. 290(6))

↓

Recovery of 3/10ths
of expenses
(sect. 24(3) and PHA
1936 sects. 291 and 293)

Note

1. The provisions of Part 12 PHA 1936 relating to appeals against, and the enforcement of, notices requiring the execution of works apply to notices in this procedure (sect. 24(2)) (page 13).

These powers have been used to secure compliance with the EC Air Quality Directive on sulphur dioxide and suspended particulates 80/779/EEC.

Smoke control areas

These may relate to the whole or part of a LA area (sect. 18(1)) and in them the emission of smoke from the chimney of a building, or a chimney serving a fixed boiler or industrial plant is generally prohibited (sects. 20(1) and (2)).

Exemptions

Exemptions from the general requirement not to emit smoke can come about in the following ways:

(a) The burning of an authorized fuel, i.e. one designated by one of the SCA (Authorized Fuels) Regulations, is a defence against proceedings (sect. 20(4) and (6)).
(b) By the use of one of a class of fireplaces exempted by a SCA (Exempted Fireplaces) Order following the SoS being satisfied that the appliance is capable of burning other than authorized fuels without producing any or a substantial quantity of smoke (sect. 21).
(c) Exemption granted by the LA in respect of a specified or class of building or specified or class of fireplaces, upon such conditions as the LA may specify (sect. 18(2)).
(d) Exemptions granted by the LA for the purpose of investigation and research (sect. 45).

Offences

Within a SCA the following are offences:

(a) emitting smoke from the chimney of a building etc. (sect. 20(1) and (2)) subject to the exemptions listed above;
(b) acquiring solid fuel, other than authorized fuel for use in a building, fixed boiler or industrial plant in a SCA, unless intended for use in an exempted building or fireplace (sect. 23(1)(a) and (b));
(c) selling by retail solid fuel, other than authorized fuel, for delivery to a building or for any fixed boiler or industrial plant in a premises in a SCA. In this case the person cannot be convicted if he had reasonable grounds for believing:
  (i) that the building was exempted or
  (ii) that the fireplace or plant for which the fuel was intended was exempted (sect. 23(1)(c) and (5)).

Upon becoming aware that an offence under sect. 20 relating to emissions of smoke in SCAs has been committed, an authorized officer of the LA must:

(a) notify the person responsible as soon as may be; and
(b) confirm the offence in writing before the end of the 4 days following the day on which the offence was committed (sect. 51).

Adaptation of fireplaces in private dwellings

The notice procedure may be used where adaptations are required to avoid contraventions of sect. 20 and these may be served on the owner or occupier of the private dwelling either between the date of making the order and its operation, or after the date of operation. The period allowed for compliance must be reasonable and not less than the appeal period, i.e. 21 days. The provisions of Part 12 of the PHA 1936 apply to the service and enforcement of these notices, but costs of carrying out work in default can be recovered only to an amount up to three-tenths of the expenditure by the LA (sect. 24(3)).

Grants

The LA is required to pay grants to owners or occupiers of private dwellings to assist with the adaptation of fireplaces where this is required to avoid contraventions of sect. 20, provided that the works are carried out to the LA's satisfaction before the order becomes operative, unless the LA has served a notice as above.

The normal grant is seven-tenths of necessary expenditure, but the LA may pay more at its discretion. Payments to occupiers other than owners for moveable cooking or heating appliances are made at seven-twentieths initially and the remainder after 2 years, provided that the appliance is still present (Sch. 2, para. 1). The LA has a discretion to make up to 100% grants for adaptations in churches, chapels or buildings used by charities (sect. 26).

Exchequer contributions towards the LA's costs ceased with orders coming into operation from 31 March 1996 other than in exceptional cases where, for example, the LA may breach 80/779/EEC and funding is not available from the LA's own resources. Otherwise LA funding requirements arising from its SCA programme will be recognized by the SoS for England by the use of the Supplementary Credit Approval System and in Wales as part of the annual Basic Credit Approval for that LA (DoE Circular 9/93).

Meaning of 'adaptations'

In relation to the payment of grants and to notices, 'adaptations' include:

(a) adapting or converting a fireplace; or
(b) replacing any fireplace by another fireplace or by some other means of heating or cooking; or
(c) altering any chimney which serves any fireplace; or

(d) providing gas ignition, electric ignition or any other special means of ignition; or

(e) carrying out any operation incidental to any of the operations (a)–(d) above, and includes works both inside and outside that dwelling.

The works must be reasonably necessary in order to make what is, in all circumstances, suitable provision for heating and cooking without contravention of sect. 20 (sect. 27).

Designated appliances

Classes of heating appliance designated by a LA for its area, or by the Minister either generally or in the area of defined LAs, which would impose undue strain on the fuel resources available, do not qualify for LA grant (Sch. 2 para. 2).

ATMOSPHERIC POLLUTION

OBTAINING INFORMATION ABOUT ATMOSPHERIC POLLUTION

References

Clean Air Act 1993 sects. 35–40 (as amended by the Pollution Prevention and Control Act 1999).
Control of Atmospheric Pollution (Research and Publicity) Regulations 1977.
Control of Atmospheric Pollution (Appeals) Regulations 1977.
Control of Atmospheric Pollution (Exempted Premises) Regulations 1977.

General provision for obtaining information etc.

A LA may obtain information about pollutants and other substances in the air from any premises, other than private dwellings or caravans in any of the following ways:

(a) issuing notices on occupiers under sect. 36; or
(b) measuring and recording itself; or
(c) entering into an arrangement for occupiers to measure and record information on behalf of the LA (sect. 35(1)).

Whilst LAs are able to obtain information under sect. 36 about processes subject to part 1 of the EPA 1990 and activities subject to regulation under sect. 2 of the PP and CA 1999, they are prevented by sect. 35(3) from undertaking any investigations into any emissions from those processes and activities unless they are the enforcing authority – see FC27. Where LAs do use this power, this can only relate to information certified by the EA to be of a kind already being supplied to them (sect. 32(6) CAA 1993 as amended

FC66 Obtaining of information about atmosphere pollution

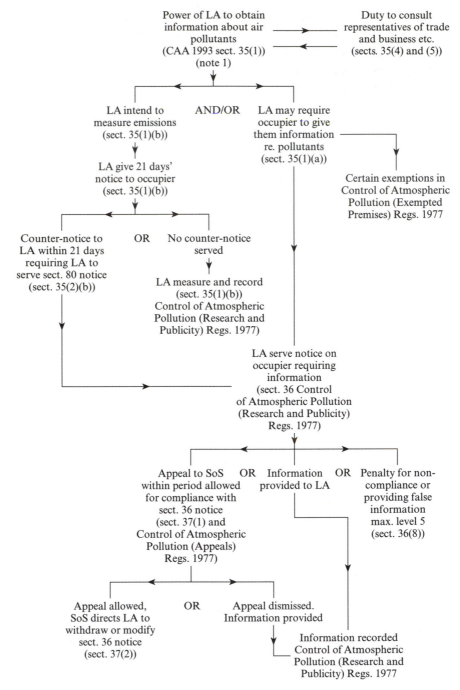

Note

1. For powers to require information about grit and dust, see CAA 1993 sect. 12.

by para. 12 of schedule 10 of the Pollution Prevention and Control (England and Wales) Regulations 2000).

The LA may also undertake, or contribute towards, investigation and research relevant to air pollution and publish information (sect. 34(1)).

Duty to consult

In exercising any of its powers to obtain information, the LA must consult from time to time, at least twice in each financial year, with trade and business and their representatives and any other persons conversant with air pollution problems or having an interest (sect. 35(4)–(6)).

Measurement by LAs

Before exercising their powers to measure pollution the LAs must, unless proceeding by agreement, serve 21 days' notice on the occupier stating:

(a) the kind of emissions in question; and
(b) the steps to be taken to measure and record.

There is no appeal against this notice but a counter-notice may be served within the 21 days requesting the LA to serve notice on them under sect. 36 requiring the occupier to provide information about air emissions (sect. 35(2)).

The LA must register the information recorded (Control of Atmospheric Pollution (Research and Publicity) Regulations 1977 para. 6). This register must be open to public inspection free of charge (sect. 38(5)).

Sect. 36 notices

These may be served on occupiers, other than occupiers of exempted or private dwellings, and require the provision of any information relating to the emission of the following pollutants and other substances from any chimney, flue or other outlet to the atmosphere:

(a) SO_2 or particulate matter derived from any combustion process where the material being heated does not contribute to the emission;
(b) any gas or particulate matter derived from any combustion process where the material being heated contributes to the emission;
(c) any gas or particulate matter derived from any non-combustion process or similar industrial activity (Control of Atmospheric Pollution (Research and Publicity) Regulations 1977 Reg. 3).

There are also more detailed requirements as to the type of information of these kinds which may be sought in the notice.

The period allowed for compliance must be at least 6 weeks. No one notice can cover information for a period exceeding 1 year and periodic information may be requested at not greater than quarterly intervals (sect. 36(4) and (5)).

Certain Crown premises are exempted from these provisions and these are listed in the Control of Atmospheric Pollution (Exempted Premises) Regulations 1977, but otherwise the procedure applies to Crown property, except that there is no power of entry (sect. 36(6)).

Appeals against sect. 36 notices

Appeal is to the SoS by the recipient of the notice or other person having an interest on the following grounds:

(a) the information would prejudice to an unreasonable degree a private interest by disclosing a trade secret;
(b) the information would be contrary to the public interest;
(c) the information is not immediately available and cannot be made so without incurring undue expenditure (sect. 37(1)).

The procedure for dealing with appeals is laid down in the Control of Atmospheric Pollution (Appeals) Regulations 1977.

Recording of information

Information obtained by the LA following sect. 36 notices must be kept in a register and adequately indexed (Control of Atmospheric Pollution (Research and Publicity) Regulations 1977 Reg. 6). This register must be open to public inspection free of charge (sect. 38(5)).

Definitions

Emissions of substances – references to the emission of substances into the atmosphere are to be construed as applying to substances in a gaseous or liquid or solid state, or any combination of those states (sect. 40(a)).

Chapter 10

CONTROL OF POLLUTION ACT 1974 (NOISE CONTROL PROVISIONS)

GENERAL PROCEDURAL PROVISIONS

The following provisions are generally applicable to procedures under the CPA 1974.

Extent

The procedures in this chapter are generally applicable in England (including the Scilly Isles), Wales and in Scotland with, in the latter case, the substitution of authorities and bodies appropriate to the structure of government in that country (CPA 1974 sects. 106 and 109).

None of the procedures in this chapter are directly applicable to Northern Ireland (CPA 1974 sect. 109(3)).

Notices

1. **Form.** Notices must be in writing (CPA 1974 sect. 105(1)).
2. **Authentication.** Notices, orders and any other documents may be signed on behalf of the authority by the proper officer and 'signature' includes a facsimile (LGA 1972 sect. 234).
3. **Service.** All notices involved in procedures in this chapter may be served in the following ways:
   (a) by either delivering it; or
   (b) by leaving it at the proper address; or
   (c) by sending it by post to the proper address; or
   (d) where after reasonable enquiry the name and address cannot be ascertained, by leaving it in the hands of a person resident or employed there, or by leaving it conspicuously affixed to some building or object on the land.
   The 'proper address' referred to is:
   (a) the last known address; or
   (b) in the case of a body corporate, the registered or principal office; or
   (c) in the case of a partnership, the principal office.

The principal office of a company registered outside the UK or of a partnership carrying on a business outside the UK shall be its principal office in the UK. If a person to be served has specified an address within the UK other than his 'proper address' as defined above, that address may be treated as his 'proper address' (LGA 1972 sect. 233).

4. **Information regarding ownership.** In order to obtain information about an owner or lessee, a LA may by notice in writing to the occupier require him to give that information. If the occupier refuses to do so or wilfully gives false information, he is liable to a maximum penalty of level 1 on the standard scale (LGA 1972 sect. 233(6)).

See also general power to obtain information regarding ownerships etc. LG(MP)A 1976 (page 10).

Power of entry

Any person authorized in writing by an authority charged with responsibilities under the CPA 1974 may enter any land or vessel at any reasonable time for the following purposes:

(a) perform any function conferred on the authority;
(b) determine whether such a function should be performed;
(c) determine whether the Act or instruments made by virtue of the Act are being complied with;
(d) carry out inspections, measurements and tests (CPA 1974 sect. 91(1)).

Admission to land or a vessel used for residential purposes, and admission with heavy equipment to any other land or vessel, cannot be demanded of right unless at least 7 days' notice of intention has been given, except in emergency or where the land or vessel is unoccupied (CPA 1974 sect. 92(3)). Where:

(a) admission has been refused; or
(b) refusal is apprehended; or
(c) the land or vessel is unoccupied; or
(d) the occupier is temporarily absent; or
(e) the case is one of emergency; or
(f) application for admission would defeat the object of entry **and** there is a reasonable ground for entry

the authority may apply to a JP for a warrant to enter, if need be by force (CPA 1974 sect. 91(2)).

An authorized person may take with him such persons and equipment as necessary (CPA 1974 sect. 92(2)).

After entering unoccupied land, the authorized person must leave it as secure against trespass as he found it (CPA 1974 sect. 92(4)). The authority is liable to pay compensation for any damage sustained in exercising its powers and responsibilities in this respect (CPA 1974 sect. 92(5)). The maximum penalty for obstruction is level 3 on the standard scale (CPA 1974 sect. 92(6)).

Information

An authority is provided with a general power to obtain information necessary to perform its functions under the CPA by the service of notice specifying the information required and giving a time period within which or time at which it must be supplied. The maximum penalty for non-compliance or wilfully giving false information is level 5 on the standard scale (CPA 1974 sect. 93).

DEFINITIONS

Best practicable means. 'Practicable' means reasonably practicable having regard among other things to local conditions and circumstances, to the current state of technical knowledge and to the financial implications. 'Means' includes design, installation, maintenance and manner and periods of operation of plant and machinery, and the design, construction and maintenance of buildings and acoustic structures. The test of best practicable means is to apply only so far as is compatible with statutory duties imposed, with safety and safe working practices and with the exigencies of any emergency or unforeseen circumstances. Regard shall be had to any relevant provisions of an approved code of practice (CPA 1974 sect. 72).

Local authority means:

(a) in England the council of a district or a London borough, the Common Council of the City of London, the Sub-Treasurer of the Inner Temple and the Under-Treasurer of the Middle Temple;
(b) in Wales, the council of a county or county borough; and
(c) in Scotland, an island or district council (CPA 1974 sect. 73(1)).

This definition includes the unitary authorities in England and Scotland as well as those in Wales.

Noise includes vibration (CPA 1974 sect. 73(1)).

Notice means a notice in writing (CPA 1974 sect. 105(1)).

Owner except in Scotland, means the person for the time being receiving the rackrent of the premises in connection with which the word is used whether on his own account or as agent or trustee for another person, or who would so receive the rackrent if the premises were let at a rackrent (CPA 1974 sect. 105(1)).

Person responsible in relation to the emission of noise, means the person to whose act, default or sufferance the noise is attributable (CPA 1974 sect. 73(1)).

Premises includes land (CPA 1974 sect. 105(1)).

NOISE FROM CONSTRUCTION SITES

References

Control of Pollution Act 1974 sects. 60 and 61.
Control of Noise (Appeals) Regulations 1975.
Control of Noise (Codes of Practice for Construction and Open Sites) (England) Order 2002.

Scope

The first of these two procedures stems from an application for consent to the LA, which, if operated in accordance with any conditions or with limitations, provides a defence against any proceedings taken by a LA following the second procedure, i.e. a LA notice specifying noise emission standards. The procedures could be operated together but only beneficially where, following application to the LA, consent has been granted without conditions or limitations, or no consent was given.

The procedures are applicable to the following works:

(a) erection, construction, alteration, repair or maintenance of buildings, structures or roads;
(b) breaking up, opening or boring under any road or adjacent land in connection with the construction, inspection, maintenance or removal of works;
(c) demolition or dredging works;
(d) any other works of engineering construction (CPA 1974 sect. 60(1)).

Criteria for consideration

In judging both applications for consent and whether or not to serve notices, LAs must have regard to the following points:

(a) BS 5228 part 1 1997, part 3 1997, part 4 1992 and part 5 1992. Noise control on construction and open sites;
(b) the need to ensure that best practicable means are employed to minimize noise;
(c) in considering the specification of any particular methods or plant etc., the interest of the applicant or recipient of the notice in regard to alternative methods etc. which might be as effective but more acceptable to them;
(d) the need to protect any persons in the locality from the effects of noise (CPA 1974 sect. 60(4)).

LA notices

1. **Person responsible.** This is the person carrying out, or going to carry out, the works and any other persons who appear to the LA to be responsible

FC67 Noise from construction sites

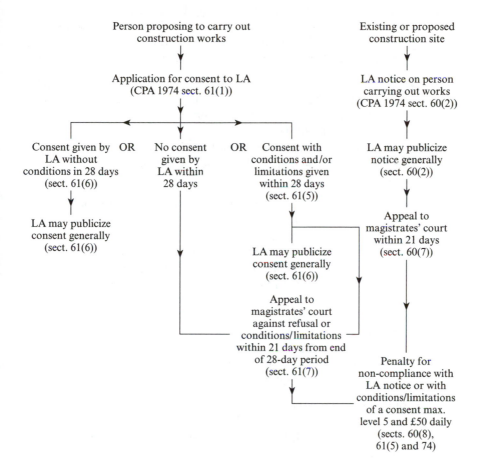

Note
1.   There are no LA powers of default following the notice procedure.

for, or have control over, those works. Notices may be served on all or any such persons (CPA 1974 sect. 60(5)).

2. **Content.** Notices may include:
    (a) specification of plant and machinery which is or is not to be used;
    (b) the hours during which the works may be carried out;
    (c) specification of noise levels from the premises, from any specified point on the premises or during specified hours;
    (d) a provision for any change of circumstances (CPA 1974 sect. 60(3)).

A notice may specify a time within which the notice should be complied with, which should not be less than the appeal period, i.e. 21 days, and may require the execution of works or other steps, but the LA has no default powers (CPA 1974 sect. 60(6)).

If the LA intends that the notice should not be suspended in the event of an appeal, this must be stated in the notice (Control of Noise (Appeals) Regulations 1975).

3. **Appeals.** The grounds of appeal are set out in the Control of Noise (Appeals) Regulations 1975 as:
    (a) the notice is not justified;
    (b) there is some material informality, defect or error in the notice;
    (c) the LA has refused unreasonably to accept compliance with alternative requirements or the requirements of the notice are unreasonable or unnecessary;
    (d) that any time period allowed for compliance is not reasonably sufficient;
    (e) the notice should have been served on some other person in substitution for, or in addition to, the appellant;
    (f) that the LA has not had regard to the matters set out in sect. 60(4).

The court may dismiss the appeal, vary or quash the notice and/or make an order as to respective responsibilities for compliance and costs.

In the event of an appeal, LA notices are suspended until the appeal has been determined or abandoned where:

    (a) the noise is caused in the performance of a statutory duty imposed on the appellant; or
    (b) compliance would involve expenditure being incurred before the hearing;

except in those cases where the LA is of the opinion that:

    (a) the noise is injurious to health; or
    (b) the noise is likely to be of limited duration so that suspension would give no practical effect to the notice; or
    (c) the expenditure to be incurred would not be disproportionate to the public benefit from compliance;

provided that this has been stated on the original LA notice (Control of Noise (Appeals) Regulations 1975).

5. **Defence.** It is a defence in any proceedings for non-compliance with a LA notice to prove that the alleged contravention amounted to the carrying out of the works in accordance with a consent under sect. 61 (CPA 1974 sect. 61(8)).

Consents

1. **Applications.** These must be made at the same time as, or later than, any application for building regulations approval and may be made by anyone intending to carry out construction works. The application must contain particulars of:
   (a) the works proposed and the methods to be used to carry them out; and
   (b) the steps proposed to minimize noise (CPA 1974 sect. 61(2) and (3)).
2. **Consent.** If the LA considers that it would not serve a notice under sect. 60 if the works are carried out in accordance with the application it must give consent and may:
   (a) attach conditions;
   (b) limit or qualify consent to allow for any changes in circumstances;
   (c) limit the duration of the consent (CPA 1974 sect. 61(4) and (5)).

   A consent under this procedure does not of itself constitute a defence against proceedings for statutory nuisance by aggrieved persons under sect. 80 EPA 1990, and this must be stated in the consent. However, it will be a defence against proceedings by a LA for failure to comply with an abatement notice under sect. 80 EPA 1990 to show that the alleged offence was covered by a notice under sect. 60 CPA or by a consent under sects. 61 or 65 CPA. It will be a defence in proceedings following sect. 60 notices to show that the works were carried out in accordance with that consent (CPA 1974 sect. 61(8) and (9) and sect. 80(9) EPA 1990).
3. **Appeals.** Grounds of appeal against no consent being granted or against attachment of conditions, limitations or qualifications are set out in the Control of Noise (Appeals) Regulations 1975 as being:
   (a) that any condition, limitation or qualification is not justified;
   (b) that there is some material informality, defect or error in connection with the consent;
   (c) that the requirements of any condition are unreasonable in character or extent, or are unnecessary;
   (d) that the time within which any conditions etc. are to be complied with is not reasonably sufficient.

Publication

In order to bring either notices under sect. 60 or consents under sect. 61 to the attention of the public who might be affected, the LA is able to publicize the contents of notices and consents in any ways which it thinks appropriate (CPA 1974 sects. 60(2) and 61(6)).

Definition

Work of engineering construction means the construction, structural altera-tion, maintenance or repair of any railway line or siding or any dock, harbour or inland navigation, tunnel, bridge, viaduct, waterworks, reservoir, pipeline, aqueduct, sewer, sewage works or gas holder (CPA 1974 sect. 73(1)).

NOISE ABATEMENT ZONES: DESIGNATION BY LAs

Reference

Control of Pollution Act 1974 sect. 63 and Sch. 1.

LA duty to inspect district

Each LA is under a statutory obligation to inspect its district from time to time to decide how to exercise its powers relating to NAZs, but this does not need to precede the designation of each zone (CPA 1974 sect. 57(b)).

Noise abatement zones

The order designating a NAZ may relate to all or part only of the LA dis-trict and may subsequently be revoked or varied by the LA. The order must detail the classes of premises which will be covered by the NAZ (CPA 1974 sect. 63).

Publication of intention

Before declaration of the NAZ, or before revoking or varying it, the LA is required to:

(a) serve notice on every owner, lessee and occupier, other than tenants of 1 month or less, of each premises in the area of a class to which the order relates; and
(b) publish a notice in the *London Gazette* and at least once in each of two successive weeks in a local newspaper.

These notices referred to in (a) and (b) above must contain:

(a) a statement that LA intends to make that order;
(b) the general effect of the order;
(c) specify where copies of the order and maps etc. may be seen by the public free of charge for at least 6 weeks from the last newspaper publication of the notice;
(d) a statement that objections may be made in writing to a LA within a 6-week period (CPA 1974 Sch. 1, paras. 1 and 2).

FC68a Noise abatement zones: designation by LAs

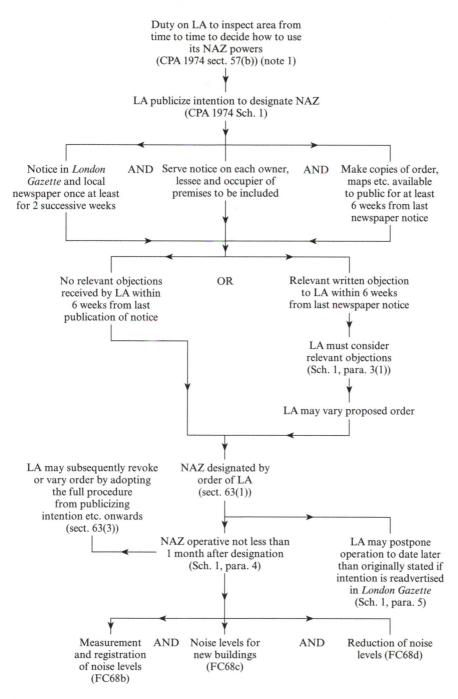

Duty on LA to inspect area from
time to time to decide how to use
its NAZ powers
(CPA 1974 sect. 57(b)) (note 1)

LA publicize intention to designate NAZ
(CPA 1974 Sch. 1)

Notice in *London Gazette* and local newspaper once at least for 2 successive weeks AND Serve notice on each owner, lessee and occupier of premises to be included AND Make copies of order, maps etc. available to public for at least 6 weeks from last newspaper notice

No relevant objections received by LA within 6 weeks from last publication of notice OR Relevant written objection to LA within 6 weeks from last newspaper notice

LA must consider
relevant objections
(Sch. 1, para. 3(1))

LA may vary proposed order

LA may subsequently revoke or vary order by adopting the full procedure from publicizing intention etc. onwards (sect. 63(3))

NAZ designated by
order of LA
(sect. 63(1))

NAZ operative not less than
1 month after designation
(Sch. 1, para. 4)

LA may postpone operation to date later than originally stated if intention is readvertised in *London Gazette* (Sch. 1, para. 5)

Measurement and registration of noise levels (FC68b) AND Noise levels for new buildings (FC68c) AND Reduction of noise levels (FC68d)

Note

1. This is not a necessary prerequisite to the declaration of each NAZ.

Objections

Before making an order designating a NAZ, the LA must consider any objections received during the 6-week period unless it is satisfied that it need not do so having regard to:
  (a) the nature of the premises to which the order will relate;
  (b) the nature of the interests of the person making the objection;
  (c) in the case of an order revoking or varying a previous order, that the substance of the objection is the same as an objection made to the initial order (CPA 1974 Sch. 1, paras. 3 and 4).

Operation

A NAZ becomes operative on a date specified in the order by the LA which, except for orders revoking or modifying previous orders, must be at least 1 month from the date on which the order was made (CPA 1974 Sch. 1, para. 4).

NOISE ABATEMENT ZONES: MEASUREMENT AND REGISTRATION OF NOISE LEVELS AND CONSENT TO EXCEED THEM

References

Control of Pollution Act 1974 sects. 64 and 65.
Control of Noise (Appeals) Regulations 1975.
Control of Noise (Measurements and Registers) Regulations 1976.

Measurement

Once a NAZ becomes operative the LA is required to measure noise levels emanating from premises within the area which are subject to the noise abatement order (CPA 1974 sect. 64(1)).

Detailed methods of measurement and calculation are laid down in the Control of Noise (Measurement and Registers) Regulations 1976.

Recording of measurements

The LA is required to record the measured levels in a register, the form and content of which is detailed in para. 6 Control of Noise (Measurement and Registers) Regulations 1976. No time period is stipulated within which these measurements must be taken. Copies of the entries are to be sent to the owner and occupier of each premises and they may appeal to the SoS (CPA 1974 sect. 64(3)).

The register must be open to public inspection free of charge and copies of entries provided at reasonable charge (CPA 1974 sect. 64(7)).

FC68b Noise abatement zones: measurement and registration of noise levels and consent to exceed them

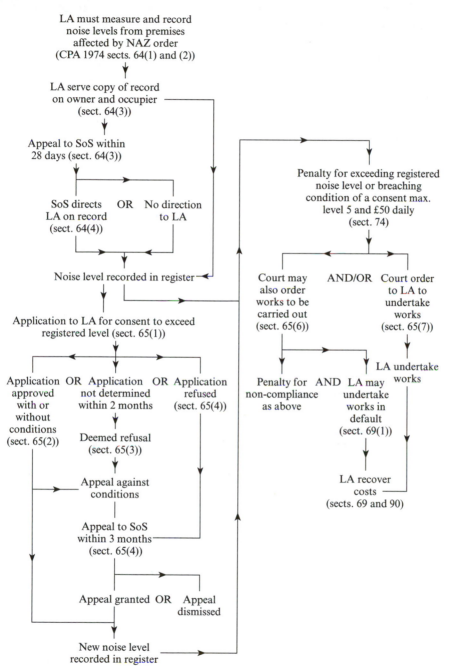

Notes
1. For noise reduction notices in NAZs, see FC68d.
2. For provisions relating to new buildings in NAZs, see FC68c.

Consents to exceed registered level

Application for such consent may be made to the LA and these may be granted, with or without conditions, or refused. If granted by the LA or subsequently on appeal, the consent must be recorded in the noise level register (CPA 1974 sect. 65(2)).

Appeals

The procedure relating to appeals to the SoS against the recorded noise level for a premises, refusal to consent to exceed the registered noise level or against conditions attached to a consent is laid down in para. 9 Control of Noise (Appeals) Regulations 1975. Within 7 days of lodging the appeal, the appellant must send to the SoS:

(a) the application made to the LA (if any);
(b) any relevant plans etc. submitted to the LA;
(c) any relevant records, consents, notices etc. issued by the LA;
(d) copies of correspondence with the LA;
(e) a plan of the premises concerned.

The SoS may require further written evidence from the appellant and from the LA and may hold a local inquiry at his discretion. No specific grounds for appeal are laid down.

Offences

An offence is committed if the registered noise level is exceeded, or a condition attached to a consent is breached or a court order to undertake works is not complied with (CPA 1974 sects. 65(5) and (6)).

No specific defences are laid down. A consent to exceed the registered noise level does not of itself provide a defence against proceedings relating to noise nuisance by aggrieved persons under sect. 82 EPA 1990, but is a defence against proceedings for failure to comply with a LA abatement notice under sect. 80 EPA 1990 (CPA 1974 sect. 65(8) and EPA 1990) sect. 80(9)(a)).

NOISE ABATEMENT ZONES: NOISE LEVEL
DETERMINATIONS FOR NEW BUILDINGS

References

Control of Pollution Act 1974 sect. 67.
Control of Noise (Appeals) Regulations 1975.

New buildings

This procedure applies in NAZs where:

(a) a new building is to be erected which will be used for a purpose classified by the noise abatement order as one to which the order will apply; and

(b) existing buildings which, when works are completed, will be used for a purpose covered by the order.

Applications to LA

LAs may decide to determine a registered noise level on their initiative but applications to them to do so may be made by:

(a) the owner or occupier; or

(b) any person negotiating to acquire an interest in the premises.

Appeals

The provisions covering appeals to the SoS are laid down in Part 3 Control of Noise (Appeals) Regulations 1975 and are the same as those relating to registration of noise levels generally in NAZs as set out on page 268.

Defences

It will be a defence in proceedings for non-compliance with a LA abatement notice under sect. 80 EPA 1990 to show that the noise did not exceed the level registered under this procedure (EPA 1990 sect. 80(9)(c)).

NOISE ABATEMENT ZONES: NOISE REDUCTION NOTICES

References

Control of Pollution Act 1974 sect. 66.
Control of Noise (Appeals) Regulations 1975.

Criteria for serving notices

Before serving a noise reduction notice the LA must be satisfied that:

(a) the premises concerned are of a class to which the noise abatement order designating the NAZ applies;

(b) the noise level emanating from the premises is not acceptable having regard to the purposes of the NAZ;

(c) a reduction in noise level is practicable at reasonable cost and would afford a public benefit (CPA 1974 sect. 66(1)).

Noise reduction notices

The notice must include:

FC68c Noise abatement zones: noise level determinations for new buildings

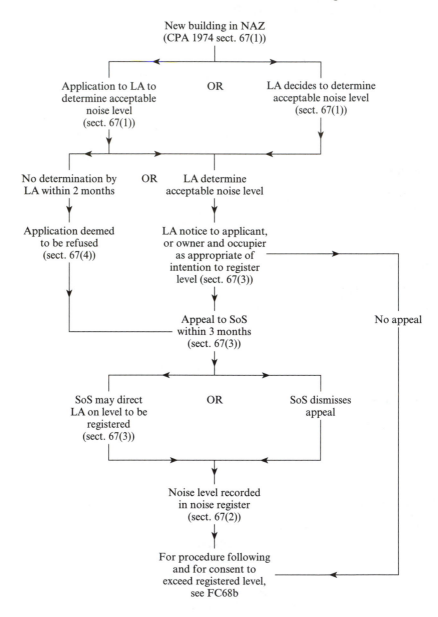

New building in NAZ
(CPA 1974 sect. 67(1))

Application to LA to determine acceptable noise level (sect. 67(1))

OR

LA decides to determine acceptable noise level (sect. 67(1))

No determination by LA within 2 months

OR

LA determine acceptable noise level

Application deemed to be refused (sect. 67(4))

LA notice to applicant, or owner and occupier as appropriate of intention to register level (sect. 67(3))

Appeal to SoS within 3 months (sect. 67(3))

No appeal

SoS may direct LA on level to be registered (sect. 67(3))

OR

SoS dismisses appeal

Noise level recorded in noise register (sect. 67(2))

For procedure following and for consent to exceed registered level, see FC68b

Notes
1. For measurement of noise levels generally in NAZs, see FC68b.
2. For noise reduction notices for new buildings, see FC68d.
3. For noise nuisances, see FC33 and 34.

FC68d Noise abatement zones: noise reduction notices

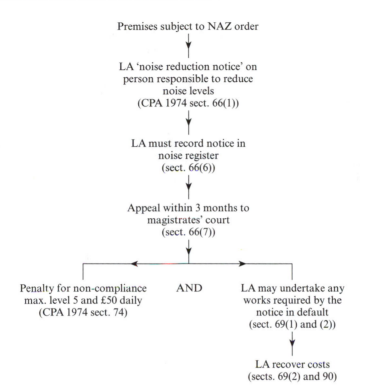

Premises subject to NAZ order

LA 'noise reduction notice' on
person responsible to reduce
noise levels
(CPA 1974 sect. 66(1))

LA must record notice in
noise register
(sect. 66(6))

Appeal within 3 months to
magistrates' court
(sect. 66(7))

Penalty for non-compliance   AND   LA may undertake any
max. level 5 and £50 daily   works required by the
(CPA 1974 sect. 74)   notice in default
                                                     (sect. 69(1) and (2))

                                                     LA recover costs
                                                     (sects. 69(2) and 90)

Notes
1. For declaration of NAZ, see FC68a.
2. For measurement and registration of noise levels in NAZs and consent to exceed, see FC68b.

(a) a requirement to reduce the level of noise to one specified in the notice;
(b) a prohibition on any subsequent increase in the level without the consent of the LA;
(c) the specification of such works as may be necessary to achieve (a) and (b) above (CPA 1974 sect. 66(2)).

In addition, the notice may specify particular times or days during which the level is to be reduced and different noise levels may be set for different times or days (CPA 1974 sect. 66(4)).

The LA must specify in the notice a time within which the noise level is to be reduced and within which any required works are to be completed. This time must be not less than 6 months or, for new buildings, 3 months (CPA 1974 sects. 66(3) and 67(5)).

A noise reduction notice may be served even if a consent under sect. 65 has been given (page 266) (CPA 1974 sect. 66(7)).

Notices are served on 'the persons responsible' (definition on page 259) (CPA 1974 sect. 66(1)).

Appeals

The grounds for appeal are set out in the Control of Noise (Appeals) Regulations 1975 as including:

(a) the notice is not justified;
(b) there is some material informality, defect or error in connection with the notice;
(c) the LA has unreasonably refused compliance with alternative requirements or the requirements of the notice are unreasonable in character or extent or are unnecessary;
(d) the time allowed is not reasonably sufficient;
(e) where the noise emanates from a trade or business, that the best practicable means have been used;
(f) the notice should have been served on some other person in addition to, or in substitution for, the appellant.

The court may vary or quash the notice or dismiss the appeal. The provisions relating to the suspension of notices pending the determination of appeal are as set out for noise from construction sites on page 262.

Defences

In proceedings for contravention of a noise reduction notice relating to a trade or business, it is a defence to prove that best practicable means have been used (CPA 1974 sect. 66(9)).

In respect of any proceedings for non-compliance with a LA abatement notice under EPA 1990 sect. 80, it will be a defence to show that the noise level at the time did not contravene the requirements of any noise reduction notice under this procedure (EPA 1990 sect. 80(9)(b)).

New buildings

These provisions relating to noise reduction notices also apply to buildings which:

(a) are to be constructed and are a class to which the noise abatement order applies; and

(b) will be of such a class when works are completed except that:

    (i) before serving a notice the LA need not be satisfied that noise reduction is practicable at reasonable cost and would afford a public benefit;

    (ii) the minimum period allowed for compliance is 3 months and not 6 months;

    (iii) the defence of best practicable means for trade or business is not available (CPA 1974 sect. 67(5)).

CONSENTS FOR THE USE OF LOUDSPEAKERS IN STREETS

References

Control of Pollution Act 1974 sect. 62.
Noise and Statutory Nuisance Act 1993 sect. 8 and Sch. 2.
DoE Circular 9/97 The Noise and Statutory Nuisance Act 1993.

Scope

This procedure allows a LA to consent to the operation of a loudspeaker in a street between the hours of 9 pm and 8 am being the period during which such operation is generally prohibited under Section 62(1) CPA 1984 (NSNA 1993 Sch. 2, para. 1).

Consents may not be granted in connection with any election or for advertising any entertainment, trade or business (Sch. 2, para. 1(2)). It is thought that LAs may wish to use a consent system, for example to allow charity events or street events to continue beyond 9 pm.

Adoption of scheme

The consent scheme operates only if the LA has adopted its provisions by resolution of the Council. Notice that such a resolution has been made must be published in a local newspaper in two consecutive weeks before becoming operative on a date specified, which cannot be less than 1 month from the date of the resolution.

The notice in the newspaper must set out the effect of the resolution and explain the procedure for applying for consents (sect. 8(1)–(4)).

FC69 Consents for the use of loudspeakers in streets

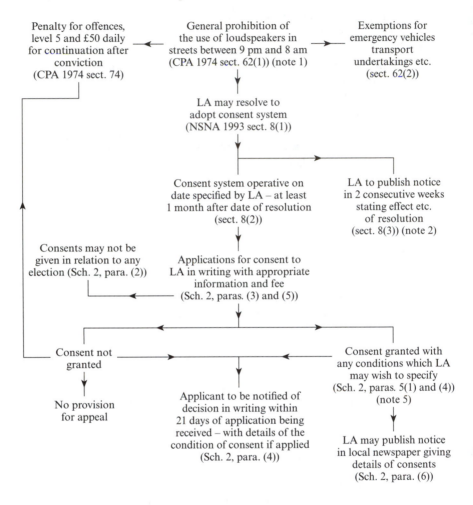

Notes
1. There is a total prohibition at any time on the use of loudspeakers in streets for advertising any entertainment, trade or business (there is an exception for advertising the presence of a vehicle selling perishable commodities e.g. ice cream, between noon and 7 pm). The consent procedure does not allow this prohibition to be breached.
2. There is no provision for the making or consideration of representations.
3. Where not otherwise indicated section and schedule numbers refer to the NSNA 1993.
4. For the control of statutory noise nuisances in streets, see FC33c.
5. Operating a loudspeaker with a consent but in breach of its conditions would appear to create an offence under CPA 1974, i.e. for breaching the general prohibition order (sect. 62(1)).

Applications

These must be made to the LA in writing and must contain the information requested by the LA to enable it to reach a decision. The LA may set a reasonable fee for each application (sect. 2, paras. 3 and 5).

Considerations

The matters to be taken into account by the LA are not specified but might reasonably include:

(a) location;
(b) adjacent activities;
(c) time;
(d) number of loudspeakers and their output power;
(e) methods of noise control;
(f) the planned route of any procession, etc.

Consents

LAs may decide not to issue a consent or issue one at their discretion, with or without conditions. There are no appeal provisions (Sch. 2, para. 1). Notification of the LAs decision must be put in writing to the applicant within 21 days of the application being made with details of any conditions which are to be attached (Sch. 2, para. 4).

The conditions attached to any consent could include the specification of noise levels and of any route to be taken but could cover any issue of legitimate concern to the LA.

The LA may (but is not obliged to) publicize details of any consents in a local newspaper (Sch. 2, para. 6).

Consents and statutory nuisance

The issue of a consent does not affect the operation of procedures to deal with statutory noise nuisances in streets (FC33c).

If during the operation of a consent such a nuisance is created the LA may use its powers under that procedure.

Definitions

Local authority see page 259.
Street means a highway and any other road, footway, square or court which is for the time being open to the public (CPA 1974 sect. 62(1)).

Chapter 11

HEALTH AND SAFETY
AT WORK ACT 1974
(with associated regulations)

GENERAL PROCEDURAL PROVISIONS

Unless otherwise indicated the following provisions are applicable to procedures in this chapter.

Extent and application

The procedures in this chapter apply in Scotland but not in Northern Ireland (sect. 84).

Enforcing authority

Sect. 18(7)(a) of HASAWA 1974 defines enforcing authority as 'the Executive (HSE) or any other authority which is by relevant statutory provision or by regulations . . . made responsible for the enforcement of any of the provisions to any extent'.

The Health and Safety (Enforcing Authority) Regulations 1998 make local authorities (as defined below) responsible for enforcement where the following are the main activities undertaken at any non-domestic premises.

1. The sale or storage of goods for retail or wholesale distribution except:
   (a) at container depots where the main activity is the storage of goods in the course of transit to or from dock premises, an airport or a railway;
   (b) where the main activity is the sale or storage for wholesale distribution of any substance or preparation dangerous for supply;
   (c) where the main activity is the sale or storage of water or sewage or their by-products or natural or town gas;

and for the purposes of this paragraph where the main activity carried on in premises is the sale and fitting of motor car tyres, exhausts, windscreens or sunroofs the main activity shall be deemed to be the sale of goods.

2. The display or demonstration of goods at an exhibition for the purposes of offer or advertisement for sale.

3. Office activities.
4. Catering services.
5. The provision of permanent or temporary residential accommodation including the provision of a site for caravans or campers.
6. Consumer services provided in a shop except dry cleaning or radio and television repairs, and in this paragraph 'consumer services' means services of a type ordinarily supplied to persons who receive them otherwise than in the course of a trade, business or other undertaking carried on by them (whether for profit or not).
7. Cleaning (wet or dry) in coin-operated units in launderettes and similar premises.
8. The use of a bath, sauna or solarium, massaging, hair transplanting, skin piercing, manicuring or other cosmetic services and therapeutic treatments, except where they are carried out under the supervision or control of a registered medical practitioner, a dentist registered under the Dentists Act 1984, a physiotherapist, an osteopath or a chiropractor.
9. The practice or presentation of the arts, sports, games, entertainment or other cultural or recreational activities except where the main activity is the exhibition of a cave to the public.
10. The hiring out of pleasure craft for use on inland waters.
11. The care, treatment, accommodation or exhibition of animals, birds or other creatures, except where the main activity is horse breeding or horse training at a stable, or is an agricultural activity or veterinary surgery.
12. The activities of an undertaker, except where the main activity is embalming or the making of coffins.
13. Church worship or religious meetings.
14. The provision of car parking facilities within the perimeter of an airport.
15. The provision of childcare, or playgroup or nursery facilities.
(sch. 1)

However, this general allocation of enforcement responsibilities to LAs does not affect any of the following activities which remain under the HSE control, even if the premises in which they take place has been allocated to the LA:

1. Any activity in a mine or quarry other than a quarry in respect of which notice of abandonment has been given under sect. 139(2) of the Mines and Quarries Act 1954.
2. Any activity in a fairground.
3. Any activity in premises occupied by a radio, television or film undertaking in which the activity of broadcasting, recording or filming is carried on, and the activity of broadcasting, recording or filming wherever carried on and, for this purpose, 'film' includes video.
4. The following activities carried on at any premises by persons who do not normally work in the premises:

  (a) construction work if:
     (i) regulation 7(1) of the Construction (Design and Management) Regulations 1994 (which requires projects which include or are intended to include construction work to be notified to the Executive) applies to the project which includes the work;
     (ii) the whole or part of the work contracted to be undertaken by the contractor at the premises is to the external fabric or other external part of a building or structure; or
     (iii) it is carried out in a physically segregated area of the premises, the activities normally carried out in that area have been suspended for the purpose of enabling the construction work to be carried out, the contractor has authority to exclude from that area persons who are not attending in connection with the carrying out of the work and the work is not the maintenance of insulation on pipes, boilers or other parts of heating or water systems or its removal from them;
  (b) the installation, maintenance or repair of any gas system, or any work in relation to a gas fitting;
  (c) the installation, maintenance or repair of electricity systems;
  (d) work with ionizing radiations except work in one or more of the categories set out in Sch. 3 to the Ionizing Radiations Regulations 1985.
5. The use of ionizing radiations for medical exposure (within the meaning of Reg. 2(1) of the Ionizing Radiations Regulations 1985).
6. Any activity in premises occupied by a radiography undertaking in which there is carried on any work with ionizing radiations.
7. Agricultural activities, and any activity at an agricultural show which involves the handling of livestock or the working of agricultural equipment.
8. Any activity on board a sea-going ship.
9. Any activity in relation to a ski slope, ski lift, ski tow or cable car.
10. Fish maggot and game breeding except in a zoo.
11. Any activity in relation to a pipeline within the meaning of Regulation 3 of the Pipelines Safety Act 1996.
12. The operation of a railway.
(sch. 2)

The 'main activity' test does not apply to the following premises all of which are allocated to the HSE:

  (a) the tunnel system within the meaning of sect. 1(7) of the Channel Tunnel Act 1987;
  (b) an offshore installation within the meaning of regulation 3 of the Offshore Installations and Pipeline Works (Management and Administration) Regulations 1995;
  (c) a building or construction site, that is to say, premises where the only activities being undertaken are construction work and activities for the purpose of or in connection with such work;

(d) the campus of a university, polytechnic, college, school or similar educational establishment;

(e) a hospital (Reg. 3(5)).

In addition all premises occupied by the following are reserved for HSE enforcement:

(a) local authority;
(b) parish and community councils;
(c) police authorities;
(d) fire authorities;
(e) organizations under the International Headquarters and Defence Organizations Act 1964 including authorities of visiting forces;
(f) United Kingdom Atomic Energy Authority;
(g) the Crown.

Where premises are mainly occupied by one of these bodies but also occupied by someone else, the HSE assumes responsibility for the whole premises (Reg. 4).

Reg. 5 allows for a transfer of enforcement responsibility between HSE and LAs in either direction by agreement.

Enforcing authority

The local authorities responsible as enforcement authorities are:

(a) in relation to England a county council so far as they are the council for an area in which there are no district councils, a district council, a London borough council, the Common Council of the City of London, the Sub-Treasurer of the Inner Temple or Under-Treasurer of the Middle Temple or the Council of the Isles of Scilly;
(b) in relation to Scotland, an islands or district council.
(c) in relation to Wales, a county or county borough council (Health and Safety (Enforcing Authority) Regulations 1998 Reg. 2(1)).

Appointment of inspectors

An enforcing authority may appoint persons having suitable qualifications as inspectors. Each appointment must be in writing and specify which of the powers given to inspectors by the Act (below) are to be exercisable. In exercising these duties the inspector must be able to produce if required his instrument of appointment or a duly authenticated copy of it (sect. 19).

Powers of inspectors

An enforcing authority may confer on an inspector all or any of the following powers in connection with his appointment:

(a) enter any premises at any reasonable time or, where a situation may be dangerous, at any time;
(b) take with him a police constable;
(c) take with him any other person duly authorized by the enforcing authority and any equipment or materials he may require;
(d) make such examination and investigation as necessary;
(e) direct that a premises or part of it to remain undisturbed for as long as reasonably necessary for examination or investigation;
(f) take such measurements, photographs or records as necessary;
(g) take samples of articles, or substances in the premises or the atmosphere in or in the vicinity of the premises;
(h) dismantle, treat or test any article or substance likely to cause danger to health or safety, if requested, in the presence of a responsible person at the premises;
(i) take possession of or detain an article or substance likely to cause danger to health or safety in order to examine it, prevent it being tampered with or to keep it available for evidence, having previously taken a sample, where possible, and left notice of his action in the premises;
(j) require any person to answer and sign a declaration of truth;
(k) require the production of, inspect and take copies of any books or documents;
(l) require such facilities or assistance as necessary;
(m) exercise any other power necessary to carry out his responsibilities (sect. 20).

Notices

(a) **Form.** Notices should be in writing.
(b) **Authentication.** Since improvement and prohibition notices are served by an inspector, they should be signed by that person.
(c) **Service.** Notices to be served by an inspector may be served by:
  (i) delivering it to the person, or leaving it at his proper address or sending it by post to him at that address;
  (ii) in the case of a body corporate, be served or given to the secretary or clerk of that body;
  (iii) in the case of a partnership, be served on or given to a partner or person having control or management of the partnership.

The proper address is:

(a) his last known address; or
(b) in the case of a body corporate on their secretary or clerk, at the registered or principal office; or
(c) in the case of a partnership, the principal office of the partnership.

If a person has specified an address in the UK, other than his proper address, to which notices should be sent, this will be acceptable as proper service. Notices to be served on an owner or occupier of any premises may be served by sending it by post to him at that premises or by addressing it to him by name but delivering it to some responsible person, resident or employee on the premises. Where, after reasonable enquiry, the name and address of the owner or coccupier cannot be found, notice may be served by addressing it to 'the owner' or 'the occupier' of the premises (describing them) and delivering it to a responsible person, resident or employee at the premises, or, if no such person is available, by fixing it, or a copy, to some conspicuous part of the premises (sect. 46).

DEFINITIONS

Unless otherwise indicated the following definitions are applicable to procedures in this chapter.

Work means work as an employee or as a self-employed person. The employee is at work throughout the time when he is in the course of his employment, but not otherwise and a self-employed person is at work throughout such time as he devotes to work as a self-employed person (sect. 52(1)).

The meaning of 'work' has been extended by regulation to include any activity involving genetic manipulation or dangerous pathogens and the duty of care under sect. 3(2) of the Act extends to such an activity undertaken by research students etc. 'Work' has also been extended to include participants on government training schemes, school age pupils on work experience and college students on 'sandwich course' external training (Health and Safety (Training for Employment) Regulations 1990).

In relation to premises outside of Great Britain but controlled under HASAWA, e.g. oil rigs, the definition of 'at work' has also been extended so that an employee or a self-employed person is deemed to be at work throughout the time that he is present at the premises (Reg. 23(2) Management of Health and Safety Regulations 1999).

Employee means an individual who works under a contract of employment.

Personal injury includes any disease and any impairment of a person's physical or mental condition.

Premises includes any place and, in particular, includes:

(a) any vehicle, vessel, aircraft or hovercraft;
(b) any installation on land (including the foreshore and other land intermittently covered by water), any offshore installation, and any other installation (whether floating, or resting on the seabed or the subsoil thereof, or resting on other land covered with water or the subsoil thereof); and
(c) any tent or moveable structure.

Substance means any natural or artificial substance including micro-organisms, whether in solid or liquid form or in the form of a gas or vapour (sect. 53).

IMPROVEMENT AND PROHIBITION NOTICES

References

Health and Safety at Work etc. Act 1974 sects. 21–24 inclusive.
Industrial Tribunal (Constitution and Rules of Procedure) Regulations 1993.
HELA Circular LAC(L)3/6/4 April 1991.
HSC Guidance Ref. HSC (G) 3 (REV) 24 March 1998.

Inspector

Improvement and prohibition notices may be served by persons appointed by an enforcing authority as an inspector under the Act and no further authority is required for them to exercise these powers (sect. 19).

Relevant statutory provisions

Both notice procedures relate to the application of 'relevant statutory provisions' which are:

 (a) Part 1 HASAWA 1974;
 (b) health and safety regulations; and
 (c) enactments and regulations made under them in force before the passing of the HASAWA 1974 and now listed in Sch. 1 of that Act (sect. 53(1)).

Enforcement procedures*

HSC Guidance contained in HSC (G) 3 (REV) issued 24 March 1998 requires inspectors when enforcing health and safety legislation to comply with the following procedures. Since the guidance was issued under sect. 18 it is the duty of LAs to act under this guidance.

1. When giving advice, explain clearly what is to be done, why, and by when. The advice should also be confirmed in writing on request, and making sure that legal requirements are clearly distinguished from best practice advice.
2. Provide an opportunity to discuss the issues before formal action is taken, for example before serving an improvement notice.
3. Explain in writing why any immediate action is considered necessary and the consequences of failing to remedy the situation.
4. Explain any rights of appeal when formal or immediate action is taken.
5. At the end of a visit, discuss with the business what further action, if any, they are going to take, and explain that they may have any advice confirmed in writing on request.
6. Provide appropriate information to employees or their representatives on matters affecting their health, safety or welfare and the action the inspector proposes to take.

* Also see HSC Statement on Enforcement Policy 2002 on page 4.

FC70 Improvement notices

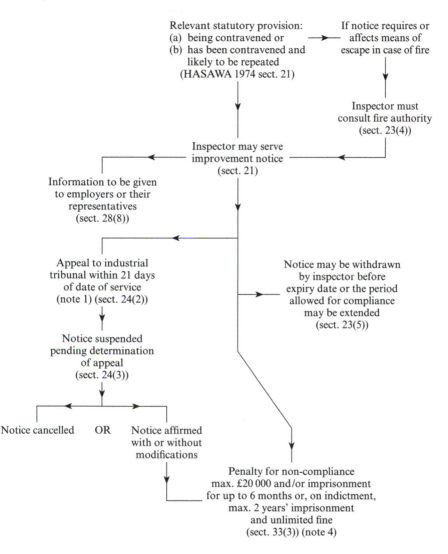

Relevant statutory provision:
(a) being contravened or
(b) has been contravened and
likely to be repeated
(HASAWA 1974 sect. 21)

If notice requires or
affects means of
escape in case of fire

Inspector must
consult fire authority
(sect. 23(4))

Inspector may serve
improvement notice
(sect. 21)

Information to be given
to employers or their
representatives
(sect. 28(8))

Appeal to industrial
tribunal within 21 days
of date of service
(note 1) (sect. 24(2))

Notice may be withdrawn
by inspector before
expiry date or the period
allowed for compliance
may be extended
(sect. 23(5))

Notice suspended
pending determination
of appeal
(sect. 24(3))

Notice cancelled   OR   Notice affirmed
with or without
modifications

Penalty for non-compliance
max. £20 000 and/or imprisonment
for up to 6 months or, on indictment,
max. 2 years' imprisonment
and unlimited fine
(sect. 33(3)) (note 4)

Notes
1. The period allowed for appeal may be extended by an industrial tribunal (Industrial Tribunals (Constitution and Rules of Procedure) Regulations 1993).
2. This procedure cannot be applied to the Crown (HASAWA 1974 sect. 48).
3. Unless the notice imposes requirements solely relating to the protection of persons at work, it must be included in a public register under the provisions of the Environment and Safety Information Act 1988.
4. The penalty on summary conviction was increased to this level by amendments to sect. 33 contained in the Offshore Safety Act 1992.

FC71 Prohibition notices

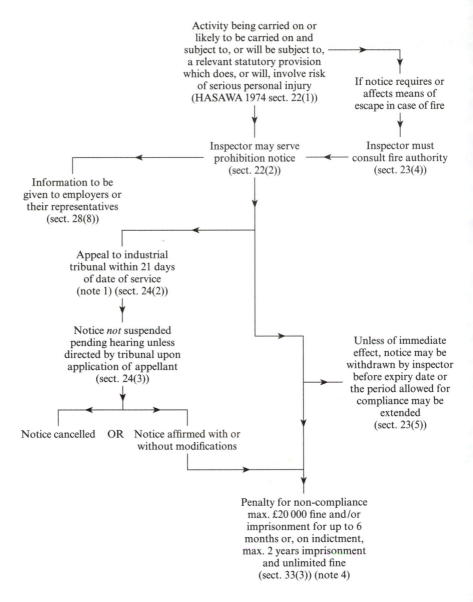

Activity being carried on or
likely to be carried on and
subject to, or will be subject to,
a relevant statutory provision
which does, or will, involve risk
of serious personal injury
(HASAWA 1974 sect. 22(1))

If notice requires or
affects means of
escape in case of fire

Inspector may serve
prohibition notice
(sect. 22(2))

Inspector must
consult fire authority
(sect. 23(4))

Information to be
given to employers or
their representatives
(sect. 28(8))

Appeal to industrial
tribunal within 21 days
of date of service
(note 1) (sect. 24(2))

Notice *not* suspended
pending hearing unless
directed by tribunal upon
application of appellant
(sect. 24(3))

Unless of immediate
effect, notice may be
withdrawn by inspector
before expiry date or
the period allowed for
compliance may be
extended
(sect. 23(5))

Notice cancelled   OR   Notice affirmed with or
without modifications

Penalty for non-compliance
max. £20 000 fine and/or
imprisonment for up to 6
months or, on indictment,
max. 2 years imprisonment
and unlimited fine
(sect. 33(3)) (note 4)

Notes
1. The period allowed for appeal may be extended by an industrial tribunal (Industrial Tribunals (Constitution and Rules of Procedure) Regulations 1993).
2. This procedure cannot be applied to the Crown (HASAWA 1974 sect. 48).
3. Unless the prohibition notice imposes restrictions solely relating to the protection of persons at work it must be included in a public register under the provisions of the Environment and Safety Information Act 1988.
4. The penalty on summary conviction was increased to this limit by amendments to sect. 33 contained in the Offshore Safety Act 1992.

Improvement notices

(a) **Scope.** Improvement notices may be served when a relevant statutory provision is either being contravened or, having been contravened, is likely to be continued or repeated (sect. 21).

(b) **Content.** The notice must include:
  (i) the inspector's opinion concerning the contravention;
  (ii) specification of the provision or provisions being contravened;
  (iii) reasons for him being of that opinion;
  (iv) a requirement to remedy the contravention; and
  (v) a period within which the contravention must be remedied which must be not less than 21 days (sect. 21).

The notice need not specify the actual measures to be taken to remedy the contravention but, if it does, it may refer to any approved code of practice and allow a choice of remedies (sect. 23(2)).

(c) **Statutory guidance.** Before serving an improvement notice the inspector should discuss with the business what the breaches are, the action which will be needed to comply and appropriate timescales for completion of the works. The business should be given the opportunity to discuss the issues with the inspector before formal action is taken and, if possible, resolve points of difference. When an inspector decides that an IN should be issued the inspector should explain what needs to be done, why, and by when.
(HSC (G) 3 (REV) 24/3/98 para. 9)

Prohibition notices

(a) **Scope.** These notices may be used by an inspector where there is a risk of serious personal injury from an activity which is either being carried on or is likely to be carried on, and to which a relevant statutory provision applies (HASAWA 1974 sect. 22(1) and (2)).

(b) **Content.** The notice must:
  (i) state the inspector's opinion relating to the risk of serious personal injury;
  (ii) specify the matters which give rise to that risk;
  (iii) where a relevant statutory provision is being contravened, specify the provision concerned and give reasons for that opinion;
  (iv) direct that the activity concerned shall not be carried on unless the matters specified in the notice have been remedied.

The notice will normally be given immediate effect if the inspector is of the opinion that there is a risk of serious personal injury. Alternatively, a period must be specified following which the notice comes into effect. There is no minimum period to be allowed and, since in the event of an appeal the notice is not automatically suspended, the period allowed may be shorter than the appeal period (sect. 22).

The notice need not specify the actual measure to be taken but if it does it may refer to any approved code of practice and allow a choice of remedies (sect. 23(2)).

(c) **Statutory guidance.** An intention to issue a prohibition notice should, where practical, be discussed with the business at the time and the views of the employees taken into account. Where the notice is issued the inspector should provide a written explanation of the reasons for the action.

(HSC (G) 3 (REV) 24/3/98 para. 10)

Standard of requirements

Unless a relevant statutory provision lays down a specific requirement, requirements in improvement notices relating to buildings must be no more onerous than that necessary to secure compliance with any building regulations in force as if the building was being newly erected (sect. 23(3)).

Person responsible

Improvement notices are to be served on the person who is contravening the relevant statutory provision or who has contravened it and is likely to continue or repeat it (sect. 21). Prohibition notices must be served on the person by or under the control of whom the particular activity is being or is about to be carried on (sect. 22(2)).

Appeals

Appeal against either notice must be made to an industrial tribunal within 21 days of the date of service although the tribunal may allow an extension of this period on application from the appellant, either before or after the expiration of the 21 days, if satisfied that it was not reasonably practicable to have brought the appeal within 21 days.

There are no specific grounds in either the Act or the regulations for lodging an appeal. The procedure to be adopted by the industrial tribunal for receiving and hearing appeals is laid down in Schedule 4 to the regulations (sect. 24 and Industrial Tribunals (Constitution and Rules of Procedure) Regulations 1993).

Where an appeal is made against an improvement notice, the notice stands suspended until the appeal is determined or the appeal is withdrawn but appeals against prohibition notices do not automatically suspend the notice. Any suspension of a prohibition notice may be ordered only by the tribunal and then only from the date of the direction (sect. 24(3)).

Defences

Where the duty imposed on a person by a relevant statutory provision is qualified by either:

(a) so far as is practicable; or

(b) as far as is reasonably practicable; or

(c) to use the best practicable means;

it is for the defendant to prove that any of those circumstances appertained (sect. 40).

SEIZURE OF DANGEROUS ARTICLES OR SUBSTANCES

Reference

Health and Safety at Work etc. Act 1974 sect. 25.
HSC Guidance HSC (G) 3 (REV) 24 March 1998.

Scope

An inspector appointed under the Act by an enforcing authority may seize any article or substance on any premises which he has power to enter if he believes that, in the circumstances in which he found it, the article or substance is a cause of imminent danger or serious personal injury to either employees or other persons (sect. 25(1)).

Customs officers are given the power to assist inspectors by seizing and detaining for up to 2 working days any imported article or substance (sect. 25A).

Sampling

Before seizing either:

(a) any article which forms part of a batch of similar articles; or

(b) any substance;

the inspector is required, where it is practicable for him to do so, to take a sample and give a portion of it, marked in such a way as to identify it, to a responsible person at the premises where it was found (sect. 25(2)).

Action following seizure

Having seized the article or substance the inspector is to render it harmless and this may be by any method including destruction (sect. 25(1)).

Having dealt appropriately with the article or substance the inspector must:

(a) prepare and sign a written report of the matter including the circumstances in which the article or substance was seized and dealt with; and

(b) give a copy of that report to a responsible person at the premises where the article or substance was found; and

FC72 Seizure of dangerous articles or substances

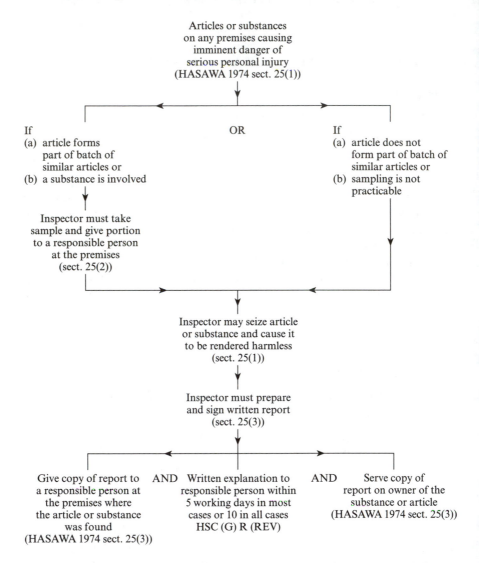

Articles or substances
on any premises causing
imminent danger of
serious personal injury
(HASAWA 1974 sect. 25(1))

OR

If
(a) article forms
part of batch of
similar articles or
(b) a substance is involved

If
(a) article does not
form part of batch of
similar articles or
(b) sampling is not
practicable

Inspector must take
sample and give portion
to a responsible person
at the premises
(sect. 25(2))

Inspector may seize article
or substance and cause it
to be rendered harmless
(sect. 25(1))

Inspector must prepare
and sign written report
(sect. 25(3))

Give copy of report to
a responsible person at
the premises where
the article or substance
was found
(HASAWA 1974 sect. 25(3))

AND

Written explanation to
responsible person within
5 working days in most
cases or 10 in all cases
HSC (G) R (REV)

AND

Serve copy of
report on owner of the
substance or article
(HASAWA 1974 sect. 25(3))

Note
1. These procedures may not be applied to Crown premises (HASAWA 1974 sect. 48).

(c) serve a copy of the notice on the owner of the article or substance where that is a different person from (b) above.

If, after reasonable enquiry, the inspector is unable to ascertain the name and address of the owner, the copy mentioned in (c) above may be served by giving it to the same person as the copy in (b) was given (sect. 25(3)).

Statutory guidance

Inspectors should send a written explanation of the action taken to the business in most cases within 5 working days and in all cases within 10 working days.
(HSC (G) 3 (REV) 24/3/98)

REPORTING OF INJURIES, DISEASES AND DANGEROUS OCCURRENCES

Reference

The Reporting of Injuries, Diseases and Dangerous Occurrences Regulations 1995.
RIDDOR Explained. HSE 31 (rev. 1) August 2000.

Scope

The following four categories of incident are to be notified or reported:

(1) Notifiable injuries:

(a) any person (i.e. at work or not) dying as a result of an accident arising out of or in connection with work (Reg. 3(1)(a));
(b) a major injury (see below) to any person at work as a result of an accident arising out of or in connection with work (Reg. 3(1)(b));
(c) a person not at work suffering any injury as a result of an accident arising out of or in connection with work and that person is taken from the scene to a hospital for treatment (Reg. 3(1)(c));
(d) a person not at work suffering a major injury as a result of an accident arising out of or in connection with work at a hospital (Reg. 3(1)(d));
(e) an injury to a person at work causing incapacity for his normal work or work of a similar nature for more than three consecutive days (excluding the day of the accident but including any days which would not have been working days) (Reg. 3(2)).

In addition, where an employee has suffered a reportable injury as a result of an accident at work and dies within one year from that cause, the employer must inform the enforcement agency in writing as soon as the death comes to his knowledge (Reg. 4).

FC73 The reporting of injuries, diseases and dangerous occurrences

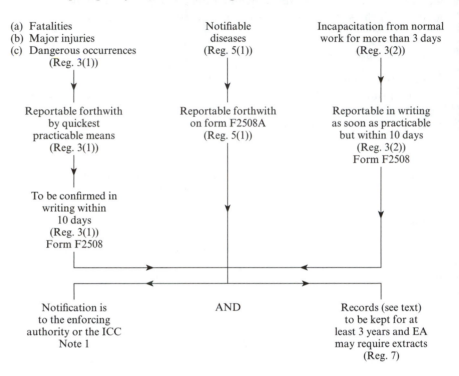

(a) Fatalities
(b) Major injuries
(c) Dangerous occurrences
    (Reg. 3(1))

Notifiable
diseases
(Reg. 5(1))

Incapacitation from normal
work for more than 3 days
(Reg. 3(2))

Reportable forthwith
by quickest
practicable means
(Reg. 3(1))

Reportable forthwith
on form F2508A
(Reg. 5(1))

Reportable in writing
as soon as practicable
but within 10 days
(Reg. 3(2))
Form F2508

To be confirmed in
writing within
10 days
(Reg. 3(1))
Form F2508

Notification is
to the enforcing
authority or the ICC
Note 1

AND

Records (see text)
to be kept for at
least 3 years and EA
may require extracts
(Reg. 7)

Note
1. There are several possible ways of making a notification – see text.

Major injuries are defined in Schedule 1 as:

1. Any fracture, other than to the fingers, thumbs or toes.
2. Any amputation.
3. Dislocation of the shoulder, hip, knee or spine.
4. Loss of sight (whether temporary or permanent).
5. A chemical or hot metal burn to the eye or any penetrating injury to the eye.
6. Any injury resulting from an electric shock or electrical burn (including any electrical burn caused by arcing or arcing products) leading to unconsciousness or requiring resuscitation or admittance to hospital for more than 24 hours.
7. Any other injury
   (a) leading to hypothermia, heat-induced illness or to unconsciousness,
   (b) requiring resuscitation, or
   (c) requiring admittance to hospital for more than 24 hours.
8. Loss of consciousness caused by asphyxia or by exposure to a harmful substance or biological agent.
9. Either of the following conditions which result from the absorption of any substance by inhalation, ingestion or through the skin
   (a) acute illness requiring medical treatment; or
   (b) loss of consciousness.
10. Acute illness which requires medical treatment where there is reason to believe that this resulted from exposure to a biological agent or its toxins or infected material.

(2) Dangerous occurrences as detailed in Schedule 2 (below) are notifiable. There are also lists detailing specific dangerous occurrences in offshore workplaces, mines, quarries and transport systems. Those which apply generally are:

Lifting machinery etc.

1. The collapse of, the overturning of, or the failure of any load-bearing part of any
   (a) lift or hoist;
   (b) crane or derrick;
   (c) mobile powered access platform;
   (d) access cradle or window-cleaning cradle;
   (e) excavator;
   (f) pile-driving frame or rig having an overall height, when operating, of more than 7 metres; or
   (g) fork lift truck.

Pressure systems

2. The failure of any closed vessel (including a boiler or boiler tube) or of any associated pipework, in which the internal pressure was above or below atmospheric pressure, where the failure has the potential to cause the death of any person.

Freight containers

3.
    (1) The failure of any freight container in any of its load-bearing parts while it is being raised, lowered or suspended.
    (2) In this paragraph, 'freight container' means a container as defined in regulation 2(1) of the Freight Containers (Safety Convention) Regulations 1984.

Overhead electric lines

4. Any unintentional incident in which plant or equipment either
    (a) comes into contact with an uninsulated overhead electric line in which the voltage exceeds 200 volts; or
    (b) causes an electrical discharge from such an electric line by coming into close proximity to it.

Electrical short circuit

5. Electrical short circuit or overload attended by fire or explosion which results in the stoppage of the plant involved for more than 24 hours or which has the potential to cause the death of any person.

Explosives

6.
    (1) Any of the following incidents involving explosives:
        (a) the unintentional explosion or ignition of explosives other than one
            (i) caused by the unintentional discharge of a weapon where, apart from that unintentional discharge, the weapon and explosives functioned as they were designed to do; or
            (ii) where a fail-safe device or safe system of work functioned so as to prevent any person from being injured in consequence of the explosion or ignition;
        (b) a misfire (other than one at a mine or quarry or inside a well or one involving a weapon) except where a fail-safe device or safe system of work functioned so as to prevent any person from being endangered in consequence of the misfire;

(c) the failure of the shots in any demolition operation to cause the intended extent of collapse or direction of fall of a building or structure;

(d) the projection of material (other than at a quarry) beyond the boundary of the site on which the explosives are being used or beyond the danger zone in circumstances such that any person was or might have been injured thereby;

(e) any injury to a person (other than at a mine or quarry or one otherwise reportable under these Regulations) involving first-aid or medical treatment resulting from the explosion or discharge of any explosives or detonator.

(2) In this paragraph 'explosives' means any explosive of a type which would, were it being transported, be assigned to Class 1 within the meaning of the Classification and Labelling of Explosives Regulations 1983 and 'danger zone' means the area from which persons have been excluded or forbidden to enter to avoid being endangered by any explosion or ignition of explosives.

Biological agents

7. Any accident or incident which resulted or could have resulted in the release or escape of a biological agent likely to cause severe human infection or illness.

Malfunction of radiation generators etc.

8.

(1) Any incident in which:

(a) the malfunction of a radiation generator or its ancillary equipment used in fixed or mobile industrial radiography, the irradiation of food or the processing of products by irradiation, causes it to fail to de-energise at the end of the intended exposure period; or

(b) the malfunction of equipment used in fixed or mobile industrial radiography or gamma irradiation causes a radioactive source to fail to return to its safe position by the normal means at the end of the intended exposure period.

(2) In this paragraph, 'radiation generator' has the same meaning as in regulation 2 of the Ionising Radiations Regulations 1985.

Breathing apparatus

9.

(1) Any incident in which breathing apparatus malfunctions:

(a) while in use, or

(b) during testing immediately prior to use in such a way that had the malfunction occurred while the apparatus was in use it would have posed a danger to the health or safety of the user.

(2) This paragraph shall not apply to breathing apparatus while it is being
   (a) used in a mine; or
   (b) maintained or tested as part of a routine maintenance procedure.

Diving operations

10. Any of the following incidents in relation to a diving operation:
   (a) the failure or the endangering of
      (i) any lifting equipment associated with the diving operation, or
      (ii) life support equipment, including control panels, hoses and breathing apparatus, which puts a driver at risk;
   (b) any damage to, or endangering of, the dive platform, or any failure of the dive platform to remain on station, which puts a diver at risk;
   (c) the trapping of a diver;
   (d) any explosion in the vicinity of a diver; or
   (e) any uncontrolled ascent or any omitted decompression which puts a diver at risk.

Collapse of scaffolding

11. The complete or partial collapse of:
   (a) any scaffold which is
      (i) more than 5 metres in height which results in a substantial part of the scaffold falling or overturning; or
      (ii) erected over or adjacent to water in circumstances such that there would be a risk of drowning to a person falling from the scaffold into the water; or
   (b) the suspension arrangements (including any outrigger) of any slung or suspended scaffold which causes a working platform or cradle to fall.

Train collisions

12. Any unintended collision of a train with any other train or vehicle, other than one reportable under Part IV of this Schedule, which caused, or might have caused, the death of, or major injury to, any person.

Wells

13. Any of the following incidents in relation to a well (other than a well sunk for the purpose of the abstraction of water):
   (a) a blow-out (that is to say an uncontrolled flow of well-fluids from a well);
   (b) the coming into operation of a blow-out prevention or diversion system to control a flow from a well where normal control procedures fail;

(c) the detection of hydrogen sulphide in the course of operations at a well or in samples of well-fluids from a well where the presence of hydrogen sulphide in the reservoir being drawn on by the well was not anticipated by the responsible person before that detection;

(d) the taking of precautionary measures additional to any contained in the original drilling programme following failure to maintain a planned minimum separation distance between wells drilled from a particular installation; or

(e) the mechanical failure of any safety critical element of a well (and for this purpose the safety critical element of a well is any part of a well whose failure would cause or contribute to, or whose purpose is to prevent or limit the effect of, the unintentional release of fluids from a well or a reservoir being drawn on by a well).

Pipelines or pipeline works

14. The following incidents in respect of a pipeline or pipeline works:

(a) the uncontrolled or accidental escape of anything from, or inrush of anything into, a pipeline which has the potential to cause the death of, major injury or damage to the health of any person or which results in the pipeline being shut down for more than 24 hours;

(b) the unintentional ignition of anything in a pipeline or of anything which, immediately before it was ignited, was in a pipeline;

(c) any damage to any part of a pipeline which has the potential to cause the death of, major injury or damage to the health of any person or which results in the pipeline being shut down for more than 24 hours;

(d) any substantial and unintentional change in the position of a pipeline requiring immediate attention to safeguard the integrity or safety of a pipeline;

(e) any unintentional change in the subsoil or seabed in the vicinity of a pipeline which has the potential to affect the integrity or safety of a pipeline;

(f) any failure of any pipeline isolation device, equipment or system which has the potential to cause the death of, major injury or damage to the health of any person or which results in the pipeline being shut down for more than 24 hours; or

(g) any failure of equipment involved with pipeline works which has the potential to cause the death of, major injury or damage to the health of any person.

Fairground equipment

15. The following incidents on fairground equipment in use or under test:

(a) the failure of any load-bearing part;

(b) the failure of any part designed to support or restrain passengers; or

(c) the derailment or the unintended collision of cars or trains.

Carriage of dangerous substances by road

16.

    (1) Any incident involving a road tanker or tank container used for the carriage of a dangerous substance in which:

        (a) the road tanker or vehicle carrying the tank container overturns (including turning onto its side);

        (b) the tank carrying the dangerous substance is seriously damaged;

        (c) there is an uncontrolled release or escape of the dangerous substance being carried; or

        (d) there is a fire involving the dangerous substance being carried.

    (2) In this paragraph, 'carriage', 'dangerous substance', 'road tanker' and 'tank container' have the same meanings as in regulation 2(1) of the Road Traffic (Carriage of Dangerous Substances in Road Tankers and Tank Containers) Regulations 1992.

17.

    (1) Any incident involving a vehicle used for the carriage of a dangerous substance, other than a vehicle to which para. 16 applies, where there is:

        (a) an uncontrolled release or escape of the dangerous substance being carried in such a quantity as to have the potential to cause the death of, or major injury to, any person; or

        (b) a fire which involves the dangerous substance being carried.

    (2) In this paragraph, 'carriage' and 'dangerous substance' have the same meaning as in regulation 2(1) of the Road Traffic (Carriage of Dangerous Substances in Packages etc.) Regulations 1992.

Collapse of building or structure

18. Any unintended collapse or partial collapse of:

    (a) any building or structure (whether above or below ground) under construction, reconstruction, alteration or demolition which involves a fall of more than 5 tonnes of material;

    (b) any floor or wall of any building (whether above or below ground) used as a place of work; or

    (c) any false-work.

Explosion or fire

19. An explosion or fire occurring in any plant or premises which results in the stoppage of that plant or as the case may be the suspension of normal work in those premises for more than 24 hours, where the explosion or fire was due to the ignition of any material.

Escape of flammable substances

20.

    (1) The sudden, uncontrolled release:

        (a) inside a building:

            (i) of 100 kilograms or more of a flammable liquid,

            (ii) of 10 kilograms or more of a flammable liquid at a temperature above its normal boiling point, or

            (iii) of 10 kilograms or more of a flammable gas; or

        (b) in the open air, of 500 kilograms or more of any of the substances referred to in sub-paragraph (a) above.

    (2) In this paragraph, 'flammable liquid' and 'flammable gas' mean respectively a liquid and a gas so classified in accordance with regulation 5(2), (3) or (5) of the Chemicals (Hazard Information and Packaging for Supply) Regulations 1994.

Escape of substances

21. The accidental release or escape of any substance in a quantity sufficient to cause the death, major injury or any other damage to the health of any person.

(3) Diseases* – where a person at work suffers from any of the listed occupational diseases and his work involves a corresponding work activity, that situation is reportable (Reg. 5(1)). To become reportable the disease must have been confirmed by a registered medical practitioner (Reg. 5(2)).

The diseases and activities are set out in Schedule 3 as follows.

Diseases	*Activities*
Conditions due to physical agents and the physical demands of work	
1. Inflammation, ulceration or malignant disease of the skin due to ionizing radiation.	
2. Malignant disease of the bones due to ionizing radiation.	Work with ionizing radiation.
3. Blood dyscrasia due to ionizing radiation.	
4. Cataract due to electromagnetic radiation.	Work involving exposure to electromagnetic radiation (including radiant heat).

* For the notification of diseases under the PH(CoD)A 1984 see page 85.

Diseases	*Activities*
5. Decompression illness.	
6. Barotrauma resulting in lung or other organ damage.	Work involving breathing gases at increased pressure (including diving).
7. Dysbaric osteonecrosis	
8. Cramp of the hand or forearm due to repetitive movements.	Work involving prolonged periods of handwriting, typing or other repetitive movements of the fingers, hand or arm.
9. Subcutaneous cellulitis of the hand (*beat hand*).	Physically demanding work causing severe or prolonged friction or pressure on the hand.
10. Bursitis or subcutaneous cellulitis arising at or about the knee due to severe or prolonged external friction or pressure at or about the knee (*beat knee*).	Physically demanding work, causing severe or prolonged friction or pressure at or about the knee.
11. Bursitis or subcutaneous cellulitis arising at or about the elbow due to severe or prolonged external friction or pressure at or about the elbow (*beat elbow*).	Physically demanding work causing severe or prolonged friction or pressure at or about the elbow.
12. Traumatic inflammation of the tendons of the hand or forearm or of the associated tendon sheaths.	Physically demanding work, frequent or repeated movements, constrained postures or extremes of extension or flexion of the hand or wrist.
13. Carpal tunnel syndrome.	Work involving the use of hand-held vibrating tools.
14. Hand-arm vibration syndrome.	Work involving:
	(a) The use of chain saws, brush cutters or hand-held or hand-fed circular saws in forestry or woodworking;
	(b) the use of hand-held rotary tools in grinding material or in sanding or polishing metal;
	(c) the holding of material being ground or metal being sanded or polished by rotary tools;
	(d) the use of hand-held percussive metal-working tools or the holding of metal being worked upon by percussive tools in connection with riveting, caulking, chipping, hammering, fettling or swaging;

Diseases	Activities
	(e) the use of hand-held powered percussive drills or hand-held powered percussive hammers in mining, quarrying or demolition, or on roads or footpaths (including road construction); or
	(f) the holding of material being worked upon by pounding machines in shoe manufacture.

Infections due to biological agents

15. Anthrax.	(a) Work involving handling infected animals, their products or packaging containing infected material; or
	(b) Work on infected sites.
16. Brucellosis.	Work involving contact with:
	(a) animals or their carcasses (including any parts thereof) infected by brucella or the untreated products of same; or
	(b) laboratory specimens or vaccines of or containing brucella.
17. (a) Avian chlamydiosis.	Work involving contact with birds infected with chlamydia psittaci, or the remains or untreated products of such birds.
(b) Ovine chlamydiosis.	Work involving contact with sheep infected with chlamydia psittaci or the remains of untreated products of such sheep.
18. Hepatitis.	Work involving contact with:
	(a) human blood or human blood products; or
	(b) any source of viral hepatitis.
19. Legionellosis.	Work on or near cooling systems which are located in the workplace and use water; or work on hot water service systems located in the workplace which are likely to be a source of contamination.
20. Leptospirosis.	(a) Work in places which are or are liable to be infested by rats, fieldmice, voles or other small mammals;

Diseases	Activities
	(b) work at dog kennels or involving the care or handling of dogs; or
	(c) work involving contact with bovine animals or their meat products or pigs or their meat products.
21. Lyme disease.	Work involving exposure to ticks (including in particular work by forestry workers, rangers, dairy farmers, game keepers and other persons engaged in countryside management).
22. Q fever.	Work involving contact with animals, their remains or their untreated products.
23. Rabies.	Work involving handling or contact with infected animals.
24. Streptococcus suis.	Work involving contact with pigs infected with streptococcus suis, or with the carcasses, products or residues of pigs so affected.
25. Tetanus.	Work involving contact with soil likely to be contaminated by animals.
26. Tuberculosis.	Work with persons, animals, human or animal remains or any other material which might be a source of infection.
27. Any infection reliably attributable to the performance of the work specified in the entry opposite hereto.	Work with micro-organisms; work with live or dead human beings in the course of providing any treatment or service or in conducting any investigation involving exposure to blood or body fluids; work with animals or any potentially infected material derived from any of the above.

Conditions due to substances

28. Poisonings by any of the following;	Any activity.
(a) acrylamide monomer;	
(b) arsenic or one of its compounds;	
(c) benzene or a homologue of benzene;	
(d) beryllium or one of its compounds;	

Diseases	Activities

(e) cadmium or one of its compounds;
(f) carbon disulphide;
(g) diethylene dioxide (dioxan);
(h) ethylene oxide,
(i) lead or one of its compounds;
(j) manganese or one of its compounds;
(k) mercury or one of its compounds;
(l) methyl bromide;
(m) nitrochlorobenzene, or a nitro- or amino- or chloro-derivative of benzene or of a homologue of benzene;
(n) oxides of nitrogen;
(o) phosphorus or one of its compounds.

29. Cancer of a bronchus or lung.

(a) Work in or about a building where nickel is produced by decomposition of a gaseous nickel compound or where any industrial process which is ancillary or incidental to that process is carried on; or

(b) work involving exposure to bis(chloromethyl) ether or any electrolytic chromium processes (excluding passivation) which involve hexavalent chromium compounds, chromate production or zinc chromate pigment manufacture.

30. Primary carcinoma of the lung where there is accompanying evidence of silicosis.

Any occupation in:
(a) glass manufacture;
(b) sandstone tunnelling or quarrying;
(c) the pottery industry;
(d) metal ore mining;
(e) slate quarrying or slate production;
(f) clay mining;
(g) the use of siliceous materials as abrasives;
(h) foundry work;
(i) granite tunnelling or quarrying; or
(j) stone cutting or masonry.

Diseases	Activities
31. Cancer of the urinary tract.	1. Work involving exposure to any of the following substances: (a) beta-naphthylamine or methylene-bis-orthochloroaniline; (b) diphenyl substituted by at least one nitro or primary amino group or by at least one nitro and primary amino group (including benzidine); (c) any of the substances mentioned in sub-paragraph (b) above if further ring substituted by halogeno, methyl or methoxy groups, but not by other groups; or (d) the salts of any of the substances mentioned in sub-paragraphs (a) to (c) above. 2. The manufacture of auramine or magenta.
32. Bladder cancer.	Work involving exposure to aluminium smelting using the Soderberg process.
33. Angiosarcoma of the liver.	(a) Work in or about machinery or apparatus used for the polymerization of vinyl chloride monomer, a process which, for the purposes of this sub-paragraph, comprises all operations up to and including the drying of the slurry produced by the polymerization and the packaging of the dried product; or (b) work in a building or structure in which any part of the process referred to in the foregoing sub-paragraph takes place.
34. Peripheral neuropathy.	Work involving the use or handling of or exposure to the fumes of or vapour containing n-hexane or methyl n-butyl ketone.
35. Chrome ulceration of; (a) the nose or throat; or (b) the skin of the hands or forearm.	Work involving exposure to chromic acid or to any other chromium compound.

Diseases	Activities
36. Folliculitis.	
37. Acne.	Work involving exposure to mineral oil, tar, pitch or arsenic.
38. Skin cancer.	

Diseases	Activities
39. Pneumoconiosis (excluding asbestosis).	1. (a) The mining, quarrying or working of silica rock or the working of dried quartzose sand, any dry deposit or residue of silica or any dry admixture containing such materials (including any activity in which any of the aforesaid operations are carried out incidentally to the mining or quarrying of other minerals or to the manufacture of articles containing crushed or ground silica rock); or

   (b) the handling of any of the materials specified in the foregoing sub-paragraph in or incidentally to any of the operations mentioned therin or substantial exposure to the dust arising from such operations.

2. The breaking, crushing or grinding of flint, the working or handling of broken, crushed or ground flint or materials containing such flint or substantial exposure to the dust arising from any of such operations.

3. Sand blasting by means of compressed air with the use of quartzose sand or crushed silica rock or flint or substantial exposure to the dust arising from such sand blasting.

4. Work in a foundry or the performance of, or substantial exposure to the dust arising from, any of the following operations:
   (a) the freeing of steel castings from adherent siliceous substance or;

Diseases	*Activities*

(b) the freeing of metal castings
   from adherent siliceous
   substance;
   (i) by blasting with an
       abrasive propelled by
       compressed air, steam or a
       wheel, or
   (ii) by the use of power-driven
        tools.

5. The manufacture of china or
   earthenware (including sanitary
   earthenware, electrical earthenware
   and earthenware tiles) and any
   activity involving substantial
   exposure to the dust arising
   therefrom.

6. The grinding of mineral graphite or
   substantial exposure to the dust
   arising from such grinding.

7. The dressing of granite or any
   igneous rock by masons, the
   crushing of such materials or
   substantial exposure to the dust
   arising from such operations.

8. The use or preparation for use of
   an abrasive wheel or substantial
   exposure to the dust arising
   therefrom.

9. (a) Work underground in any mine
      in which one of the objects of
      the mining operation is the
      getting of any material;
   (b) the working or handling above
       ground at any coal or tin mine
       of any materials extracted
       therefrom or any operation
       incidental thereto;
   (c) the trimming of coal in any
       ship, barge, lighter, dock or
       harbour or at any wharf or
       quay; or
   (d) the sawing, splitting or dressing
       of slate or any operation
       incidental thereto.

10. The manufacture or work
    incidental to the manufacture of
    carbon electrodes by an industrial
    undertaking for use in the

Diseases	Activities
	electrolytic extraction of aluminium from aluminium oxide and any activity involving substantial exposure to the dust therefrom.
	11. Boiler scaling or substantial exposure to the dust arising therefrom.
40. Byssinosis.	The spinning or manipulation of raw or waste cotton or flax or the weaving of cotton or flax, carried out in each case in a room in a factory, together with any other work carried out in such a room.
41. Mesothelioma.	(a) The working or handling of asbestos or any admixture of asbestos;
42. Lung cancer.	(b) the manufacture or repair of asbestos textiles or other articles containing or composed of asbestos;
43. Asbestosis.	(c) the cleaning of any machinery or plant used in any of the foregoing operations and of any chambers, fixtures and appliances for the collection of asbestos dust; or
	(d) substantial exposure to the dust arising from any of the foregoing operations.
44. Cancer of the nasal cavity or associated air sinuses.	1. (a) Work in or about a building where wooden furniture is manufactured;
	(b) work in a building used for the manufacture of footwear or components of footwear made wholly or partly of leather or fibre board; or
	(c) work at a place used wholly or mainly for the repair of footwear made wholly or partly of leather or fibre board.
	2. Work in or about a factory building where nickel is produced by decomposition of a gaseous nickel compound or in any process which is ancillary or incidental thereto.

Diseases	Activities
45. Occupational dermatitis.	Work involving exposure to any of the following agents: (a) epoxy resin systems; (b) formaldehyde and its resins; (c) metalworking fluids; (d) chromate (hexavalent and derived from trivalent chromium); (e) cement, plaster or concrete; (f) acrylates and methacrylates; (g) colophony (rosin) and its modified products; (h) glutaraldehyde; (i) mercaptobenzothiazole, thiurams, substituted paraphenylene-diamines and related rubber processing chemicals; (j) biocides, anti-bacterials, preservatives or disinfectants; (k) organic solvents; (l) antibiotics and other pharmaceuticals and therapeutic agents; (m) strong acids, strong alkalis, strong solutions (e.g. brine) and oxidizing agents including domestic bleach or reducing agents; (n) hairdressing products including in particular dyes, shampoos, bleaches and permanent waving solutions; (o) soaps and detergents; (p) plants and plant-derived material including in particular the daffodil, tulip and chrysanthemum families, the parsley family (carrots, parsnips, parsley and celery), garlic and onion, hardwoods and the pine family; (q) fish, shell-fish or meat; (r) sugar or flour; or (s) any other known irritant or sensitizing agent including in particular any chemical bearing the warning 'may cause sensitization by skin contact' or 'irritating to the skin'.
46. Extrinsic alveolitis (including farmer's lung).	Exposure to moulds, fungal spores or heterologous proteins during work in:

Diseases	Activities
	(a) agriculture, horticulture, forestry, cultivation of edible fungi or malt-working;
	(b) loading, unloading or handling mouldy vegetable matter or edible fungi whilst same is being stored;
	(c) caring for handling birds; or
	(d) handling bagasse.
47. Occupational asthma.	Work involving exposure to any of the following agents:
	(a) isocyanates;
	(b) platinum salts;
	(c) fumes or dust arising from the manufacture, transport or use of hardening agents (including epoxy resin curing agents) based on phthalic anhydride, tetrachlorophthalic anhydride, trimellitic anhydride or triethylene-tetramine;
	(d) fumes arising from the use of rosin as a soldering flux;
	(e) proteolytic enzymes;
	(f) animals including insects and other arthropods used for the purposes of research or education or in laboratories;
	(g) dusts arising from the sowing, cultivation, harvesting, drying, handling, milling, transport or storage of barley, oats, rye, wheat or maize or the handling, milling, transport or storage of meal or flour made therefrom;
	(h) antibiotics;
	(i) cimetidine;
	(j) wood dust;
	(k) ispaghula;
	(l) castor bean dust;
	(m) ipecacuanha;
	(n) azodicarbonamide;
	(o) animals including insects and other arthropods (whether in their larval forms or not) used for the purposes of pest control or fruit cultivation or the larval forms of animals used for the purposes of

Diseases	*Activities*
	research or education or in laboratories;
	(p) glutaraldehyde;
	(q) persulphate salts or henna;
	(r) crustaceans or fish or products arising from these in the food processing industry;
	(s) reactive dyes;
	(t) soya bean;
	(u) tea dust;
	(v) green coffee bean dust;
	(w) fumes from stainless steel welding;
	(x) any other sensitizing agent, including in particular any chemical bearing the warning 'may cause sensitization by inhalation'.

(4) Gas incidents – death or major injury arising out of gas being distributed, filled, imported or supplied are notifiable to HSE (Reg. 6).

Exemptions

Road traffic accidents causing death or injury are not notifiable under these provisions unless they involved the exposure of substances carried on the vehicle or another work activity, e.g. roadworks adjoining the carriageway. None of the requirements relates to the Armed Forces (Reg. 10).

Notifications

The responsible person must notify deaths, major injuries and dangerous occurrences without delay.

Notifications of other reportable injuries are also to be reported as soon as practicable and in any event within ten days of the accident (Reg. 3(2)).

As from 1 April 2001 all of these incidents may be notified to a new Incident Contact Centre (ICC) in any of the following ways:

1. by telephone on Monday to Friday from 8.30 am to 5.00 pm to 0845 300 9923
2. by fax to 0845 300 9924
3. by internet at www.riddor.gov.uk
4. by e-mail to riddor@natbrit.com
5. by post to Incident Contact Centre, Caerphilly Business Park, Caerphilly, CF83 3GG.

Written notifications and confirmations are to be made on form F2508A for occupation-related diseases or F2508 for all other incidents.

Notification to the local enforcing agency is also acceptable by phone and confirmation by the appropriate form and these reports will be sent onto the ICC for processing.

Records

In respect of deaths, major injuries and dangerous occurrences the following particulars must be kept by the responsible person:

1. Date and time of the accident or dangerous occurrence.
2. In the event of an accident suffered by a person at work, the following particulars of that person:
   (a) full name;
   (b) occupation;
   (c) nature of injury.
3. In the event of an accident suffered by a person not at work, the following particulars of that person (unless they are not known and it is not reasonably practicable to ascertain them):
   (a) full name;
   (b) status (for example 'passenger', 'customer', 'visitor' or 'bystander');
   (c) nature of injury.
4. Place where the accident or dangerous occurrence happened.
5. A brief description of the circumstances in which the accident or dangerous occurrence happened.
6. The date on which the event was first reported to the relevant enforcing authority.

In relation to notifiable diseases the following information must be recorded:

1. Date of diagnosis of the disease.
2. Name of the person affected.
3. Occupation of the person affected.
4. Name or nature of the disease.
5. The date on which the disease was first reported to the relevant enforcing authority.
6. The method by which the disease was reported.

The records must be kept either at the place where the work to which it relates is carried on or at the usual place of business of the responsible person and must be kept for at least three years.

Enforcing authorities may require extracts from the records to be sent to them (Reg. 7).

Defences

In proceedings for offences under these provisions it will be a defence to prove that the person was not aware of the event and had taken all reasonable steps to have such events brought to his attention (Reg. 11).

Definitions

Accident includes:

(a) an act of non-consensual physical violence done to a person at work; and

(b) an act of suicide which occurs on, or in the course of the operation of, a relevant transport system.

Disease includes a medical condition.

Responsible person means:

(a) in the case of:

    (i) a mine, the manager of that mine;

    (ii) a quarry, the owner of that quarry;

    (iii) a closed tip, the owner of the mine or quarry with which that tip is associated;

    (iv) an offshore installation (otherwise than in the case of a disease reportable under regulation 5), the duty holder for the purposes of the Offshore Installations and Pipeline Works (Management and Administration) Regulations 1995 provided that for the purposes of this provision regulation 3(2)(c) of those Regulations shall be deemed not to apply;

    (v) a dangerous occurrence at a pipeline (being an incident to which paragraph 14(a)–(f) of Part I of Schedule 2 applies), the owner of that pipeline;

    (vi) a dangerous occurrence at a well, the person appointed by a concession owner to execute any function organizing or supervising any operation to be carried out by the well, or, where no such person has been appointed, the concession owner (and for this purpose 'concession owner' means the person who at any time has the right to exploit or explore mineral resources in any area, or to store gas in any area and to recover gas so stored if, at any time, the well is, or is to be, used in the exercise of that right);

    (vii) a diving operation (otherwise than in the case of a disease reportable under regulation 5), the diving contractor;

    (viii) a vehicle to which paragraph 16 or 17 of Part I of Schedule 2 applies, the operator of the vehicle;

(b) (where sub-paragraph (a) above does not apply) in the case of the death of or other injury to an employee reportable under regulation 3 or of a disease suffered by an employee reportable under regulation 5, his employer; and

(c) in any other case, the person for the time being having control of the premises in connection with the carrying on by him of any trade, business or other undertaking (whether for profit or not) at which, or in connection with the work at which, the accident or dangerous occurrence reportable under regulation 3, or case of disease reportable under regulation 5, happened.

(Reg. 2)

NOTIFICATION OF COOLING TOWERS AND EVAPORATIVE CONDENSERS

Reference

The Notification of Cooling Towers and Evaporative Condensers Regulations 1992.
Approved Code of Practice and Guidance. Legionnaires' disease: the control of legionella bacteria in water systems. HSE 2001.

Scope

Notification is required to be made to the LA in whose area the premises are situated in respect of all cooling towers and evaporative condensers except:

(a) where it contains no water exposed to air;
(b) its water supply is not connected; and
(c) its electrical supply is not connected (Reg. 2 and 3(1)),

and where the device is situated in premises other than domestic premises, which are used for, or in connection with, the carrying on of a trade, business or undertaking (whether for profit or not).

The purpose of the notification is to allow the LA to monitor the environmental control of the device with particular reference to the avoidance of legionnaire's disease and to have an awareness of the location of them in order to investigate outbreaks of such disease.

Person responsible

Notification is to be made by each person who has, to any extent, control of the premises and must be made before the device is situated there. Where the premises fall to any extent under the control of the manufacturer of the device, e.g. during installation, that person is required to effect the notification (Reg. 3(1) and (2)).

Notification

This is to be done in writing using the form prescribed by the HSE for the purpose. The detail to be supplied is:

(a) The address of the premises where the notifiable device is to be situated.
(b) The name, address and telephone number of a person who has, to any extent, control of the premises referred to in (a) above.
(c) The number of notifiable devices at the premises referred to in (a) above.
(d) The location on the premises of each notifiable device referred to in (c) above.

FC74 Notification of cooling towers and evaporative condensers

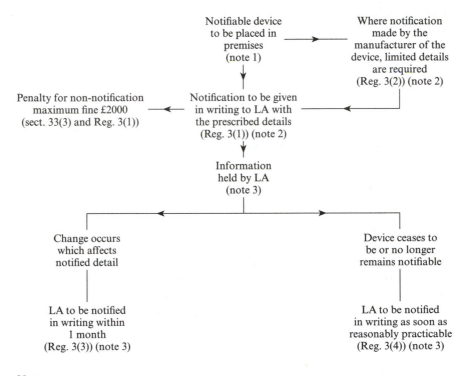

Notes
1.   Notification must be made before the device is placed in the premises otherwise an offence is committed (Reg. 3(1)).
2.   The form to be used is prescribed by the HSE.
3.   LAs should make available to the HSE information from notifications relating to equipment on premises for which the HSE has enforcement responsibility.

Where the notification is being made by the manufacturer, the details required relate only to (a) and (b) above (Reg. 3(1) and (2)).

Subsequent change

Any changes to the notified details must be notified to the LA by either the person having control of the premises or the manufacturer in writing within 1 month (Reg. 3(3)).

The LA must be informed in writing as soon as reasonably practicable where the device ceases to be used, other than for seasonal shutdown (Reg. 3(4)).

Definitions

Cooling tower means a device whose main purpose is to cool water by direct contact between that water and a stream of air.

Evaporative condenser means a device whose main purpose is to cool a fluid by passing that fluid through a heat exchanger which is itself cooled by contact with water passing through a stream of air.

Heat exchanger means a device for transferring heat between fluids which are not in direct contact with one another.

Local authority means:

(a) in relation to England and Wales, a district council, a London borough council, the Common Council of the City of London, the Sub-Treasurer of the Inner Temple or the Under-Treasurer of the Middle Temple or the Council of the Isles of Scilly;

(b) in relation to Scotland, an island or district council.

NB. This definition includes unitary authorities in England, Wales and Scotland (Reg. 2).

Chapter 12

FOOD SAFETY ACT 1990 AND OTHER FOOD CONTROL LEGISLATION

GENERAL PROCEDURAL PROVISIONS

The following general provisions relate to the procedures included in this chapter unless otherwise indicated.

Extent

Unless indicated, these procedures apply in England, Wales and in Scotland (with the application of a different court system), the Isles of Scilly and the Channel Islands but not in Northern Ireland (sects. 57 and 60(5)).

Crown premises

The provisions of the Act and of regulations etc. made under it apply to Crown premises and to persons in the service of the Crown, except that the power of entry may be limited by order of the SoS in the interests of national security. Contraventions by the Crown are not criminal offences unless declared to be so by the High Court (sect. 54).

CoP No. 13 gives guidance on the inspection of Crown premises by enforcement officers.

Food authorities

Local authorities operating the enforcement procedures in this chapter are known as 'food authorities', and are:

(i) in England:
   (a) London borough councils, district councils, and non-metropolitan county councils;
   (b) the Common Council of the City of London;
   (c) the appropriate Treasurer of the Inner and Middle Temples;
   (d) the Council of the Isles of Scilly;
(ii) in Wales, the council of each county or county borough, i.e. the unitary authorities;
(iii) in Scotland, the unitary authorities;
(iv) port health authorities to which functions under the Act have been assigned (sect. 5(1) and (2) as amended).

The effect of this is that within London boroughs, metropolitan districts, unitary authorities in non-metropolitan areas of England and in Wales and Scotland all enforcement is carried out by that unitary authority, but within non-metropolitan areas of England outside of the unitary authorities enforcement and the operation of the procedures in this chapter is a concurrent function of district councils and the county council. However, Ministers are able to assign responsibilities to one or other of these authorities (sect. 6) and, under the Food Safety (Enforcement Authority) (England and Wales) Order 1990, emergency orders under sect. 12 (FC79) are operated by district councils, and enforcement of sect. 15 regarding the description and labelling of food is assigned to county councils. In these areas, all other functions are exercised concurrently, but a statutory code of practice (No. 1 Responsibilities for Enforcement of the FSA 1990) has been issued by the Ministers under sect. 40 and, in general terms, this requires matters of food safety involving immediate risks to health, e.g. contamination by micro-organisms and their toxins, to be dealt with by the district and by the county council where there is no such risk. Consumer protection matters are to be dealt with by the counties, including sect. 14 relating to food not of the nature, substance or quality demanded except that districts should deal with all foreign bodies and mould complaints. So far as regulations made under the Act are concerned, the division of responsibility for their enforcement is indicated in each statutory instrument (sect. 6(4)).

Throughout this chapter the term 'food authority' is used.

Notices

(a) **Form.** All notices must be in writing (sect. 49(1)) and the Ministers may prescribe forms of notice etc. These are identified in the appropriate procedure (sect. 49(2)).

(b) **Authentication.** Notices must be signed (includes a facsimile signature) by the proper officer or any officer so authorized in writing (sect. 49(3)).

(c) **Service.** This may be effected by any of the following methods:
   (i) delivering to the person.
   (ii) in the case of an authorized officer of the enforcement authority, by leaving it, or sending it in a prepaid letter addressed to him at his office;
   (iii) in the case of any other person, by leaving it, or sending it in a pre-paid letter addressed to him, at his usual or last known residence;
   (iv) for an incorporated company or body, by delivering it to its secretary or clerk at its registered or principal office, or by sending it in a prepaid letter addressed to him at that office;
   (v) in the case of an owner or occupier of a premises, if it is not practicable after reasonable enquiries to ascertain the name and address of the person, or if the premises are unoccupied,

by addressing it to the person concerned by the designation of 'owner' or 'occupier' of the premises (naming them) and:

(a) delivering it to some person on the premises; or

(b) if there is no person on the premises to whom it can be given, by affixing it, or a copy of it, to some conspicuous part of the premises (sect. 50 as amended by Sch. 16, para. 18 of the D and CO Act 1994).

Methods of service set out in sect. 233 of the LGA 1972 (page 8) may also be used.

Authorized officer

This is a person (whether or not an officer of the enforcement authority) authorized in writing by the authority either generally or specially to act in matters arising under the FSA. (sect. 5(6)).

Code of Practice No. 19: Qualifications and experience of authorized officers (revised 2000) limits the LAs' appointment of authorized officers to those that are suitably qualified, experienced and competent to carry out the tasks and duties they are being authorized to perform. The CoP sets out these requirements in detail relating to the particular types of premises and food standards enforcement to be carried out and requirements for other specific procedures are set out in the text at the appropriate point.

Power of entry

An AO of an enforcement authority (food authority), upon producing, if required, a duly authenticated document showing his authority, has a right of entry at all reasonable hours to:

(a) any premises within the authority's area, or any business premises inside or outside of the authority's area, to see if there is or has been contraventions of the Act or of regulations etc. made under it or evidence of such contraventions; and

(b) any premises for the performance by the authority of its functions under the Act.

Having secured entry, an AO may:

(a) inspect any records relating to a food business including computerized records;

(b) seize and detain records which may be required in evidence;

(c) require computerized records to be produced in a form which may be taken away.

Admission to a premises used only as a private dwelling cannot be demanded unless 24 hours' notice of intended entry has been given to the occupier.

If a JP is satisfied on sworn information in writing that:

(a) there is reasonable ground for entry; and
(b) admission has been refused or is apprehended and that notice of intention to apply for a warrant has been given to the occupier; or
(c) the premises are unoccupied or the occupier is temporarily absent; or
(d) that an application for admission would defeat the object of entry;

he may issue a warrant authorizing the authorized officer to enter the premises, if need be by reasonable force. A warrant continues in force for 1 month. The AO may take with him such other persons who may be necessary and must leave any unoccupied premises as effectively secured against trespass as he found them.

The disclosure of any trade secret by any person using the powers of those provisions is an offence unless the disclosure was made in the performance of his duty.

Persons obstructing an AO in his execution of the Act are liable to fines not exceeding level 5 on the standard scale and/or imprisonment not exceeding 3 months (sects. 32, 33 and 35).

Defences

In any proceedings for offences in relation to procedures dealing with improvement notices, prohibition orders, emergency prohibition notices and orders and the detention and seizure of food, it will be a defence to prove that the defendant took all reasonable precautions and exercised all due diligence to avoid the commission of the offence by himself or by a person under his control (sect. 21(1)).

Time limit for prosecution

Prosecution for offences, other than those involving obstruction of officers, must be commenced before:

(a) 3 years from the commission of the offence; or
(b) 1 year from the discovery;

whichever is the earlier (sect. 34).

Enforcement procedures for food safety

The role of the FSA including its' relationship with the food authorities is described on page 3 and overlies the procedures in this chapter. The FSA has taken over most of the functions previously exercised by MAFF in these procedures (see the Food Standards Act 1999 (Transitional and Consequential Provisions and Savings) (England and Wales) Regulations 2000). This is indicated where necessary in each procedure.

The following statutory codes of practice issued under sect. 40 of the FSA 1990 have a particular and general relevance to these procedures:–

CoP No. 2 Legal procedures – this sets out various requirements in relation to the legal aspects of enforcement and includes a list of factors to be considered before determining whether or not an offence should result in prosecution

CoP No. 9 (rev. 2000) Food Hygiene Inspections – this deals with the purposes and objectives of food hygiene inspections, hazard rating of premises, scheduling of inspections, file records, inspection techniques, the quality of inspections and the monitoring of the inspection programme.

CoP 19 (revised 2000) Qualifications and experience of authorized officers – see above. In addition there are several CoPs which relate to particular procedures and these are noted within each of them.

DEFINITIONS

The following definitions are applicable to all FSA 1990 procedures in this chapter.

Animal means any creature other than a bird or fish (sect. 53(1)).

Article does not include a live animal or bird, or a live fish which is not used for human consumption while it is alive (sect. 53(1)).

Authorized officer In relation to an enforcement authority means any person (whether or not an officer of the authority) who is authorized by it in writing, either generally or specially, to act in matters arising under this Act; but if regulations made by the Ministers so provide, no person shall be so authorized unless he has such qualifications as may be prescribed by the regulations (sect. 5(6)).

Business includes the undertaking of a canteen, club, school, hospital or institution, whether carried on for profit or not, and any undertaking or activity carried on by a public or LA (sect. 1(3)).

Commercial operation, in relation to any food or contact material, means any of the following, namely:

(a) selling, possessing for sale and offering, exposing or advertising for sale;
(b) consigning, delivering or serving by way of sale;
(c) preparing for sale or presenting, labelling or wrapping for the purpose of sale;
(d) storing or transporting for the purpose of sale;
(e) importing and exporting;

and, in relation to any food source, means deriving food from it for the purpose of sale, or for purposes connected with sale (sect. 1(3)).

Fish includes crustaceans and molluscs (sect. 53(1)).

Food includes:

(a) drink;

(b) articles and substances of no nutritional value which are used for human consumption;

(c) chewing gum and other products of a like nature and use; and

(d) articles and substances used as ingredients in the preparation of food or anything falling within this subsection;

but does not include:

(a) live animals or birds, or live fish which are not used for human consumption while they are alive;

(b) fodder or feeding stuffs for animals, birds or fish;

(c) controlled drugs within the meaning of the Misuse of Drugs Act 1971; or

(d) subject to such exceptions as may be specified in an order made by the Ministers:

  (i) medicinal products within the meaning of the Medicines Act 1968 in respect of which product licences within the meaning of that Act are for the time being in force; or

  (ii) other articles or substances in respect of which such licences are for the time being in force in pursuance of orders under sect. 104 or 105 of that Act (application of Act to other articles and substances) (sect. 1(1) and (2)).

NB. Through this definition water is deemed to be food and is subject to the Act. However, water as supplied to any premises by a water undertaker or through a private supply is not subject to this Act but is dealt with in the Water Industry Act 1991 and regulations made under it (Chapter 4) (sects. 55 and 56).

Food business means any business in the course of which commercial operations with respect to food or food sources are carried out (sect. 1(3)).

Food premises means any premises used for the purposes of a food business (sect. 1(3)).

Human consumption includes use in the preparation of food for human consumption (sect. 53(1)).

Premises includes any place, any vehicle, stall or moveable structure and, for such purposes as may be specified in an order made by the Ministers, any ship or aircraft of a description so specified (sect. 1(3)).

Preparation, in relation to food, includes manufacture and any form of processing or treatment, and 'preparation for sale' includes packaging, and 'prepare for sale' shall be construed accordingly (sect. 53(1)).

Proprietor, in relation to a food business, means the person by whom that business is carried on (sect. 53(1)).

Sale:

(a) the supply of food, otherwise than on sale, in the course of a business; and

(b) any other thing which is done with respect to food and is specified in an order made by the Ministers:

shall be deemed to be a sale of the food, and references to purchasers and purchasing shall be construed accordingly.

This Act shall apply:

(a) in relation to any food which is offered as a prize or reward or given away in connection with any entertainment to which the public are admitted whether on payment of money or not, as if the food were, or had been, exposed for sale by each person concerned in the organization of the entertainment;

(b) in relation to any food which, for the purpose of advertisement or in furtherance of any trade or business, is offered as a prize or reward or given away, as if the food were, or had been, exposed for sale by the person offering or giving away the food; and

(c) in relation to any food which is exposed or deposited in any premises for the purpose of being so offered or given away as mentioned in paragraph (a) or (b) above, as if the food were, or had been, exposed for sale by the occupier of the premises;

and in this subsection 'entertainment' includes any social gathering, amusement, exhibition, performance, game, sport or trial of skill (sect. 2).

GENERAL ENFORCEMENT PROCEDURES

The procedures in this section apply to all food premises in the chapter including those which have separate approval or registration procedures.

DETENTION AND SEIZURE OF FOOD

References

Food Safety Act 1990 sect. 9.
Detention of Food (Prescribed Forms) Regulations 1990.
The Meat (Enhanced Enforcement Powers) Regulations 2000.
Code of Practice No. 4: Inspection, Detention and Seizure of Suspected Food (under review).
Code of Practice No. 19: Qualifications and experience of authorised officers (revised 2000).

Power of inspection of food

An AO of a FA may at all reasonable times inspect any food intended for human consumption which:

FC75 Detention and seizure of food

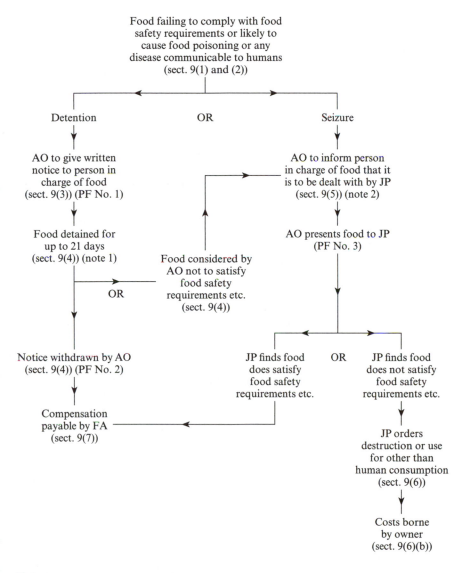

Food failing to comply with food safety requirements or likely to cause food poisoning or any disease communicable to humans (sect. 9(1) and (2))

Detention | OR | Seizure

AO to give written notice to person in charge of food (sect. 9(3)) (PF No. 1)

AO to inform person in charge of food that it is to be dealt with by JP (sect. 9(5)) (note 2)

Food detained for up to 21 days (sect. 9(4)) (note 1)

AO presents food to JP (PF No. 3)

Food considered by AO not to satisfy food safety requirements etc. (sect. 9(4))

OR

Notice withdrawn by AO (sect. 9(4)) (PF No. 2)

JP finds food does satisfy food safety requirements etc.

OR

JP finds food does not satisfy food safety requirements etc.

Compensation payable by FA (sect. 9(7))

JP orders destruction or use for other than human consumption (sect. 9(6))

Costs borne by owner (sect. 9(6)(b))

Notes
1. It is an offence to breach the requirements of the notice (sect. 9(3)).
2. Persons liable to prosecution have a right to attend the hearing and present evidence and witnesses (sect. 9(5)(a)).
3. The prescribed forms are detailed in the Detention of Food (Prescribed Forms) Regulation 1990.

(a) has been sold;
(b) is offered or exposed for sale;
(c) is in the possession of, has been deposited with or consigned to any person for the purpose of sale or preparation for sale (sect. 9(1)).

Detention and seizure of food

Where either:

(a) on inspection food is found to fail to comply with food safety requirements; or
(b) it appears to the AO that any food is likely to cause food poisoning or any disease communicable to humans;

the AO may either:

(i) detain the food; or
(ii) seize it and remove it in order to have it dealt with by a JP (sect. 9(1)–(3)).

Food is deemed not to comply with food safety requirements if:

(a) it is unfit for human consumption; or
(b) it has been rendered injurious to health by the addition of articles or substances, by the use of an article or substance as an ingredient in its preparation, by the abstraction of any constituent or by the application of any process or treatment; or
(c) it is so contaminated (by extraneous matters or otherwise) that it would not be reasonable to expect it to be used for human consumption in that state (sect. 8(2)).

The Meat (Enhanced Enforcement Powers) Regulations 2000 have extended the use of sect. 9 powers dealing with the inspection and seizure of food to include illegally produced meat which does not have the required health mark.

CoP 19 requires that the procedures dealing with the inspection, detention and seizure of food under sect. 9 should only be exercised by appropriately qualified officers i.e. environmental health officers and, where appropriate, official veterinary surgeons and in relation to meat only officers qualified under the Authorised Officers (Meat Inspection) Regulations 1987.

Detention

The detention of the food is effected by a notice (form prescribed by the Detention of Food (Prescribed Forms) Regulations 1990) to be served on the person in charge of the food requiring that:

(a) the food is not to be used for human consumption; and
(b) it is not to be removed except to a place specified in the notice.

Failure to comply with the notice is an offence for which the penalty on indictment is an unlimited fine and/or imprisonment for up to 2 years or on summary conviction to a fine up to the statutory maximum and/or imprisonment for not exceeding 6 months (sects 9(3) and 35(2)).

Having served such notice the AO is required to determine whether or not the food complies with food safety requirements as soon as is reasonably practicable and in any event within 21 days. If satisfied of compliance, the food is released and the notice withdrawn (PF No. 2); if satisfied of non-compliance the food must be seized (sect. 9(4)).

Seizure

Seizure of food, with or without prior detention, is effected by the AO removing it to have it dealt with by a JP. The person in charge of the food must be informed (PF No. 3) and any person who may be liable to prosecution for the condition of the food may attend before the JP and is entitled to be heard and call witnesses (sect. 9(5)).

If the JP considers on the basis of the evidence that the food fails to comply with food safety requirements or is likely to cause food poisoning or a human communicable disease he must condemn the food and order it to be destroyed or disposed of so as to prevent it being used for human consumption. In this case the owner of the food is required to defray the costs incurred in the destruction or disposal of the food (sect. 9(6)).

The CoP requires that, if possible, food which is seized should be presented to a JP within *2* days and, where the food is highly perishable, as soon as possible.

Compensation

If the JP is not satisfied that the food fails to comply, the FA is required to compensate the owner for any depreciation in its value. This is also the case if a detention notice is withdrawn by the AO (sect. 9(7)).

Defences

See under 'General procedural provisions' (page 317).

SAMPLING OF FOOD FOR ANALYSIS AND EXAMINATION

References

Food Safety Act 1990 sects. 29–30.
Food Safety (Sampling and Qualifications) Regulations 1990.
Code of Practice No. 7: Sampling For Analysis or Examination (revised 2000).
Code of Practice No. 19: Qualifications and Experience of Authorised Officers (revised 2000).

FC76 Sampling of food for analysis and examination

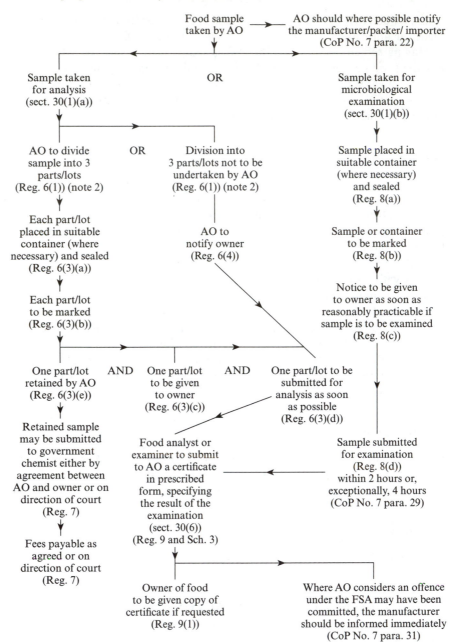

Food sample taken by AO → AO should where possible notify the manufacturer/packer/ importer (CoP No. 7 para. 22)

OR

Sample taken for analysis (sect. 30(1)(a))

Sample taken for microbiological examination (sect. 30(1)(b))

AO to divide sample into 3 parts/lots (Reg. 6(1)) (note 2) **OR** Division into 3 parts/lots not to be undertaken by AO (Reg. 6(1)) (note 2)

Sample placed in suitable container (where necessary) and sealed (Reg. 8(a))

Each part/lot placed in suitable container (where necessary) and sealed (Reg. 6(3)(a))

AO to notify owner (Reg. 6(4))

Sample or container to be marked (Reg. 8(b))

Each part/lot to be marked (Reg. 6(3)(b))

Notice to be given to owner as soon as reasonably practicable if sample is to be examined (Reg. 8(c))

One part/lot retained by AO (Reg. 6(3)(e)) **AND** One part/lot to be given to owner (Reg. 6(3)(c)) **AND** One part/lot to be submitted for analysis as soon as possible (Reg. 6(3)(d))

Retained sample may be submitted to government chemist either by agreement between AO and owner or on direction of court (Reg. 7)

Food analyst or examiner to submit to AO a certificate in prescribed form, specifying the result of the examination (sect. 30(6)) (Reg. 9 and Sch. 3)

Sample submitted for examination (Reg. 8(d)) within 2 hours or, exceptionally, 4 hours (CoP No. 7 para. 29)

Fees payable as agreed or on direction of court (Reg. 7)

Owner of food to be given copy of certificate if requested (Reg. 9(1))

Where AO considers an offence under the FSA may have been committed, the manufacturer should be informed immediately (CoP No. 7 para. 31)

Notes

1. Regulation numbers refer to the Food Safety (Sampling and Qualifications) Regulations 1990.
2. Division into three parts or lots is not required if this is not reasonably practicable or may impede analysis (Reg. 6(1)).
3. This sampling procedure does not apply to the enforcement of certain food regulations (page 326).
4. Sometimes there are difficulties in establishing who the 'owner' is in certain multiple retailer situations – in these cases the registered head office of the company should be notified.

Powers of sampling

An AO, properly trained in the appropriate techniques and competent to carry out the duties, has the power to:

(a) purchase samples of food or any substance capable of being used in the preparation of food;
(b) take (i.e. without payment) such samples of food and substances which either;
(i) appear to be intended for sale or have been sold; or
(ii) are found in any premises which the AO is entitled to enter;
(c) take a sample from a food source or of any contact material found on any such premises; and
(d) take a sample of any article or substance found at such premises which may be required as evidence in proceedings under the Act (sect. 29).

Analysis and examination

The FSA 1990 recognizes that the microbiological examination of food is a separate activity from its analysis. Analysis is defined as:

'. . . includes microbiological assay and any technique for establishing the composition of the food' (sect. 53(1)).

Examination is defined as:

'a microbiological examination' (sect. 28(2)).

Samples taken by the AO are to be submitted either to a public analyst for analysis or, if for examination, to a food examiner. The necessary qualifications for each are specified in the Food Safety (Sampling and Qualifications) Regulations 1990 (sect. 30(1) and Regs 3 and 4). CoP 7 requires that samples for chemical analysis must be submitted to a laboratory accredited for that purpose and which appears on the list of official food control laboratories.

FAs, other than remaining two-tier district councils in England, who are not generally involved in matters of food composition, are required to appoint one or more public analysts for their areas (sect. 27).

FAs including district councils may provide their own facilities for microbiological examination (sect. 28) or, alternatively, may submit samples to a food examiner.

Analyst's and examiner's certificates

Following analysis or examination, the analyst or examiner must submit a certificate to the AO specifying the result (sect. 30(6)). Each certificate must be signed (sect. 30(7)). The form of certificate to be used is specified in Sch. 3 Food Safety (Sampling and Qualifications) Regulations 1990.

In any legal proceedings taken following the analysis or examination, the certificate or a copy of it is sufficient evidence of the facts stated therein unless either party wishes the food analyst or examiner to be called as a witness (sect. 30(8)).

Sampling procedure

Code of Practice 7 asks FAs to prepare and publish their sampling strategy on an annual basis in consultation with the food examiner and public analyst.

Following sampling, the sample must be dealt with in accordance with the Food Safety (Sampling and Qualifications) Regulations 1990 and the procedure is indicated in FC76. More detailed guidance will be found in the CoP No. 7: Sampling for Analysis or Examination (revised 2000).

This procedure is *not* applicable to sampling under the following food regulations in which particular sampling procedures are laid down:

The Milk-based Drinks (Scotland) Regulations 1983
The Cream (Heat Treatment) (Scotland) Regulations 1983
Poultry Meat (Water Content) Regulations 1984 and (Scotland) Regulation 1993
Natural Mineral Waters Regulations 1985
The Materials and Articles in Contact with Food Regulations 1987
Milk and Dairies (Semi-skimmed and Skimmed Milk) (Heat Treatment) (Scotland) Regulations 1988
Milk (Special Designations) (Scotland) Order 1958
The Animals, Meat and Meat Products (Examinations for Residues and Maximum Residue Limits) Regulations 1991
The Plastic Materials and Articles in Contact with Food Regulations 1992

The sampling procedure detailed in FC76 covers the sampling of foods in all other situations.

Neither the regulations nor the CoP deals with samples taken informally except that the CoP recommends that all samples produced for microbiological examination should be taken in accordance with the requirements of both.

Division of samples

Samples intended for analysis (but not for examination) are to be divided into three parts unless either:

(a) this is not reasonably practicable; or
(b) it is likely to impede analysis (Reg. 6(1) and (4)).

If the sample itself consists of sealed containers and the AO believes that the opening of them would impede proper analysis, he may proceed by dividing the sample into parts by putting the containers into three lots (Reg. 6(2)).

Definitions (also page 318)

Container includes any basket, pail, tray, package or receptacle of any kind, whether open or closed (sect. 53(1)).

Food source means any growing crop or live animal, bird or fish from which food is intended to be derived (whether by harvesting, slaughtering, milking, collecting eggs or otherwise) (sect. 1(3)).

IMPROVEMENT NOTICES

References

Food Safety Act 1990 sects. 10 and 37–39.
Food Safety (Improvement and Prohibition – Prescribed Forms) Regulations 1991.
Code of Practice No. 5: The Use of Improvement Notices (revised April 1994).
Code of Practice No. 19: Qualifications and Experience of Authorised Officers (revised 2000).

Scope

An improvement notice may be served by an AO (whether or not an officer of the authority) where there are reasonable grounds for believing that the proprietor of a food business is failing to comply with any regulations:

(a) for requiring, prohibiting, or regulating the use of any process or treatment in the preparation of food; or
(b) dealing with hygienic conditions and practices connected with food or food sources at a commercial operation (sect. 10(1) and (3)).

In the main the regulations comprise the 'horizontal' Food Safety (General Food Hygiene) Regulations 1995 and the 'vertical' series which include the Dairy Products (Hygiene) Regulations 1995, the Meat Products (Hygiene) Regulations 1994 and the Food Safety (Fishery Products and Live Shellfish) (Hygiene) Regulations 1998. The CoP No. 5 gives guidance on the circumstances in which an improvement notice should be served (paras. 11–19).

Issue of improvement notices

The CoP No. 19 indicates the types of officers who should be authorized by enforcing authorities to sign improvement notices. These are environmental health officers or veterinary surgeons and those officers holding the Higher Certificate in Food Premises Inspection. In relation to the lower risk premise

FC77 Improvement notices

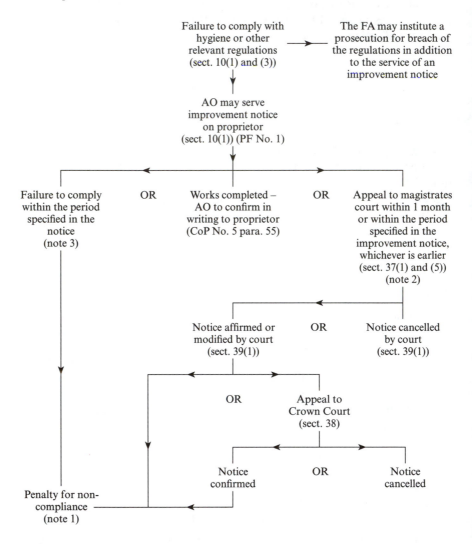

Failure to comply with hygiene or other relevant regulations (sect. 10(1) and (3)) → The FA may institute a prosecution for breach of the regulations in addition to the service of an improvement notice

AO may serve improvement notice on proprietor (sect. 10(1)) (PF No. 1)

Failure to comply within the period specified in the notice (note 3) — OR — Works completed – AO to confirm in writing to proprietor (CoP No. 5 para. 55) — OR — Appeal to magistrates court within 1 month or within the period specified in the improvement notice, whichever is earlier (sect. 37(1) and (5)) (note 2)

Notice affirmed or modified by court (sect. 39(1)) — OR — Notice cancelled by court (sect. 39(1))

OR — Appeal to Crown Court (sect. 38)

Notice confirmed — OR — Notice cancelled

Penalty for non-compliance (note 1)

Notes

1. The penalty for each improvement notice not complied with is, on indictment, an unlimited fine and/or imprisonment for up to 2 years and, on summary conviction, a fine not exceeding level 5 and/or imprisonment for up to 6 months.
2. An appeal suspends the period for compliance until that appeal has been determined (sect. 39(2)).
3. Although there is no provision for the AO to extend the period allowed the CoP considers that it would not be unreasonable to do so in appropriate circumstances (para. 41). This should be done in writing.

categories C to F detailed in C of P No. 9, holders of the Ordinary Certificate may also serve improvement notices for those categories only.

Minded-to notices

The regulation enabling sections of the Deregulation and Contracting Out Act 1994, which made provision for this early intimation of the intent to issue an improvement notice, were repealed by the Regulatory Reform Act 2001. Such notices are therefore not now necessary.

Person responsible

Improvement notices are to be served on the proprietor of the food business (sect. 10(1)). If it is not possible to name him, the notice may be addressed to 'the owner' of the premises and left there (sect. 50(2)) (also 'General procedural provisions' on page 315).

Wherever possible, a copy of the notice should be sent to the person within the organization responsible for taking action (CoP para. 26).

Content

The improvement notice should:

(a) state the AO's grounds for believing there is a failure to comply;
(b) specify the matters of non-compliance;
(c) specify the measures required to secure compliance; and
(d) require compliance within a specified period of not less than 14 days (sect. 10(1)).

CoP No. 5 requires that the wording of the notice should be clear and easily understood (para. 30) and must include details of the regulation contravened and the reason for the opinion that there has been a contravention (para. 32). The time period allowed must be realistic and wherever possible should be set after discussion with the proprietor (para. 36).

Appeals

Any person aggrieved by the AO's decision to serve the improvement notice may appeal to the magistrates' court within 1 month or within the period specified by the AO in the notice for requiring compliance if that is less than 1 month (sect. 37(1) and (5)). An appeal suspends the period of compliance until the appeal has been determined (sect. 39(2)).

There is a subsequent appeal to the County Court upon dismissal of an appeal by a magistrates' court by any aggrieved person, including the AO. The CoP asks that full information about the rights of appeal and the necessary procedures should be included in notes attached to the notice (para. 50).

Defences

See under 'General procedural provisions' on page 317.

PROHIBITION ORDERS

References

Food Safety Act 1990 sects. 11, 37 and 38.
Food Safety (Improvement and Prohibition – Prescribed Forms) Regulations 1991.
Code of Practice No. 6: Prohibition Procedures (under review).

Prohibition orders

These orders may be made by a court upon the conviction of the proprietor of a food business for offences against certain regulations made under the Act. There is no requirement for application by the FA although they may bring the procedure to the attention of the court (sect. 11(1)).

Relevant regulations

The offences must relate to non-compliance with regulations made under Part 2 of the Act and which make provisions:

    (a) for requiring, prohibiting, or regulating the use of any process or treatment in the preparation of food; or

    (b) dealing with hygienic conditions and practices connected with food or food sources at a commercial operation (sects. 10(3)) and 11(1)).

(Note to 'Relevant Regulations' on page 327).

Health risk condition

Before making an order, the court must be satisfied that there is a risk of injury to health from:

    (a) the use of any process or treatment;

    (b) the construction of any premises or the use of any equipment; or

    (c) the condition of the premises or equipment (sect. 11(2)).

Where these conditions exist and the proprietor is convicted of an offence under appropriate regulations, the court must make a prohibition order (sect. 11(1)), although the prohibition of a person from the management of a business is discretionary (sect. 11(4)).

Types of prohibition

The type of prohibition specified in the order must be appropriate to the circumstances and may either:

FC78 Prohibition orders

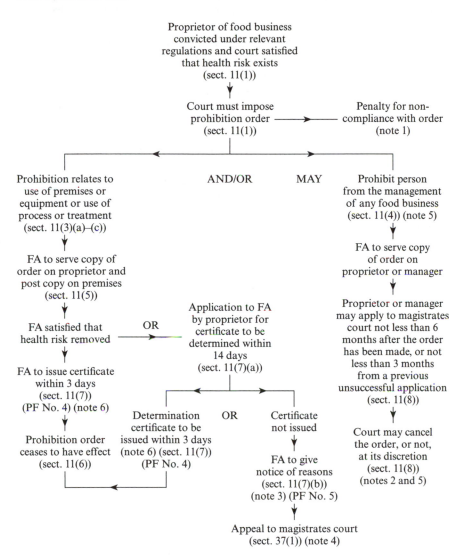

Notes

1. The penalty for non-compliance with the order is a fine up to the statutory maximum and/or up to 6 months' imprisonment, or, on indictment, an unlimited fine and/or up to 2 years' imprisonment (sect. 35).
2. There is no appeal against these determinations by the court.
3. No time period is indicated within which those reasons must be given.
4. This appeal must be made within 1 month from the notice of the decision and there is a subsequent appeal to the Crown Court (sects. 37(5) and 38).
5. FA should notify other FAs through the Chartered Institute of Environmental Health (CoP No. 6 para. 61).
6. Saturdays, Sundays and bank holidays are not taken into account in relation to periods specified (sect. 53(5)).

(a) prohibit a particular process or treatment; or
(b) prohibit the use of (close down) the premises or prohibit it being used for particular types of food business; or
(c) prohibit the use of equipment at any food business or at a specified business; or
(d) in the case of a proprietor or manager of a food business convicted on hygiene offences, prohibit his participation in the management of any food business or businesses of a specified description or class (sect. 11(3) and (4)).

Guidance on the appropriate use of each prescribed prohibition is given in CoP No. 6. FAs are regularly informed by LACOTS and CIEH of those persons who have been prohibited from managing a food business under sect. 11(4). To enable this to happen, FAs are asked to notify those bodies of the details of persons prohibited in their areas and when such prohibition is cancelled (CoP No. 6 para. 61).

Determination

Orders prohibiting a person from participating in the management of a food business are determined by the court on the application of the person concerned, whereas all others are dealt with by the FA either on its own initiative or on application from the proprietor of the business. Decisions of the FA not to issue a certificate that it is satisfied that the health risk condition is no longer fulfilled are subject to appeal to the magistrates' court (sect. 11(6) and (7)), and details of this should be given with the certificate of continuing risk to health (CoP No. 6 para. 91).

Defences

See under 'General procedural provisions' on page 317.

Definition (also page 318)

Manager, in relation to a food business, means any person who is entrusted by the proprietor with the day-to-day running of the business, or any part of the business (sect. 11(11)).

EMERGENCY PROHIBITION NOTICES AND ORDERS

References

Food Safety Act 1990 sects. 12, 37 and 38.
Food Safety (Improvement and Prohibition – Prescribed Forms) Regulations 1991.
Code of Practice No. 6: Prohibition Procedures (under review).

FC79 Emergency prohibition notices and orders

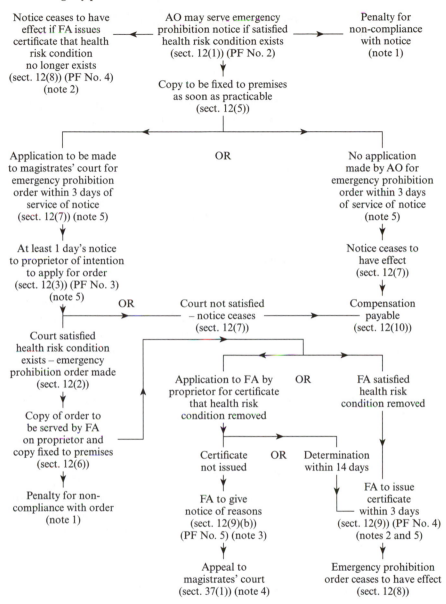

Notice ceases to have effect if FA issues certificate that health risk condition no longer exists (sect. 12(8)) (PF No. 4) (note 2)

AO may serve emergency prohibition notice if satisfied health risk condition exists (sect. 12(1)) (PF No. 2)

Penalty for non-compliance with notice (note 1)

Copy to be fixed to premises as soon as practicable (sect. 12(5))

Application to be made to magistrates' court for emergency prohibition order within 3 days of service of notice (sect. 12(7)) (note 5)

OR

No application made by AO for emergency prohibition order within 3 days of service of notice (note 5)

At least 1 day's notice to proprietor of intention to apply for order (sect. 12(3)) (PF No. 3) (note 5)

Notice ceases to have effect (sect. 12(7))

OR

Court not satisfied – notice ceases (sect. 12(7))

Compensation payable (sect. 12(10))

Court satisfied health risk condition exists – emergency prohibition order made (sect. 12(2))

Application to FA by proprietor for certificate that health risk condition removed

OR

FA satisfied health risk condition removed

Copy of order to be served by FA on proprietor and copy fixed to premises (sect. 12(6))

Certificate not issued

OR

Determination within 14 days

Penalty for non-compliance with order (note 1)

FA to give notice of reasons (sect. 12(9)(b)) (PF No. 5) (note 3)

FA to issue certificate within 3 days (sect. 12(9)) (PF No. 4) (notes 2 and 5)

Appeal to magistrates' court (sect. 37(1)) (note 4)

Emergency prohibition order ceases to have effect (sect. 12(8))

Notes

1. The penalty for non-compliance with the emergency prohibition notice and the order is a fine up to the statutory maximum and/or up to 6 months' imprisonment or, on indictment, an unlimited fine and/or up to 2 years' imprisonment (sect. 35).
2. The certificate must be issued within 3 days of the FA being satisfied that the health risk condition is no longer fulfilled (sect. 12(9)).
3. No time period is indicated within which these reasons must be given.
4. The appeal must be made within one month of the notice of the decision and there is a subsequent appeal to the Crown Court (sects. 37(5) and 38).
5. Saturdays, Sundays and bank holidays are not taken into account in relation to periods specified (sect. 53(5)).

Code of Practice No. 19: Qualifications and experience of authorised officers (revised 2000).

Scope

This procedure allows an AO to instigate emergency action to secure a prohibition order where there is **imminent** risk of injury to health in respect of any food business (sect. 12(4)).

The parameters of the health risk condition to exist before the action is taken are the same as for prohibition orders (FC78) except that the risk of injury to health must be **imminent** (sect. 12(4)). This emergency procedure does not require court conviction before it can be activated. Guidance on 'imminent risk to health' is given in CoP No. 6.

Emergency prohibition notices

This is served by the AO (whether or not an officer of the authority) on the proprietor of the food business where satisfied that there is an imminent risk to health arising from any of the defined health risk conditions (page 330). It immediately imposes such appropriate prohibition as specified in the notice. These are those identified on page 332 in relation to prohibition orders, but cannot include the prohibition from management of a business (sect. 12(1)).

CoP No. 19 states that emergency prohibition notices should only be signed by environmental health officers who have had 2 years post qualification experience in food safety matters and are currently involved in food enforcement.

Emergency prohibition orders

The notice ceases to have effect after 3 days (or before if the conditions are removed), unless the AO applies to the magistrates' court for the issue of an emergency prohibition order. The proprietor must be given at least 1 day's notice of the intention to apply for the order (sect. 12(2)).

Types of prohibition by emergency notices and orders

These are the same as for normal prohibition notices (page 332) except that they cannot include the prohibition from the management of food businesses.

Enforcement of emergency prohibition notices and orders

The use and enforcement of these notices and orders may only be undertaken by district councils including the unitary authorities in England, Wales and Scotland (Food Safety (Enforcement Authority) (England and Wales) Order 1990).

Determination of emergency notices and orders

This is effected by the FA when satisfied that the health risk condition has been removed and may be undertaken on its own initiative in relation to notices or on its own initiative or upon application in relation to orders (sect. 12(7) and (9)).

Certificates by the FA that the health risk condition has been removed must be issued within 3 days of the determination and determinations following application by the proprietor in relation to an order must be made within 14 days (sect. 12(9)).

Compensation

Where either:

(a) the emergency prohibition notice is not followed by an application for an order within 3 days; or
(b) upon the hearing of such an application, the court is not satisfied that the imminent health risk condition exists;

compensation is payable by the FA in respect of any loss suffered by reason of the interim compliance with the notice (sect. 12(10)).

Defences

See under 'General procedural provisions' on page 317.

FOOD PREMISES

REGISTRATION OF FOOD PREMISES

References

Food Premises (Registration) Regulations 1991.
Food Premises (Registration) Amendment Regulations 1993 and 1997.
Food Premises (Registration) (Welsh Form Applications) Regulations 1993.
Code of Practice No. 11: Enforcement of the Food Premises (Registration) Regulations (under review).

Scope

All food businesses (definition on page 319) are required to be registered with the LA, with the exceptions specified below (Reg. 2).

Exemptions from registration requirement

Registration is not required in the following cases.

FC80 Registration of Food Premises

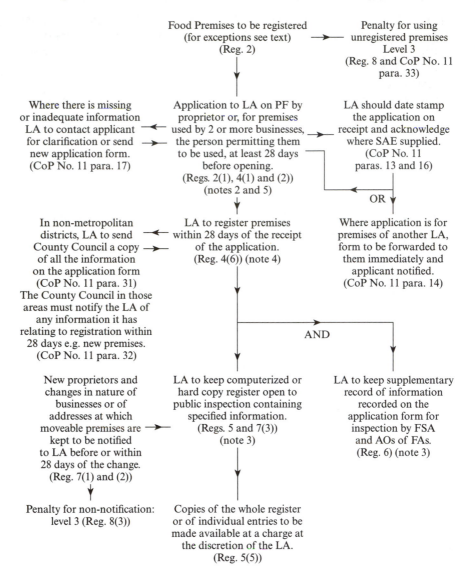

Food Premises to be registered
(for exceptions see text)
(Reg. 2)

Penalty for using
unregistered premises
Level 3
(Reg. 8 and CoP No. 11
para. 33)

Where there is missing
or inadequate information
LA to contact applicant
for clarification or send
new application form.
(CoP No. 11 para. 17)

Application to LA on PF by
proprietor or, for premises
used by 2 or more businesses,
the person permitting them
to be used, at least 28 days
before opening.
(Regs. 2(1), 4(1) and (2))
(notes 2 and 5)

LA should date stamp
the application on
receipt and acknowledge
where SAE supplied.
(CoP No. 11
paras. 13 and 16)

In non-metropolitan
districts, LA to send
County Council a copy
of all the information
on the application form
(CoP No. 11 para. 31)
The County Council in those
areas must notify the LA of
any information it has
relating to registration within
28 days e.g. new premises.
(CoP No. 11 para. 32)

LA to register premises
within 28 days of the receipt
of the application.
(Reg. 4(6)) (note 4)

Where application is for
premises of another LA,
form to be forwarded to
them immediately and
applicant notified.
(CoP No. 11 para. 14)

OR

AND

New proprietors and
changes in nature of
businesses or of
addresses at which
moveable premises are
kept to be notified
to LA before or within
28 days of the change.
(Reg. 7(1) and (2))

LA to keep computerized or
hard copy register open to
public inspection containing
specified information.
(Regs. 5 and 7(3))
(note 3)

LA to keep supplementary
record of information
recorded on the
application form for
inspection by FSA
and AOs of FAs.
(Reg. 6) (note 3)

Penalty for non-notification:
level 3 (Reg. 8(3))

Copies of the whole register
or of individual entries to be
made available at a charge at
the discretion of the LA.
(Reg. 5(5))

Notes

1. Regulation numbers refer to the Food Premises (Registration) Regulations 1991.
2. The form of application is prescribed in Schedule 4 to the regulations.
3. The LA may not alter the register or supplementary record unless the information came from the proprietor or has been notified to that person 28 days before (Reg. 7(4)).
4. Registration does not require periodic renewal.
5. The application form should be retained by the LA (in origin form or microfilmed) until there is a new application or details of the entry change or following an inspection, whichever is the earlier (CoP No. 11 para. 20).

1. Places used *only* for the following activities and which are *already* approved or licensed for that purpose.
   (a) Slaughterhouse.
   (b) Poultry meat slaughterhouses and cutting premises.
   (c) Meat export cutting premises, cold stores and transhipment centres.
   (d) Meat product plants approved for export to another country in the EEC.
   (e) Production holding or dairy establishment.
   (f) Wild game processing facility.
   (g) Premises producing minced meat or meat preparations for export to another state
   (Reg. 3(2)).
   Premises which are subject to licensing under the Food Safety (General Food Hygiene) (Butchers' Shops) Amendment Regulations 2000 – see FC81 – are still required to be registered under this procedure.
2. Premises used irregularly or only occasionally i.e. premises used for less than 5 days in any 5 consecutive weeks (this will exempt village fetes, car boot sales, markets held only irregularly, some sporting events, etc.). The 5 days do not have to be consecutive and thus any premises used regularly once a week will be included (Reg. 2(1) and (2)).
3. Premises at which the only commercial operations carried on in relation to food or food sources are at:
   (a) Places where game including deer is killed in sport (e.g. grouse moors).
   (b) Places where fish is taken for food (but not processed).
   (c) Places only harvesting, cleaning, storing and packing crops (e.g. vineyards, arable farms) except where the crops are packed on those premises in the form in which they are to be sold to the ultimate consumer – thus farms which harvest, clean and pack vegetables in cellophane wrapping in which they will ultimately be sold retail are included in the scope of registration.
   (d) Places where honey is harvested.
   (e) Places where eggs are produced or packed.
   (f) Retail sale by automatic vending machine.
   (g) The supply of beverages, biscuits, crisps, confectionery or similar products, ancillary to the business whose principal activity is not food.
   (h) The supply of food at a religious ceremony (Reg. 3(1)).
4. Exemptions for some food businesses run from domestic premises:
   (a) Domestic premises where the person resident is not the owner of the food business (e.g. where the resident is a volunteer preparing food for another business as in 'meals on wheels' *except* where the food business involves peeling shrimps or prawns).
   (b) Domestic premises used for the production of honey or subsequent preparation storage, bottling or sale (whether wholesale or retail) of honey.

(c) Domestic premises where crops are produced, cleaned, stored, packed and sold (whether wholesale or retail).
(d) Domestic premises where bed and breakfast accommodation is provided in not more than 3 bedrooms.
(e) Domestic premises where the proprietor lives on the premises and the only commercial undertaking relates to the preparation of food for sale from a market stall run by W1 Country Markets Ltd (Reg. 3(4) as amended 1997).

5. Exemptions for some vehicles and stalls etc.:
   (a) Private motor cars.
   (b) Aircraft.
   (c) Ships, unless they are permanently moored or used for pleasure excursions in inland or coastal waters only.
   (d) Food vehicles normally based outside Great Britain.
   (e) Vehicles and stalls kept at or used from premises which are themselves registered or which are exempt (e.g. food trolleys used within a hospital, forklift trucks in food warehouses, delivery vans run from a registered baker's shop).
   (f) Market stalls provided by the controller of the market, although the market itself must be registered.
   (g) Tents and marquees, awnings and similar structures (not including stalls) Regs. 1(3) and 2.

6. Childminders: the sale of food ancillary to acting as a childminder – this applies only to domestic premises where the registered childminder resides. Childminding on other premises, or where the childminder is caring for more than 6 children, does not benefit from this exemption (1993 Amendment Regulations).

7. Other exemptions:
   (a) Places run by voluntary or charitable organizations and used only by those types of organizations, if no food (except tea, coffee, sugar or biscuits or similar dry products) is stored on the premises (e.g. some village and church halls) (Reg. 3(3)).
   (b) Crown premises where a certificate of exemption has been issued for reasons of national security.
   (c) places supplying food or drink in the course of religious ceremonies (Reg. 3(1)).
   (d) Places where food is sold, stored or prepared for use in the event of an emergency or national disaster (Reg. 2).

Local authority

It is the LA and not the FA which is responsible for the registration procedure.

The LAs are

(a) Unitary and District Councils in England
(b) London borough councils

(c) The Common Council in the City of London
(d) Unitary authorities in Scotland and Wales
(e) The Council of the Isles of Scilly
(f) Port Health Authorities, or Port Local Authorities in Scotland, to whom the function of a food authority have been assigned (Reg. 9).

Registration of moveable premises is to be undertaken by the LA in whose area it is normally kept or garaged (Reg. 4(3)).

Applications

The form of application is prescribed in Schedule 4 to the regulations. They are normally to be made by the proprietor of the premises except that for premises used by 2 or more food businesses, application is to be made by the person permitting the premises to be used (Reg. 4). Forms in the Welsh language are also specified.

Application for registration must be made at least 28 days before the food business first operates (Reg. 2(1)). All premises existing before the regulations became operative had to be registered before 1 May 1992 (Regs. 2(1)(c) and 4).

Registration

Registration is mandatory on the LA since the procedure is an informatory one only, bringing to the attention of the LA premises which will require assessment for enforcement attention. The LA must effect the registration within 28 days of receipt of the application (Reg. 4). No charge may be made by the LA and the registration does not require periodic renewal.

The particulars to be kept in the register, which is open to public inspection at no cost, are specified and contain particulars of the name, nature and address of the business. Where the premises is moveable, the address where it is ordinarily kept should also be shown (Reg. 5). Copies of the register or of individual entries may be made available by the LA at a charge. Hard copy registers should be no more than 2 months out of date (CoP No. 11 para. 26).

Supervision

The ongoing food safety surveillance of premises registered under this procedure is a duty of the FA applying the requirements of the Food Safety (General Food Hygiene) Regulations 1995 (Reg. 8 of the 1995 regulations).

Supplementary record

In addition to registration details above, the LA is also required to keep a record of all the other information shown on the application form. This is to be made available to other FAs and to persons acting on behalf of the

FSA but is not available to the public (Reg. 6). However, the proprietor of any food business should be given access to any information supplied by him on the registration form (CoP No. 11 para. 8).

Alterations to register

The LA must be notified by a proprietor, before or within 28 days of the change, if he takes over an existing business or if the nature of the business changes and the LA must reflect these changes in its register. The LA must not change the registration relating to information which it has obtained other than from the proprietor unless it gives the owner of the business 28 days notice in writing (Reg. 7).

MEAT AND MEAT PRODUCTS

LICENSING OF BUTCHERS' SHOPS

References

The Food Safety (General Food Hygiene) Regulations 1995.
The Food Safety (General Food Hygiene) (Butchers' Shops) Amendment Regulations 2000.
Guidance Notes on the Food Safety (General Food Hygiene) (Butchers' Shops) Amendment Regulations 2000. Food Standards Agency. May 2000 (FSA GN).
LAC 11 00 4 Butchers' Shops Licensing: clarification of some points of interpretation and application LACOTS August 2000.
LAC 12 00 9 Licensing of Butchers' Shops and the Home Authority Principle LACOTS September 2000.

Scope

It is an offence to use premises as a butchers' shop without a current licence issued by the appropriate FA (Sch. 1A para. 2(a)).

A **butchers' shop** is defined as 'the premises of a food business in or from which both:

(a) commercial operations are carried out in relation to unwrapped meat, and
(b) raw meat and ready to eat food are both placed on the market for sale or supply; other than catering premises' (Sch. 1A para. 1).

The definitions of meat and raw meat taken together cover fresh meat which is not ready to eat and includes red meat, poultry, rabbit and wild and farmed game meat (FSA GN para. 10). Premises requiring a licence therefore include retail butchers' shops, mobile shops, market stalls and some on-farm shops.

FC81 Licensing of Butchers' Shops

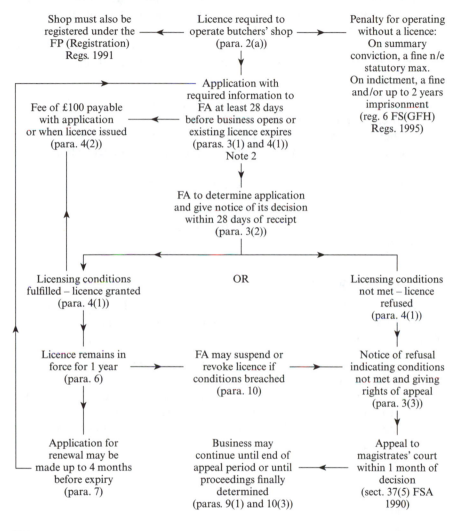

Notes

1. Unless stated otherwise, all reference numbers refer to schedule 1A of the FS(GFH) Regs. 1995.

2. A form of application is contained in appendix 1 of schedule 1A but does not have to be used.

Extent

This procedure is based on regulations that apply in England.

A similar scheme has been introduced in Wales through the Food Safety (General Food Hygiene) (Butchers' Shops) (Amendment) (Wales) Regulations 2000 which covers not only butchers' shops but also other premises selling both unwrapped raw meat and ready-to-eat foods.

In Scotland provision is made through the Food Safety (General Food Hygiene) (Butchers' Shops) Amendment (Scotland) Regulations 2000 with the main differences being:

(a) Strict physical separation of raw meat and ready to eat foods is acceptable as an alternative to HACCP.

(b) There is no provision for mixed business premises. Any shop coming within the definition of a butchers' shop requires a licence that covers the whole premises.

(c) The requirements for training are slightly different.

In Northern Ireland the licensing of butchers' shops has been implemented through the Food Safety (General Food Hygiene) (Amendment) Regulations (Northern Ireland) 2001.

Mixed business premises

Mixed business premises on which less than a half of the floor area is devoted to goods other than meat are included but where that proportion is greater than half, licensing only applies:

(a) where both unwrapped raw meat and ready to eat food are handled, stored or displayed in the same area,

(b) where unwrapped raw meat is handled, stored or displayed in one area and ready to eat food in an adjoining area, or

(c) where the same food worker is working on both unwrapped meat and ready to eat food or the same equipment is being used for both (Sch. 1A para. 5(6) and FSA GN para. 6).

Registration

In addition to these licensing requirements butchers' shops must also be registered under the Food Premises (Registration) Regulations 1991. This procedure is dealt with in FC 80.

Exemptions

These licensing requirements do not apply to catering establishments or to activities carried out in food premises subject to product specific food hygiene regulations listed in Reg. 3(2) of the Food Safety (General Food

Hygiene) Regulations 1995 as amended e.g. meat products establishments (FSAGN para. 20).

Food Authority

This is defined on page 314. Licensing is undertaken by the FA responsible for the area in which the shop is situated or the FA that has registered the shop under Reg. 2 of the Food Premises (Registration) Regulations. In the case of moveable premises, e.g. mobile shops and market stalls, the appropriate FA is that for the area where the shop is ordinarily kept (Sch. 1A para. 1).

Applications

An application for a licence must be made in writing to the FA at least 28 days before business commences or the existing licence expires and may be made by any person. An existing licence holder may apply up to 4 months before the its expiry. A form of application is contained in Appendix 1 to the FSA GN but does not have to be used so long as the application contains the following information:

(a) address of applicant
(b) location of premises or place where moveable premises is kept
(c) information required by the FA to determine if the conditions of licensing are met. This might include:
   (i) a list of products handled and sold
   (ii) the areas to be covered by the licensed operation
   (iii) information on the existing or proposed commercial operations, staff training, HACCP control procedures and number of staff (Sch. 1A paras. 4(1), 5(7) and 7 and FSA GN paras. 21–27).

Fees

A fee of £100 is payable to the FA on receipt of the licence. If payment is made with the application it is refundable in the event of refusal. Where the fee is not paid after receipt of a licence it is recoverable by the FA as a civil debt (Sch. 1A paras. 4(2) and (3)).

Determination of applications

The FA must make its decision and notify the applicant in writing within 28 days of receiving the application (Sch. 1A para. 3(2)).
   A licence must be issued if the following conditions are satisfied:

(a) the application has been properly made (see above);
(b) there will be compliance with both the Food Safety) General Food Hygiene) Regulations 1995 and with the Food Safety (Temperature Control) Regulations 1995;

(c) food handlers are trained in food hygiene to standards that enable them to perform their duties to ensure that the shop meets food safety requirements;

(d) at least one person is sufficiently trained in food hygiene to supervise the activities of the business and ensure that it meets food safety requirements and that the relevant HACCP procedures are met;

(e) HACCP procedures are in place.

(Sch. 1A paras. 4(1) and 5(1))

Licences

The form of licence is not specified but a suggested format is contained in Appendix 2 of the FSA GN.

Licences remain in force for 1 year. In the case of new licences this is from the day of issue and, for the renewal of licences, from the date on which the existing licence expires (Sch. 1A para. 6).

Refusals

If the FA is not satisfied that the necessary conditions are fulfilled (see 'Determination of applications' above) they must refuse the application. The notice of refusal must be sent within 28 days of receipt and specify the conditions on which it is not satisfied. This notice must also indicate the rights of appeal (Sch. 1A para. 3(2) and (3)).

Where the application refused is in relation to an existing licence, the refusal does not take effect for 1 month or, in the event of an appeal, until those proceedings are finally determined (Sch. 1A para. 9(1)).

Enforcement

During the duration of a licence, in addition to ensuring general compliance with the food hygiene regulations the applicant, proprietor or manager must make available to the FA:

(a) the HACCP procedures and, for at least 12 months, a record of how these have been applied;

(b) records of training undertaken by each individual, kept until that person ceases to work in the shop.

These records may either be in writing or in electronic form.

In respect of mixed business premises, e.g. supermarkets, where over half of the floor area is used for goods other than meat, the requirements to keep records applies only to those areas specified in (a)–(d) under 'Mixed business premises' above (Sch. 1A para. 5(4)(5) and (6)).

Material changes

The licence holder must notify the FA of any material change which may reduce the safety of the food sold or supplied from the shop including a change of layout, ownership or operation (Sch. 1A para. 8).

Suspension/revocation

The FA may suspend or revoke a licence without prior notice where any of the licence conditions as outlined in 'Determination of applications' and 'Enforcement' above are breached (Sch. 1A para. 10).

The FSA advises that the FA should make a decision about the use of these steps in the light of relevant codes of practice on enforcement and the nature and seriousness of the breach. Revocation or suspensions are not seen as the first option where other forms of enforcement, e.g. improvement notices, are likely to secure compliance within an appropriate timescale. (FSA GN paras. 50–56)

Notices of revocation and suspension must indicate the conditions that have been breached but do not become effective for 1 month i.e. the period allowed for appeal or, in the event of an appeal, until those proceedings have been finally determined (Sch. 1A para. 10(3)).

Appeals

Appeals against the refusal, suspension and revocation of licences are made to the magistrates' court under sect. 37(5) of the FSA 1990 and must be entered within 1 month of the notification of decision by the FA (FSA 1990 sect. 37).

Offences

The penalty for operating a butchers' shop without a licence is:

(a) on summary conviction, a fine not exceeding the statutory maximum, or
(b) on conviction on indictment, a fine or imprisonment for not exceeding 2 years or both (Reg. 6 the FS (GFH) Regs. 1995, Reg. 4 of the 2000 Regs. and sch. 1A para. 2).

Definitions

Catering premises means premises, or parts of premises, which are used solely for the purposes of a restaurant, canteen, club, public house, school, hospital or similar establishment (including a vehicle or a fixed mobile stall) where, in the course of a business:

(a) food is prepared for delivery to the ultimate consumer, and
(b) no food is prepared or supplied with a view to it being subject to further treatment or processing after it has left the premises.

HACCP procedures mean procedures critical to ensuring food safety by Hazard Analysis and Critical Control Points systems and which are based on the following principles:

(a) analysis of the potential food hazards in a food business operation;
(b) identification of the points in those operations where food hazards may occur;

(c) deciding which of the points identified are critical to ensuring food safety ('critical points');

(d) identification and implementation of effective control and monitoring procedures (including critical limits), at those critical points;

(e) verification to confirm that the Hazard Analysis and Critical Control Points system is working effectively;

(f) review of the analysis of food hazards, the critical points and the control and monitoring procedures periodically, and whenever the food business' operations change; and

(g) documentation of all procedures appropriate to the effective application of the principles listed in (a) to (f), including documentation which identifies the persons who have undertaken training in accordance with paragraph 5(1)(b) or (c).

Meat means fresh meat within the meaning of regulation 2(1) of the Fresh Meat (Hygiene and Inspection) Regulations 1995 or regulation 2(1) of the Poultry Meat, Farmed Game Bird Meat and Rabbit Meat (Hygiene and Inspection) Regulations 1995 and wild game meat as defined in regulation 2(1) of the Wild Game Meat (Hygiene and Inspection) Regulations 1995.

Raw meat means meat which is not ready to eat food.

Ready to eat food means any food for consumption without further treatment or processing.

Unwrapped means neither wrapped so as to prevent the passage of microorganisms nor enclosed in a receptacle which prevents the passage of such organisms.

APPROVAL OF MEAT PRODUCTS PREMISES

References

The Meat Products (Hygiene) Regulations 1994 (as amended).
Code of Practice No. 17: Enforcement of the Meat Products (Hygiene) Regulations 1994.
Guidance notes to be read in conjunction with CoP No. 17 (MAFF/DoH February 1996 and MAFF 1999).
LAC 26 99 6 LACOTS.

Scope

The regulations generally control hygiene and food safety in premises used for the manufacture and handling of meat products and the primary manufacture of certain other products of animal origin intended for human consumption e.g. rendered animal fat. Examples of meat products falling within this definition are cooked meats, meat pies, bacon and sandwiches containing cooked meats. Also included are meat-based prepared meats such as roast meat dishes with vegetables and cottage pies.

FC82 Approval of meat products premises

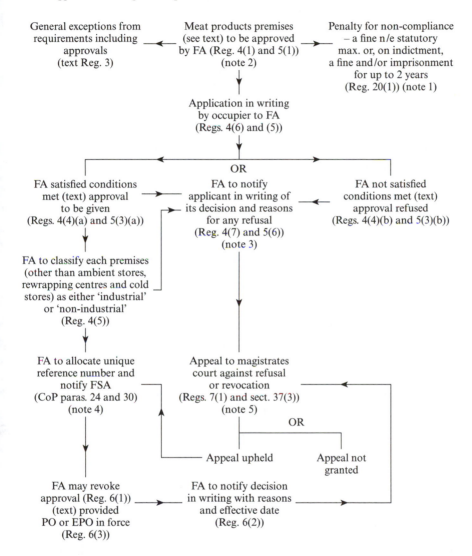

General exceptions from requirements including approvals (text Reg. 3)

Meat products premises (see text) to be approved by FA (Reg. 4(1) and 5(1)) (note 2)

Penalty for non-compliance – a fine n/e statutory max. or, on indictment, a fine and/or imprisonment for up to 2 years (Reg. 20(1)) (note 1)

Application in writing by occupier to FA (Regs. 4(6) and (5))

OR

FA satisfied conditions met (text) approval to be given (Regs. 4(4)(a) and 5(3)(a))

FA to notify applicant in writing of its decision and reasons for any refusal (Reg. 4(7) and 5(6)) (note 3)

FA not satisfied conditions met (text) approval refused (Regs. 4(4)(b) and 5(3)(b))

FA to classify each premises (other than ambient stores, rewrapping centres and cold stores) as either 'industrial' or 'non-industrial' (Reg. 4(5))

FA to allocate unique reference number and notify FSA (CoP paras. 24 and 30) (note 4)

Appeal to magistrates court against refusal or revocation (Regs. 7(1) and sect. 37(3)) (note 5)

OR

Appeal upheld

Appeal not granted

FA may revoke approval (Reg. 6(1)) (text) provided PO or EPO in force (Reg. 6(3))

FA to notify decision in writing with reasons and effective date (Reg. 6(2))

Notes
1. No prosecution may commence later than 3 years from the commission of the offence or 1 year from its discovery whichever is the earlier (Reg. 20(2)).
2. This flow chart combines the procedures for all meat products premises including ambient stores, rewrapping centres and cold stores (text).
3. No time period for the decision is specified in the regulations. The CoP asks FAs to deal promptly with applications (para. 24).
4. FSA will keep a central list of approved premises (CoP para. 30).
5. Revocations do not take effect until the appeal is determined (Reg. 7(3)).

The regulations do not cover fresh meat, poultry, farmed game or rabbit meat or minced meat which are covered by other regulations. This procedure deals with meat products premises, ambient stores, rewrapping centres and cold stores (see definitions below) which must be approved for the handling of meat products by FA (Regs. 4(1) and 5(1)).

Extent

This procedure applies in England, Wales and Scotland with special provisions relating to Northern Ireland (Reg. 23).

Exemptions

The regulations, including the requirement for approval, do not apply to premises handling meat products which are solely for sale to the final consumer (Reg. 3) – this includes caterers and take-away premises, but butchers etc. do not come within the scope if they supply only to the public or catering premises direct (CoP para. 13 and guidance notes). Premises dealing with other products of animal origin (see definitions below) do not require approval but must comply with Schedule 5.

Approval authority

This is the FA except in the following cases where the approval role is undertaken by the FSA:

(a) premises where meat processing operations are carried on and which share a common curtilage with a slaughterhouse;
(b) cold stores storing both meat products and fresh meat; and
(c) combined cutting and meat products plants where any of the cut fresh meat is dispatched off-site (Reg. 2(1) and CoP para. 7).

Applications for approval

These must be made by the occupier (or intending occupier) in writing (Regs. 4(6) and 5(5)). No form or details are specified.

Considerations

For meat products premises other than ambient stores, rewrapping centres and cold stores the FA must issue an approval if satisfied that:

(a) for the handling of meat products there is compliance with Schedule 1 and parts 1, 2, 3, 5 and 7 of Schedule 2 – these deal with general conditions for approval of premises, general condition of hygiene (Schedule 1) and special conditions relating to particular premises, for raw materials, wrapping and labelling and storage and transport (Schedule 2);

(b) for the handling of meat products containing only a small percentage of meat (10% or less by weight), there is compliance as in (a) above except Part 1 of Schedule 2, i.e. these premises do not need to comply with the special conditions of approval dealing with certain structural requirements;

(c) for the pasteurization and sterilization of meat products in hermetically sealed containers, there is compliance as in (a) above and with the special requirements of Part 8 of Schedule 2 which deal with the hygiene of the meat treatment and canning process; and

(d) for the manufacture of meat-based prepared meals, there is compliance as in (a) above and with Part 9 of Schedule 2 which deals with temperature control requirements for such products (Reg. 4(4)).

A lesser standard is required for 'non-industrial' premises – see 'classification' below.

In relation to ambient stores, the FA needs only to be satisfied as to compliance with Part 1 of Schedule 3 (which sets out a basic structural requirement only) and for rewrapping centres compliance only with Part 2 of Schedule 3 is required (these are special requirements for such centres).

For cold stores, approval must be given under these regulations if they are already licensed as a cold store or for the storage of fresh meat or poultry **or** if, although not licensed, they comply with Part 3 of Schedule 3 which sets out particular structural and handling requirements for such premises (Reg. 5(3) and (4)).

Approvals

If the FA is satisfied that the plant complies with the requirements identified in the section above it must give approval (Regs. 4(4)(a) and 5(3)(a)).

The FA must notify its approval in writing to the applicant (Regs. 4(7) and 5(6)).

The FA must allocate to each premises a unique reference number known as the 'approval code' (CoP para. 24). This approval code must be notified to the FSA who will maintain a central list of all approved meat products premises (and will notify other Member States (CoP para. 30).

There is no provision for the FA to attach conditions to any approval; approval is subject to continuing compliance with the regulations. However, see 'classification' below.

Refusals

If the FA is not satisfied that the relevant considerations are fulfilled, it must refuse the application (Regs. 4(4)(b) and 5(3)(b)). It must notify the applicant of the decision and give reasons for the refusal (Regs. 4(7) and 5(6)). There is provision for appeal to the magistrates' court against such a decision

(Reg. 7(1) and sect. 37(2)) and FAs should bring this to the applicant's attention (CoP para. 29). The regulations do not specify any time periods within which notification must be given.

Classification

FAs must classify each premises to be approved as either 'industrial' or non-industrial' (Definitions below) using the weekly output from premises of finished meat products. The FA is able to classify a premises as 'industrial' even where the output is below the 7.5 tonnes threshold where the nature of the product and the production process are such that an unacceptable risk to health would otherwise exist (Reg. 4(5) and CoP paras. 21–23).

The applicant must be informed of the classification by the FA (CoP para. 23). The effect of classification as a 'non-industrial meat products plant' is to lower the hygiene requirements and such premises must only comply with Schedule 1 (general conditions of hygiene) with certain minor exceptions and with Parts 2, 3, 5, 7 and Part 8 where appropriate of Schedule 2 (Reg. 4(5)).

Supervision

The FA is responsible for the continuing supervision of premises which it has approved (Reg. 19) and the particular issues of concern which the AOs must address are listed in Part 4 of Schedule 2. These include cleanliness, the occupiers' checks, the microbiological and hygiene conditions of the products, the efficacy of any meat treatment processes and the use of the health mark. The AO may take samples, make any other necessary checks and has full access at all reasonable times (Part 4 Schedule 2).

The specific duties of occupiers of approved meat products plants include:

1. general compliance with the regulations;
2. carrying out checks to ensure that:
   (a) critical points relative to the process are identified (and accepted by FA) and monitored;
   (b) samples are undertaken to check cleaning and disinfection;
3. ensuring that records of these activities are kept and made available to the AO;
4. controlling the health marking process and providing a commercial document bearing the approval code to accompany each consignment;
5. notifying the FA when samples or other information reveals a health risk;
6. in the event of an imminent health risk, withdraw from the market appropriate meat products and hold under the supervision of the FA until destruction, reprocessing or use for other than human consumption;
7. ensuring adequate training for staff in hygiene (Reg. 13).

Health mark

Meat products, with some specified exceptions, must carry a health mark in the required form as set out in Part 6 of Schedule 2. It is the responsibility of the occupier to see that the correct health mark is applied (Reg. 13(1) and of the FA to satisfy itself that this requirement is being met (Schedule 2 part 4 para. 1(a) (vi)).

Revocation

In exercising its supervisory role over meat products plants FAs have available the general enforcement powers of the FSA 1990 (FCs 75–79) but may also revoke an approval where either:

(a) there is an obvious failure to comply with the regulations;
(b) there are obstacles to an adequate inspection; or
(c) the premises no longer fall within the regulations; **and**

there is in force a prohibition notice or an emergency prohibition order under sects. 11 or 12 of the 1990 Act.

The FA is not obliged to give any prior warning of its intention to revoke or to hear any representations but must notify the occupier in writing with its reasons (Reg. 6). There is an appeal (Reg. 7).

Definitions

Ambient store means any premises, not being part of approved meat products premises, which store unpackaged meat products under non-refrigerated conditions;

Cold store means any premises, not being part of approved meat products premises, which store unpackaged meat products under refrigerated conditions.

Final consumer means a person who buys meat products or other products of animal origin:

(a) for his own consumption;
(b) for direct transport to, and consumption on, premises either in his ownership or under his personal supervision or in the ownership or under the personal supervision of a person employed by him; or
(c) for direct transport to premises used for handling meat products either in his ownership or under his personal supervision or in the ownership or under the personal supervision of a person employed by him for sale as a ready-cooked take-away meat product for consumption off those premises.

Handling means manufacturing, preparing, processing, packaging, wrapping or rewrapping.

Industrial meat products premises means meat products premises whose production exceeds 7.5 tonnes of finished meat products per week or 1 tonne per week in the case of premises producing foie gras, or a lower level of production where a special hygiene direction is given.

Meat-based prepared meal means a wrapped meat product (excluding sandwiches or products made with pastry, pasta or dough) in which meat has been mixed with other foodstuffs before, during or after cooking and requires refrigeration for preservation.

Meat products means products for human consumption prepared from or with meat which has undergone treatment such that the cut surface shows that the product no longer has the characteristics of fresh meat, but not:

(a) meat which has undergone only cold treatment;
(b) minced meat;
(c) mechanically recovered meat;
(d) meat preparations.

Non-industrial meat products premises means meat products premises whose production does not exceed 7.5 tonnes of finished meat products per week or 1 tonne per week in the case of premises producing foie gras, or a lower level of production where a special hygiene direction is given.

Other products of animal origin means the following products intended for human consumption:

(a) meat extracts;
(b) rendered animal fat, fat derived from rendering meat, including bones;
(c) greaves: the protein-containing residue of rendering, after partial separation of fat and water;
(d) meat powder, powdered rind, salted or dried blood, plasma;
(e) stomachs, bladders and intestines, cleaned, salted or dried, and/or heated.

Rewrapping centre means premises where any of the following operations is carried out:

(a) meat products are unwrapped, sliced or cut and subsequently rewrapped prior to despatch;
(b) unpackaged meat products from different manufacturers are assembled into batches for despatch (Reg. 2(1)).

APPROVAL OF MINCED MEAT AND MEAT PREPARATIONS PREMISES

References

The Minced Meat and Meat Preparations (Hygiene) Regulations 1995. Guidance Notes on the Enforcement of the Minced Meat and Meat Preparations (Hygiene) Regulations 1995 (MAFF/DOH).

FC83 Approval of minced meat and meat preparation premises by FAs

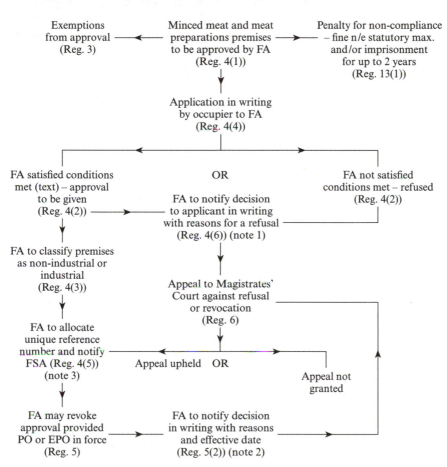

Notes
1. No time period is specified within which the application must be dealt with.
2. Revocations are suspended until the appeal is finally determined or withdrawn (Reg. 6(3)).
3. FSA keep a central register of all approvals.

Scope

The regulations lay down the structural, hygiene and supervision standards for premises engaged in the production (see definitions below) of minced meat and meat preparations (see definitions below). Those premises whose products are for the UK and EEC markets are required to have the approval of the relevant FA (Reg. 4). Products only for the UK market do not need to be approved, but are covered by the Food Premises (Registration) Regulations 1991 and the Food Safety (General Food Hygiene) Regulations 1995.

Meat products (meat which has undergone processing such that the cut surface shows that the product no longer has the characteristics of fresh meat) do not fall within the scope of these regulations but see FC82 dealing with the approval of meat products premises.

Extent

The provisions apply in England, Wales and Scotland.

Exemptions

The regulations, including the requirement for approval, do not apply to:

(a) any premises all of whose production of minced meat and meat preparation is intended for sale to the final consumer (see below);

(b) the production of mechanically recovered meat (MRM); and

(c) the production or sale of minced meat intended to be used as a raw material for the production of sausage meat destined for use in a meat product.

In relation to (a) above, sales to the final consumer include:

(i) sales direct from a sales point on the premises to the consumer (or from a local sales point owned by the operator if the premises do not have an attached sales point);

(ii) sales to caterers; and

(iii) sales to take-away premises for sale as ready cooked take-away. (Reg. 3).

Approval authority

Enforcement is by FAs who approve premises, supervise and enforce the regulations except in the following cases:

(a) combined slaughtering and minced meat/meat preparations premises;

(b) combined cutting premises with minced meat/meat preparations premises; and

(c) combined licensed cold stores and minced meat/meat preparations premises.

In these three cases enforcement is by the FSA through the NMHS (Reg. 2(1) and (4)).

Those premises for which the FA is the enforcement authority are known as 'independent premises'.

Applications for approval

These must be made to the FA in writing by the occupier (or intending occupier) (Reg. 4(4)). No form or particular details are specified.

Considerations

In relation to independent premises (where the FA is the approving authority), the FA must approve the application provided it complies with Schedule 3 of the Regulations. Schedule 3 generally applies the requirements in Schedule 1 of the Meat Products (Hygiene) Regulations 1994 and some additional points to premises to be approved under these regulations.

Premises licensed under the Fresh Meat Regulations or approved under the Meat Products Regulations must comply with Schedules 1 and 2 of the Minced Meat etc. Regulations (Reg. 4(2)).

In the case of premises classified as 'non-industrial' (see below), the same standards apply, but there are derogations from the requirements relating to hand-operable taps near to work stations, lockers instead of changing rooms and the manufacture and wrapping of meat in the same room (Reg. 4(3)).

Approvals

The FA must indicate its approval in writing (Reg. 4(6)) and allocate a unique identification reference (approved Code) (Reg. 4(5)). The approval code must be notified to FSA who will maintain a central list and notify other Member States. There is no provision for the FA to attach conditions to approvals, which are subject to continuing compliance with the regulations.

Refusals

If the FA is not satisfied that the relevant considerations (see above) are met, it must refuse the application (Reg. 4(2)). The applicant must be notified of this decision in writing, giving reasons for the refusal to grant approval (Reg. 4(6)). The FA should draw the applicants attention to the provisions for appeals. The regulations do not specify any period within which decisions on applications must be made by the FA.

Classification

The FA should classify any appropriate premises as 'non-industrial' (see definitions below) and this should be made clear to the applicant (Reg. 4(3)).

Supervision

FAs are responsible for the continuing supervision of premises which they have approved (Reg. 12(4)) and this role must be performed in accordance with Schedule 6 of the Regulations. This includes:

(a) checks on:
  (i) cleanliness of premises, equipment and staff;
  (ii) the occupiers own supervision;
  (iii) microbiological and hygienic condition of the minced meat and meat preparations as in Schedule 11;
  (iv) health marking;
  (v) storage and transport conditions;
(b) taking samples for laboratory testing;
(c) any other checks considered necessary.

Where the meat is to be consigned to another EC state, supervision by the FA must be at least once a day while production is being carried out (Schedule 6, para. 3(b)).

The specific duties of occupiers of approved premises are:

1. General compliance with the regulations as laid out in Reg. 7.
2. Identifying critical points to the satisfaction of the FA.
3. Establishing monitoring of the critical points.
4. Undertaking sampling.
5. Keeping records of sampling activities available to a AO.
6. Controlling the health marking process.
7. Notifying the FA where sampling reveals a health risk.
8. In the event of an imminent health risk, withdraw from the market appropriate products to be held under the supervision of a AO.
9. Ensure hygiene training for staff.
10. Label products to identify temperature requirements.
11. Undertake microbiological testing (Reg. 11).

Health marking

Minced meat and meat preparations despatched to Member States must be health marked in accordance with Schedule 7 (Reg. 7(e)).

Revocation

In exercising its supervisory role the FA has available the general enforcement powers of the FSA 1990 (FCs 75 and 79) but may also revoke its approval where either:

(a) the premises no longer comply with these regulations;
(b) adequate health inspection is being hampered;
(c) the business which required approval is no longer there or has become exempt.

In respect of (a) and (b), revocation may not be effected unless there is already in force either a prohibition order under Section 11 or an emergency prohibition order under Section 12 of the FSA 1990 (Reg. 5).

The FA must give written notice of revocation to an occupier giving the reasons and the date on which it is to take effect (Reg. 5(2)).

Appeals

Appeals against either a refusal to approve the premises or a revocation of an approval may be made to a magistrates' court. In the case of an appeal against a revocation, this does not take effect until the appeal has been finally disposed of or abandoned (Reg. 6).

Definitions

Final consumer means a person who buys minced meat or meat preparations:

(a) otherwise than for the purpose of resale;
(b) for direct transport to, and consumption on, premises either in his ownership or under his personal supervision or in the ownership or under the personal supervision of a person employed by him; or
(c) for direct transport to premises either in his ownership or under his personal supervision or in the ownership or under the personal supervision of a person employed by him for sale as ready-cooked take-away food for consumption off the premises.

Meat preparation means meat to which foodstuffs, seasonings or additives have been added or which has undergone a treatment insufficient to modify its internal cellular structure and alter its characteristics.

Mechanically recovered meat means meat which:

(a) comes from residual meat on bones apart from:
　(i) the bones of the head; and
　(ii) the extremities of the limbs below the carpal and tarsal joints and, in the case of swine, the coccygeal vertebrae;
(b) has been obtained by mechanical means; and
(c) has been passed through a fine mesh such that its cellular structure has been broken down and it flows in puree form.

Minced meat means meat which has been minced into fragments or passed through a spiral screw and includes such meat to which not more than 1% salt has been added.

Non-industrial premises means any establishment whose total production of meat products and meat preparations does not exceed 7.5 tonnes per week and which does not produce minced meat.

Production means manufacturing, preparing, processing, packaging, wrapping or rewrapping and 'produce' has a corresponding meaning (Reg. 2).

SHELLFISH AND FISHERY PRODUCTS

GENERAL PROCEDURES

The following applies to all of the procedures dealt with in this section.

References

The Food Safety (Fishery Products and Live Shellfish) (Hygiene) Regulations 1998 as amended 1999.
Code of Practice No. 14: The Enforcement of the Food Safety (Live Bivalve Molluscs and other Shellfish) Regulations 1992 (under review).
Code of Practice No. 15: The Enforcement of the Food Safety (Fishery Products) Regulations 1992 (under review).
NB. Although drafted in relation to the earlier 1992 regulations replaced in 1998, these codes are still applicable but are to be read in the context of the 1998 regulations.

Extent

These procedures apply in England, Wales and Scotland but not in Northern Ireland.

Charges

The 1998 regulations contain provisions that harmonize hygiene inspection charges for fishery products directly landed at UK ports by third country vessels. The Food Safety (Fishery Products and Live Shellfish) (Hygiene) Amendment (No. 2) Regulations 1999 introduced hygiene inspection charges for fishery products landed by EU vessels and fishery products entering preparation and processing plants.

DEFINITIONS

Auction or wholesale market means any premises where the display and sale by wholesale of fishery products, but no other activities associated with the production and placing on the market of fishery products, takes place.

Bivalve molluscs means filter – feeding lamellibranch molluscs.

Designated bivalve production area means an area of seawaters or brackish waters designated by the Food Standards Agency in accordance with Reg. 3(1) as an area from which live bivalve molluscs may be taken.

Dispatch centre means any on-shore or offshore installation for the reception, conditioning, washing, cleaning, grading and wrapping of live shellfish fit for human consumption.

Establishment means, with regard to the production of fishery products for human consumption, any premises where fishery products are prepared, processed, frozen, chilled, packaged or stored other than:

(a) cold stores where only the handling of wrapped products takes place or
(b) auction or wholesale markets where only display and sale by wholesale takes place.

Factory vessel means any vessel on which any fishery products undergo one or more of the following operations – filleting, slicing, skinning, mincing, freezing or processing – followed by packaging but the following are not deemed to be factory vessels:

(a) fishing vessels in which only shrimps and molluscs are cooked on board; and
(b) fishing vessels on board which only freezing is carried out.

Fishery products means:

(a) all seawater or freshwater animals including their roes; and
(b) parts of such animals, except in circumstances where they:
  (i) are combined (in whatever way) with other foodstuffs, and
  (ii) comprise less than 10% of the total weight of the combined foodstuffs,

but excluding aquatic mammals, frogs and aquatic animals covered by Community acts other than the Fishery Products Directive, and parts of such mammals, frogs and aquatic animals.

Laying means a foreshore, bed, pond, pit, ledge, float or similar place, including a relaying area, where live shellfish are liable to be gathered, harvested or deposited.

Placing on the market means with regard to:

(a) fishery products, the holding or displaying for sale, offering for sale, selling, delivering or any other form of placing on the market in the European Community, except for:
  (i) retail sales (which includes retail sales to catering businesses and sales by catering businesses), and
  (ii) direct transfers on local markets of small quantities by fishermen to retailers or consumers in the circumstances specified in regulation 41; and
(b) live shellfish, the holding or displaying for sale, offering for sale, selling, delivering or any other form of placing on the market of live shellfish for human consumption either raw or for the purposes of processing in the European Community, except for direct transfers on local markets of small quantities by coastal fishermen to retailers or consumers in the circumstances specified in regulation 20.

Private laying means a laying where live shellfish are usually harvested or deposited by the owner or by the tenant of the laying.

Production area means any sea, estuarine or lagoon area containing natural deposits of shellfish or sites used for the cultivation of shellfish (including relaying areas) from which live shellfish are taken.

Purification centre means an establishment with tanks fed by naturally clean seawater or seawater which has been cleaned by appropriate treatment, in which live bivalve molluscs are placed for the time necessary to remove microbiological contamination, so making them fit for human consumption.

Registration document means a document for the identification of live shellfish during transport from a production area to a dispatch centre, purification centre, relaying area or processing plant which:

(a) conforms with the standard registration document contained in a measure intended to give effect to the requirements of point 6 of Chapter II of the Live Bivalve Molluscs Directive that has been made in accordance with the procedure laid down in Article 12 of the Live Bivalve Molluscs Directive; and

(b) is issued by:
  (i) a competent authority within the meaning of the Live Bivalves Directive; or
  (ii) where the document meets the requirements of that Directive and any law in force in the Channel Islands or the Isle of Man, a competent authority in those islands or that Isle.

Relaying means an operation whereby live bivalve molluscs are transferred to a sea, lagoon or estuarine area in order to remove contamination, but does not include a transfer to an area more suitable for further growth or fattening.

Relaying area means any sea, lagoon or estuarine area with boundaries clearly marked and indicated by buoys, posts or any other fixed means which is used exclusively for the natural purification of live bivalve molluscs.

Shellfish means only bivalve molluscs, echinoderms, tunicates and marine gastropods (Reg. 2(1)).

REGISTRATION OF CERTAIN FISHING VESSELS

Scope

Fishing vessels on board which only shrimps and molluscs are processed by cooking as part of a food business are exempt from approval (below) but are required to be registered with the FA for the area in which the vessel is based. (Reg. 21(1) and (2)). Registration is, however, not necessary where such processing is to be supplemented subsequently by cooking (Reg. 21(1)).

Notifications

Before the vessel begins to operate the operator must notify the FA in writing with the following information:

FC84 Registration of certain fishing vessels

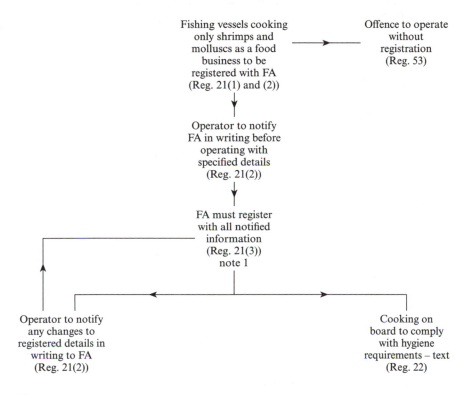

Fishing vessels cooking
only shrimps and
molluscs as a food
business to be
registered with FA
(Reg. 21(1) and (2))

Offence to operate
without
registration
(Reg. 53)

Operator to notify
FA in writing before
operating with
specified details
(Reg. 21(2))

FA must register
with all notified
information
(Reg. 21(3))
note 1

Operator to notify
any changes to
registered details in
writing to FA
(Reg. 21(2))

Cooking on
board to comply
with hygiene
requirements – text
(Reg. 22)

Note
1.   There is no power for a FA to refuse to register provided that all of the required information has been given by the proprietor.

(a) the name of the vessel
(b) the usual place of landing
(c) the name and address of the owner of the vessel.

Any subsequent changes to these details must be notified to the FA in writing (Reg. 21(2)).

Penalties

Not complying with these requirements is an offence for which the penalty of either:

(a) on summary conviction, a fine not exceeding the statutory maximum, or
(b) on indictment, an unlimited fine or imprisonment for not exceeding 2 years or both (Reg. 53).

Registration

The FA must register each vessel notified to it and there is no discretion to refuse to do so. The FA must maintain a register of all of the information notified to it (Reg. 21(3)).

Hygiene requirements

Any cooking on board the vessel must be in accordance with:

(a) para. 5 of sect. 1 of chapter 3 of schedule 3 (disinfection of instruments etc.), and
(b) para. 7 of sect. 4 of chapter 4 of schedule 3 (cooking processes).

REGISTRATION OF AUCTION AND WHOLESALE MARKETS

Scope

Auction and wholesale markets (definitions page 358) where no preparation, processing or chilling takes place are exempt from approval (below) but are required to register with the FA for the area in which the premises are situated (Reg. 25).

Penalties

The penalty for operating without registration is:

(a) on summary conviction, a fine not exceeding the statutory maximum, and
(b) on indictment, an unlimited fine and/or imprisonment for up to 2 years (Reg. 53).

FC85 Registration of auction and wholesale markets for fishery products

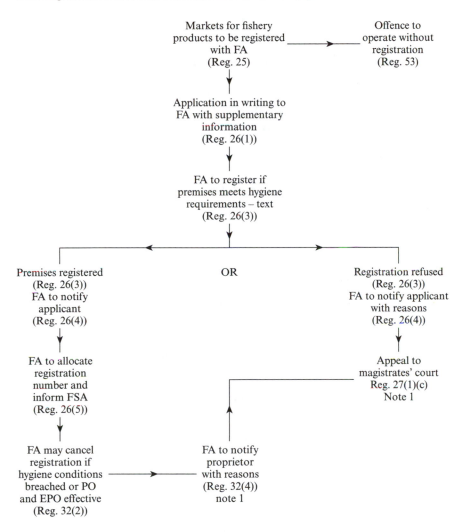

Note

1. On appeal against the FA decision to cancel the registration the cancellation is effective either after the time for appeal has elapsed or the appeal has been determined (Reg. 32(4)).

Applications

Applications must be in writing and be accompanied by sufficient supplementary information as necessary to enable the FA to make its decision (Reg. 26(1)).

Registration

The FA can only register the premises if it meets the relevant hygiene conditions specified in chapters 2 and 3 of schedule 3. The FA's decision must be notified to the applicant in writing with, in the case of refusal, its reasons (Reg. 26(4)). Upon registration the FA must designate the premises with a unique registration number which has to be notified to the FSA (Reg. 26(5)).

Appeals

There is a right of appeal to the magistrates' court against a refusal to register (Reg. 27(1)(c)).

Cancellation of registration

The FA may cancel the registration if, following an inspection or enquiry, they are satisfied that:

(a) there has been a serious (on animal or public health grounds) and manifest breach of the applicable provisions of chapter 2 or 3 of schedule 3;
(b) the proprietor is either unable or not prepared to remedy the breach; and
(c) there is either a prohibition order (FC87) or emergency prohibition order (FC89) in force and, as a consequence of that order, commercial operations with regard to fishery products may not be carried out (Reg. 32(2)).

The FA must notify its decision to the proprietor and set out the reasons for cancellation. The cancellation is effective either after the period for appeal has elapsed or after the determination of an appeal (Reg. 32(4)). There is an appeal to a magistrates' court against cancellation (Reg. 32(3)).

APPROVAL OF FISHERY PRODUCTS ESTABLISHMENTS AND FACTORY VESSELS

Scope

Fishery products establishments (definitions) and factory vessels (definitions, page 359) may not be operated unless they have received the approval of the relevant FA. For fishery products establishments this is the FA for the area in which the premises is situated and for factory vessels the FA for the area at which the vessel usually lands fishery products in Great Britain (Reg. 23(1)).

FC86 Approval of fishery products establishments and factory vessels

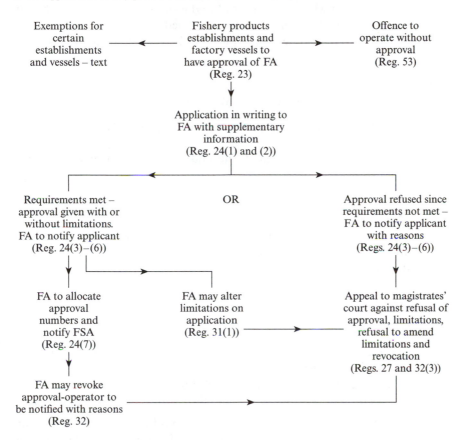

Exemptions for certain establishments and vessels – text

Fishery products establishments and factory vessels to have approval of FA (Reg. 23)

Offence to operate without approval (Reg. 53)

Application in writing to FA with supplementary information (Reg. 24(1) and (2))

OR

Requirements met – approval given with or without limitations. FA to notify applicant (Reg. 24(3)–(6))

Approval refused since requirements not met – FA to notify applicant with reasons (Regs. 24(3)–(6))

FA to allocate approval numbers and notify FSA (Reg. 24(7))

FA may alter limitations on application (Reg. 31(1))

Appeal to magistrates' court against refusal of approval, limitations, refusal to amend limitations and revocation (Regs. 27 and 32(3))

FA may revoke approval-operator to be notified with reasons (Reg. 32)

Cold stores where only the handling of wrapped products takes place and auction or wholesale markets where only display and sale by wholesale takes place are exempt from approval but are required to be registered with the FA (see FC85).

Factory vessels in which only shrimps and molluscs are cooked on board and those on board which only freezing is carried out are also exempt from approval although the former are required to be registered (see FC84).

Penalties

Offences of operating without approvals are subject to:

(a) on summary conviction, a fine not exceeding the statutory maximum, and
(b) on indictment, an unlimited fine and/or imprisonment for up to 2 years (Reg. 53).

Applications

These are to be made to the relevant FA in writing with supplementary information, documents, plans and diagrams sufficient to enable the FA to determine the application (Reg. 24(2)).

Approvals

The FA may only give approval if:

(a) in the case of a factory vessel, the conditions in chapters 1 and 2 of schedule 3 are met, and
(b) in the case of fishery products establishments, the conditions of chapters 3 and 4 of schedule 3 are complied with (Reg. 24(3) and (4)).

The FA may grant its approval subject to limitations as to the particular activities, methods of operation and intensity of activity (Reg. 24(5)). The FA may later amend these upon application (Reg. 31(1)).

The FA must notify its decision to the applicant in writing and include the reasons for any refusal or specifying any limitations (Reg. 24(6)).

Upon approval the FA is to allocate a unique reference number for the establishment or vessel and notify the FSA of this (Reg. 24(7)).

Revocation

The FA may revoke the approval where, after inspection or inquiry, it is satisfied that:

(a) there has been a serious (on animal or public health grounds) and manifest breach of the conditions/limitations or of the applicable hygiene requirements in schedule 3;
(b) the proprietor is either unable or not prepared to remedy the breach; and

(c) there is in force either a prohibition order or emergency prohibition order under sect. 11 of the FSA1990 (Reg. 32 (1)).

Appeals

Appeals may be made to the magistrates' court against refusal to approve an application, the imposition of conditions or refusal to alter them and against revocation (Reg. 27(1) and 32(3)). In the event of an appeal against the imposition of limitations or revocation the FA's decision does not become effective until the period for appeal has elapsed or, in the event of an appeal, it has been determined (Reg. 27(2) and 32(3)).

DESIGNATION OF SHELLFISH PRODUCTION AREAS

Scope

Sea or brackish waters from which live bivalve molluscs are taken commercially are classified by the FSA according to the microbiological quality of the shellfish derived form the water (Reg. 3 (1)). Designation is usually undertaken annually and environmental health departments are notified of the classifications by the FSA. The FAs play an important role in this process by providing scientific data based on their sampling programmes to the FSA.

Other shellfish and scallops that are not farmed are not subject to the classification process and may be taken from any area unless there is a prohibition order in place. They are, however, subject to the end-product standard in schedule 2.

Classifications

The classification standards are set out in detail in chapter 1 of schedule 2 and include:–

Class	Microbiological standard	Control
A	<230 *E. Coli*/100g flesh <300 faecal coliforms/100g No *Salmonella*/25g	May go for direct consumption if satisfying sch. 2 chapter 4
B	<4600 *E. Coli*/100g <6000 faecal coliforms/100g in 90% of the samples	To be purified in approved centres
C	<60 000 faecal coliforms/100g	To be relayed for at least 2 months with or without purification

FC87 Designation of production, relaying and prohibition areas

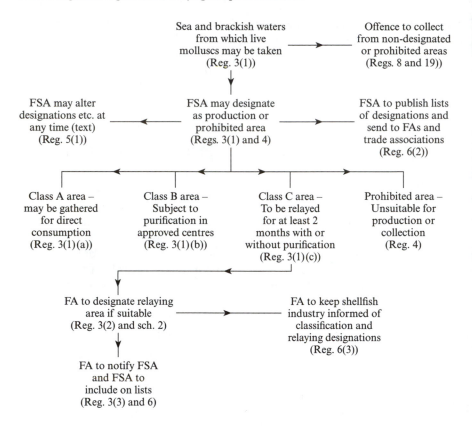

Sea and brackish waters from which live molluscs may be taken (Reg. 3(1))

Offence to collect from non-designated or prohibited areas (Regs. 8 and 19))

FSA may alter designations etc. at any time (text) (Reg. 5(1))

FSA may designate as production or prohibited area (Regs. 3(1) and 4)

FSA to publish lists of designations and send to FAs and trade associations (Reg. 6(2))

Class A area – may be gathered for direct consumption (Reg. 3(1)(a))

Class B area – Subject to purification in approved centres (Reg. 3(1)(b))

Class C area – To be relayed for at least 2 months with or without purification (Reg. 3(1)(c))

Prohibited area – Unsuitable for production or collection (Reg. 4)

FA to designate relaying area if suitable (Reg. 3(2) and sch. 2)

FA to keep shellfish industry informed of classification and relaying designations (Reg. 6(3))

FA to notify FSA and FSA to include on lists (Reg. 3(3) and 6)

All shellfish must meet the standard set out in schedule 2 before being placed on the market for human consumption (Reg. 19).

Prohibited Areas

The FSA may designate any area as being unsuitable, for health reasons, for the production or collection of:

(a) live bivalve molluscs
(b) live echinoderms, tunicates and marine gastropods
(c) live shellfish (Reg. 4).

Collecting from such areas is an offence (Reg. 8).

FAs have the power to make temporary prohibitions in certain circumstances and these are dealt with in FC 89.

Variations etc.

The FSA may at any time in appropriate circumstances alter its, designations in any of the following ways:

(a) vary the boundaries of the area
(b) impose limitations, restrictions or conditions
(c) alter the classification
(d) revoke the designation where the area is no longer suitable
(e) vary or revoke a designation where the area has been made a prohibited area (Reg. 5).

Relaying areas

These are areas designated by FAs from either class A or B areas in order to provide for the treatment of shellfish from class C waters (Reg. 3(2)). The conditions for relaying are specified in schedule 2 and are supervised by the FA.

The FA must notify the FSA of designated relaying areas (Reg. 3(3)). The FSA must publish lists of these and keep all FAs and trade associations informed (Reg. 6(2)). FAs are required to keep informed the industry in its area (Reg. 6(3)).

APPROVAL OF SHELLFISH DISPATCH AND PURIFICATION CENTRES

Scope

Any premises to be used as a dispatch or purification centre (definitions, page 358) for shellfish requires the approval of the FA (Reg. 10).

FC88 Approval of shellfish dispatch and purification centres

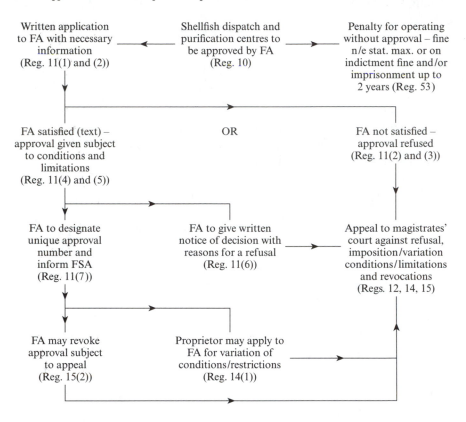

Written application to FA with necessary information (Reg. 11(1) and (2))

Shellfish dispatch and purification centres to be approved by FA (Reg. 10)

Penalty for operating without approval – fine n/e stat. max. or on indictment fine and/or imprisonment up to 2 years (Reg. 53)

FA satisfied (text) – approval given subject to conditions and limitations (Reg. 11(4) and (5))

OR

FA not satisfied – approval refused (Reg. 11(2) and (3))

FA to designate unique approval number and inform FSA (Reg. 11(7))

FA to give written notice of decision with reasons for a refusal (Reg. 11(6))

Appeal to magistrates' court against refusal, imposition/variation conditions/limitations and revocations (Regs. 12, 14, 15)

FA may revoke approval subject to appeal (Reg. 15(2))

Proprietor may apply to FA for variation of conditions/restrictions (Reg. 14(1))

Applications

These are to be made in writing and be accompanied by such supplementary information, documents, plans and diagrams as necessary to enable the FA to determine the application (Reg. 11(1) and (2)).

Requirements to be satisfied

Before giving approval to a purification centre the FA must be satisfied that:

(a) the hygiene requirements of sections 1, 2, and 3 of chapter 4 of schedule 2 are met, and
(b) any additional conditions which may have been specified by the FSA are also met (Reg. 11(4)).

Before approving a dispatch centre the FA must be satisfied that it meets the conditions set out in sections 1, 2 and 4 of chapter 4 of schedule 2 (Reg. 11(3)).

Approvals

Approvals may be subject to limitation as to the particular activities, methods of operation, intensity of use and, for purification centres, to any FSA specified conditions (Reg. 11(5)).

Decisions must be notified in writing, with the reasons for any refusal, and are subject to appeal to a magistrates' court (Regs. 11(6) and 12).

The FA must allocate each approved centre a unique reference number and the FSA notified of this (Reg. 11(7)).

Variations/revocation of conditions etc.

The terms or limitations of approvals may be varied by the FA on application but where this relates to conditions directed by the FSA this may be done only with their consent. The FA's decision is subject to appeal to the magistrates' court and, if this relates to a FSA direction, the FA must notify them of the appeal.

The court may affirm the FA's decision, vary the condition or limitation, remit the matter to the FA with its opinion or make such other order as it thinks fit (Reg. 14).

Revocation of approvals

Approvals may be revoked by the FA where:

(a) there has been a serious and manifest breach of conditions or limitations (on animal or public health grounds) or of the applicable provisions of Chapter IV of Schedule 2;

(b) the proprietor is either unable or not prepared to remedy the breach; and

(c) there is in force in relation to that centre a prohibition order under sect. 11 if the FSA 1990 or an emergency prohibition order under sect. 12.

and as a consequence of the order, commercial operations with regard to placing live shellfish on the market may not be carried out.

The FA must notify its decision in writing to the proprietor with its reasons and there is provision for appeal to a magistrates' court. A revocation takes effect immediately after the time allowed for appeal or, if an appeal is brought, after that appeal is determined (Reg. 15).

TEMPORARY PROHIBITION ORDERS

Scope

A FA may make a temporary prohibition order (TPO) which prohibits the collection of live shellfish from an area where it is satisfied that consumption is likely to cause a risk to public health (Reg. 13(1)).

The Secretary of State has powers under the Food and Environment Act 1985 to prevent the human consumption of food rendered unsuitable for such – see for example various Food Protection (Emergency Prohibition) (Paralytic Shellfish Poisoning) Orders.

Considerations

TPOs may be used to deal with microbiological problems from sudden pollution, chemical pollution or toxin producing planktons. Before making a TPO the FA should seek advice from the CCDC, PHLS and any other specialists and discuss the situation with the FSA (CoP no. 14 paras. 34 and 36).

Temporary prohibition orders

When made these orders last for 28 days although they may be revoked earlier if there is no longer a risk to public health (Reg. 7(2)).

In the case of private layings (definitions, page 359) owners and tenants must be given written notice together with the reasons for making the order and indicating the proposed conditions and restrictions (Reg. 7(3)(a)).

For other production areas the FA must erect notices in the vicinity of the laying and take other steps necessary to bring the TPO to the attention of those affected (Reg. 7(3)(b)).

FC89 Temporary prohibition orders

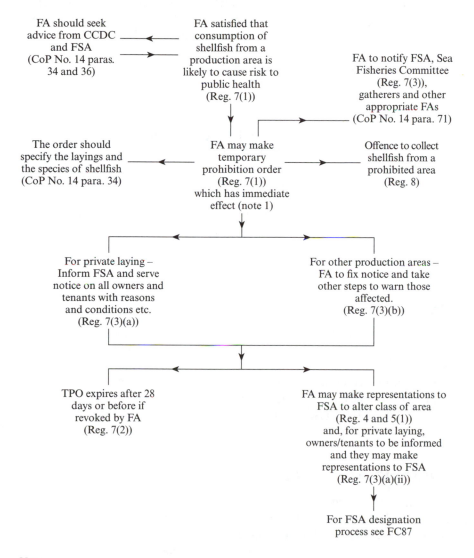

FA should seek advice from CCDC and FSA (CoP No. 14 paras. 34 and 36)

FA satisfied that consumption of shellfish from a production area is likely to cause risk to public health (Reg. 7(1))

FA to notify FSA, Sea Fisheries Committee (Reg. 7(3)), gatherers and other appropriate FAs (CoP No. 14 para. 71)

The order should specify the layings and the species of shellfish (CoP No. 14 para. 34)

FA may make temporary prohibition order (Reg. 7(1)) which has immediate effect (note 1)

Offence to collect shellfish from a prohibited area (Reg. 8)

For private laying – Inform FSA and serve notice on all owners and tenants with reasons and conditions etc. (Reg. 7(3)(a))

For other production areas – FA to fix notice and take other steps to warn those affected. (Reg. 7(3)(b))

TPO expires after 28 days or before if revoked by FA (Reg. 7(2))

FA may make representations to FSA to alter class of area (Reg. 4 and 5(1)) and, for private laying, owners/tenants to be informed and they may make representations to FSA (Reg. 7(3)(a)(ii))

For FSA designation process see FC87

Note

1. TPOs may not be made if one has been made for the same area in the preceding 28 days unless the FSA consents (Reg. 7(4)).

Designation of production areas

If the FA believes that prohibition of collection is likely to be required for in excess of 28 days it may make representations to the FSA. On doing so, it must notify owners/tenants of any private laying, who may make representations (Reg. 4 and 5(1)). In such cases the FA should liase with the FSA as early as possible and take any additional samples requested by them to aid a decision (CoP 14 para. 37).

The designation process to be undertaken by the FSA to declare a prohibition area is shown at FC87.

REGISTRATION DOCUMENTS FOR
THE MOVEMENT OF SHELLFISH

Scope

A registration document is required to accompany and identify each batch of live shellfish during transport from the production area to the dispatch centre, purification centre, relaying area or processing plant (Sch. 2 chapter 2 para. 6A(1)).

Registration documents

These are to be of the form of the standard document specified in point 6 of chapter 2 of the Live Bivalve Molluscs Directive 97/492/EEC. They are issued a FA for the harvesting area on the request of the gatherer. The gatherer must retain the document for at least 12 months as must the receiving centre which must also record the date of receipt of the consignment.

The issuing FA must number each document sequentially and keep a register of them (Sch. 2 chapter 2 para. 6A).

Permanent transport authorization

These may be issued to a gatherer by the FA where it is satisfied that the gatherer will comply with the requirements concerning gathering and handling without each batch needing to receive a registration document. Such authorization may be withdrawn at any time (Sch. 2 chapter 2 para. 6A(9)).

Healthmarks

These are required to provide an ability to trace consignments of live shellfish from the final business user back to the original, approved dispatch centre. The form of healthmark is set out in sch. 2 chapter 10 and conditions for how the mark is to be applied are also specified there.

FC90 Registration documents for the movement of live shellfish

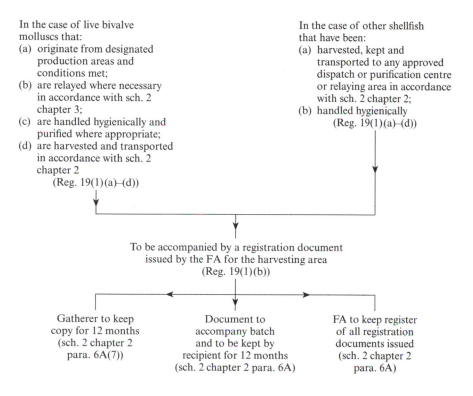

In the case of live bivalve molluscs that:
(a) originate from designated production areas and conditions met;
(b) are relayed where necessary in accordance with sch. 2 chapter 3;
(c) are handled hygienically and purified where appropriate;
(d) are harvested and transported in accordance with sch. 2 chapter 2
     (Reg. 19(1)(a)–(d))

In the case of other shellfish that have been:
(a) harvested, kept and transported to any approved dispatch or purification centre or relaying area in accordance with sch. 2 chapter 2;
(b) handled hygienically
     (Reg. 19(1)(a)–(d))

To be accompanied by a registration document issued by the FA for the harvesting area
(Reg. 19(1)(b))

Gatherer to keep copy for 12 months (sch. 2 chapter 2 para. 6A(7))

Document to accompany batch and to be kept by recipient for 12 months (sch. 2 chapter 2 para. 6A)

FA to keep register of all registration documents issued (sch. 2 chapter 2 para. 6A)

MILK AND MILK PRODUCTS

APPROVAL OF DAIRY ESTABLISHMENTS

References

The Dairy Products (Hygiene) Regulations 1995 as amended 1996.
Code of Practice No. 18: Enforcement of the Dairy Products (Hygiene) Regulations 1995 (under review).
Guidance Notes on the Dairy Products (Hygiene) Regulations 1995 – MAFF/DoH October 1995.
The Food Standards Act 1999 (Transitional and Consequential Provisions and Savings) (England and Wales) Regulations 2000.

Scope

The Regulations generally cover the production and placing on the market of raw milk, heat treated milk and milk-based products (including dairy ice-cream) from cows, sheep, goats and buffaloes.

This procedure deals with dairy establishments (definition on page 380) which may not be used for the handling of dairy products (definition on page 380) unless the premises have been approved by the FA (Reg. 6(1)).

Extent

This procedure applies in England and Wales. Similar provisions have been introduced for Scotland (see the Dairy Products (Hygiene) (Scotland) Regulations 1995) and Northern Ireland.

Exemptions

The regulations generally, including the requirement for approval, do not apply to:

1. establishments producing or handling products:
   (a) for consumption only by the occupier;
   (b) for supply otherwise than on sale;
   (c) any person engaged in the production or handling of dairy products as in (a) or (b) above (Reg. 3(1)); and
2. any catering establishments or shop premises handling or selling dairy products except that the provisions relating to raw cow's milk in Reg. 12 and to the heat treatment of cream and ice-cream in Schedule 6 to apply (Reg. 3(2)–(3)).

In addition, the requirement to obtain an approval does not apply to any processing establishment comprising part of a farm premises selling or handling milk-based products from raw milk produced on the farm or raw

FC91 Approval of dairy establishments

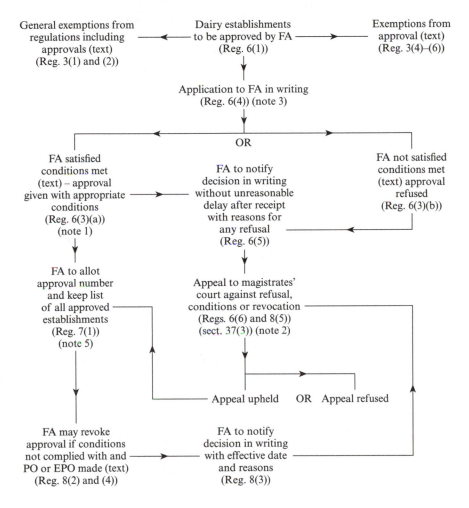

General exemptions from regulations including approvals (text) (Reg. 3(1) and (2)) ←——— Dairy establishments to be approved by FA (Reg. 6(1)) ———→ Exemptions from approval (text) (Reg. 3(4)–(6))

Application to FA in writing (Reg. 6(4)) (note 3)

OR

FA satisfied conditions met (text) – approval given with appropriate conditions (Reg. 6(3)(a)) (note 1)

FA to notify decision in writing without unreasonable delay after receipt with reasons for any refusal (Reg. 6(5))

FA not satisfied conditions met (text) approval refused (Reg. 6(3)(b))

FA to allot approval number and keep list of all approved establishments (Reg. 7(1)) (note 5)

Appeal to magistrates' court against refusal, conditions or revocation (Regs. 6(6) and 8(5)) (sect. 37(3)) (note 2)

Appeal upheld   OR   Appeal refused

FA may revoke approval if conditions not complied with and PO or EPO made (text) (Reg. 8(2) and (4))

FA to notify decision in writing with effective date and reasons (Reg. 8(3))

Notes
1. Approval may be given as a 'limited production' establishment (page 378) (Reg. 6(9)).
2. For appeals against refusal for existing premises and revocations, the business may continue until the appeal is determined (Regs. 6(10) and 8(7)).
3. The regulations do not specify the form of application or the details to accompany it.
4. For the registration of production holdings by the FSA, see page 380.
5. FAs are to notify the FSA who will keep a central list (CoP para. 32).

milk wrapped in that establishment where the activity consists exclusively of dealing with sales direct to the ultimate consumer, although the requirements relating to general conditions of hygiene in Schedule 2, Part 3, and the relevant standards still apply and are enforced by FAs (Reg. 3(4)–(6) and CoP No. 18 paras 13–15).

Premises exempt from these regulations but handling dairy products (including ice-cream and heat-treated cream) are subject to the Food Safety (General Food Hygiene) Regulations 1995.

Applications for approval

These must be made in writing to the FA by the occupier but no form or details are specified (Reg. 6(4)).

Considerations

The FA must issue an approval for the establishment if satisfied that:

    (a) it conforms with the appropriate requirements of Schedule 2 which identifies both general conditions of hygiene to be achieved in all dairy establishments and particular standards for different types of premises; and

    (b) the arrangements at the establishment will be such as to ensure full compliance with the duties placed upon occupiers by Reg. 13(1) which relate to hygiene control, sampling and keeping records and co-operation with the FA (Reg. 6(3)).

Detailed interpretation of the schedules to the regulations is given in the Guidance Notes.

Approval

Each approved establishment must be given an approval number and a list of all approved establishments is to be maintained (Reg. 7(1)).

Notification of each approval must be sent to the FSA who maintain a central list (CoP No. 18 para. 32).

Conditions

Each approval must be subject to a condition that any alterations must comply with Schedule 2 (Reg. 6(8)).

The approval may designate the establishment as of 'limited production' in which case full compliance with the requirements of parts 1 and 2 of Schedule 2 is not required. These relate to size, layout and structure and special conditions for treatment and processing establishments (Reg. 6(9)). **Limited production** means production by means of a separate circuit within any dairy establishment in which heat-treated milk is manufactured and the annual usage of milk for the production of heat-treated drinking milk by such separate circuit does not exceed 2 million litres.

There is an appeal against any condition imposed by the FA (Reg. 6(6)).

Refusals

If the FA is not satisfied that the required standards are met it must issue a refusal against which there is an appeal (Reg. 6(3)(b) and (6)).

Supervision (See CoP No. 18 paras. 46–63)

The FA is responsible for the supervision of the dairy establishment (Reg. 16(1) and (4)) and has available the general enforcement powers under the FSA (FCs 75–79).

The specific duties of occupiers of approved dairy establishments are laid down in Regulation 13 and include:

1. General compliance with the regulations.
2. Carrying out checks to ensure that:
    (a) critical points are identified and monitoring and control of these points are established,
    (b) tests are undertaken to detect residues and added water,
    (c) samples are taken to check cleaning and disinfection methods.
3. Keep records of (2) above and produce these for authorized officers.
4. Administer and control the health mark.
5. Inform FA when serious health risk revealed.
6. Furnish assistance and information to the FA.
7. In the event of an immediate human health risk, withdraw from the market appropriate dairy products and hold under the supervision of the FA until used for other than human consumption, reprocessing or destruction.
8. Staff training (Reg. 13).

Health mark

Dairy products must carry a health mark consisting of an oval surround containing an indication of its UK origin, the approval number of the establishment and the letters 'ECC' (Schedule 10, Part 2, para. 4).

It is the responsibility of the occupier to ensure that the health mark is properly applied (Reg. 13(1)(e)) and of the FA to ensure that these requirements are met (Reg. 16(1) and (4)).

Revocation

In addition to the use of its general enforcement powers, the FA may revoke an approval where either:

(a) there is an obvious failure to comply with the regulations;
(b) there are obstacles to an adequate inspection of the dairy establishment; or
(c) the premises have ceased to be used as a dairy establishment or have become exempt under Reg. 3 (Reg. 8(2)).

The FA is not obliged by the legislation to give any prior warning of its intention to revoke or to hear any representation but must notify its decision in writing with reasons and there is an appeal (Reg. 8(3)).

Revocation should not take place unless the establishment has been made subject to a Prohibition Order or Emergency Prohibition Order and has effectively ceased to trade as a result (Reg. 8(4) and CoP para. 31).

Production Holdings

These premises, defined by Reg. 2(1) as 'premises at which one or more milk-producing cows, ewes, goats or buffaloes are kept', need to be registered by the FSA through a procedure which does not involve FAs (Reg. 4(1)). Such production holdings need to comply with the hygiene standards set out in Schedule 1 and provisions concerned with raw milk in paras. 1 and 2 of Schedule 7.

If the Minister is not satisfied about such compliance, registration must be refused (Reg. 4(3)).

Definitions

Catering establishment means a restaurant, canteen, club, public house, school, hospital, institution, or similar establishment (including a vehicle or a fixed or mobile stall) where, in the course of a business, food is prepared for delivery to the ultimate consumer for immediate consumption without further preparation.

Dairy establishment means any undertaking handling dairy products and is either:

(a) a standardization centre, or any one of the following undertakings operating alone or in combination;
(b) a treatment establishment;
(c) a processing establishment; or
(d) a collection centre.

Dairy product means milk or any milk-based product.

Milk means the milk of cows, ewes, goats or buffaloes intended for human consumption.

Milk-based product means:

(a) a milk product exclusively derived from milk to which other substances necessary for their manufacture may have been added, provided that those substances do not replace in part or in whole any milk constituent; and

(b) a composite milk product where no part replaces or is intended to replace any milk constituent and of which milk or a milk product is an essential part either in terms of quantity or for characterization of the product

intended for human consumption.

(For further clarification of this definition see CoP No. 18 paras. 11 and 12).

Ultimate consumer means any person who buys otherwise than:

(a) for the purpose of resale;
(b) for the purpose of a catering establishment; or
(c) for the purposes of a manufacturing business (Reg. 2(1)).

MILK TREATMENT ORDERS

References

Milk and Dairies (General) Regulations 1959 Reg. 20.
Food Safety Act 1990 (Consequential Modifications) (England and Wales) Order 1990.
The Food Standards Act 1999 (Transitional and Consequential Provisions and Savings) (England and Wales) Regulations 2000.
NB. This procedure was **not** replaced by the Dairy Products (Hygiene) Regulations 1995 and therefore continues to be available to FAs.

Scope

This procedure may be used by a proper officer of a FA when either:

(a) satisfied on evidence that milk (definition on page 383) supplied within the district has caused disease or is infected with a disease communicable to man; or
(b) there are reasonable grounds for suspecting that this is the case but he does not have the supporting evidence (Reg. 20(1) and (2)).

The approval of dairy establishments by FAs is dealt with in FC91.

Extent

These regulations apply in England at Wales but not in Scotland where there are separate regulations.

Notice

The proper officer's notice must be in writing and is to be served on the occupier of the premises supplying the milk in question.

FC92 Milk treatment orders

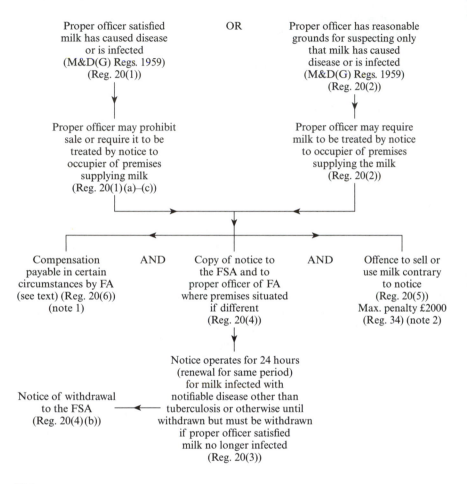

Proper officer satisfied
milk has caused disease
or is infected
(M&D(G) Regs. 1959)
(Reg. 20(1))

OR

Proper officer has reasonable
grounds for suspecting only
that milk has caused
disease or is infected
(M&D(G) Regs. 1959)
(Reg. 20(2))

Proper officer may prohibit
sale or require it to be
treated by notice to
occupier of premises
supplying milk
(Reg. 20(1)(a)–(c))

Proper officer may require
milk to be treated by notice
to occupier of premises
supplying the milk
(Reg. 20(2))

Compensation
payable in certain
circumstances by FA
(see text) (Reg. 20(6))
(note 1)

AND

Copy of notice to
the FSA and to
proper officer of FA
where premises situated
if different
(Reg. 20(4))

AND

Offence to sell or
use milk contrary
to notice
(Reg. 20(5))
Max. penalty £2000
(Reg. 34) (note 2)

Notice of withdrawal
to the FSA
(Reg. 20(4)(b))

Notice operates for 24 hours
(renewal for same period)
for milk infected with
notifiable disease other than
tuberculosis or otherwise until
withdrawn but must be withdrawn
if proper officer satisfied
milk no longer infected
(Reg. 20(3))

Notes
1. Disputes about the right to or amount of compensation are to be settled by arbitration (Reg. 20(7)).
2. See text under 'offences' for additional actions by the FA.

The notice must state the evidence or grounds for suspicion upon which the decision to take the action has been based. It may either:

(a) if the premises are in the proper officer's district, totally prohibit sale or use for manufacture of products for human consumption; or

(b) if the premises are in another district, prohibit sale within the district; or

(c) prohibit sale or use, in and/or outside the district as the case may be, unless it is treated in such a way as to render it safe for consumption, e.g. heat treatment of raw milk.

If the notice is based upon suspicion only, the notice may only specify treatment and not prohibit sale or use (Reg. 20(1) and (2)).

Operation of notice

For milk which is infected or suspected of being infected with a notifiable disease other than tuberculosis the notice is only operative for 24 hours but is renewable for further 24-hour periods. In all other cases the notice operates until it is withdrawn and it must be withdrawn when the proper officer is satisfied that the milk is no longer likely to cause disease (Reg. 20(3)).

Offences

It is an offence to sell or use milk contrary to the terms of a milk treatment order (Reg. 20(5)) and there is a penalty of a maximum £2000 (Reg. 34).

In the circumstances of such a breach, the FA may also wish to consider the use of a prohibition notice under the FSA (see FC78) and review its approval of the dairy establishment involved (FC91).

Compensation

The FA whose proper officer served the notice is required to compensate anyone sustaining damage or loss unless:

(a) disease was in fact caused or the milk was shown to be infected; or

(b) there were reasonable grounds for the suspicion.

Compensation is payable despite (a) and (b) if the notice was not withdrawn as soon as the circumstances required. Disputes about compensation are dealt with through arbitration (Reg. 20(6) and (7)).

Definitions

Milk means cows' milk intended for sale or sold for human consumption or intended for manufacture into products for sale for human consumption and includes cream, skimmed milk and separated milk as well as standardized whole milk and non-standardized whole milk (Reg. 2).

NB. It should be noted that this is a narrower definition of 'milk' than that contained in the Dairy Products (Hygiene) Regulations, i.e. Milk

Treatment Orders cannot be used to deal with milk from ewes, goats and buffaloes.

Notifiable disease means food poisoning, gastroenteritis and a disease notifiable under the Public Health (Control of Disease) Act 1984 (PH CoD)A 1984 sect. 10).

EGGS

APPROVAL OF EGG PRODUCTS ESTABLISHMENTS

References

The Egg Products Regulations 1993.
The Food Standards Act 1999 (Transitional and Consequential Provisions and Savings) (England and Wales) Regulations 2000.

Scope

The manufacture of, or heat treatment to, any egg products for human consumption may only be carried out in establishments approved by the FA (Reg. 5(1)).

Extent

The procedure applies in England, Wales and, with modifications, in Scotland.

Definitions

An **egg product establishment** is one where 'egg products are treated, handled or obtained for the purpose of sale for human consumption' whilst **egg products** are defined as meaning 'products obtained from eggs, their various components or mixtures thereof, after removal of the shell and outer membranes, intended for human consumption, and includes such products when partially supplemented by other foodstuffs and additives and such products when liquid, concentrated, crystallised, frozen, quick-frozen, coagulated or dried but does not include finished foodstuffs'.

Eggs means an egg laid by a hen, duck, goose, turkey, guinea fowl or quail (Reg. 2(1)).

Food safety control

The 1993 regulations provide a control system which includes:

(a) approval of plants (dealt with in this procedure);
(b) health control and supervision of plants;

FC93 Approval of egg product establishments

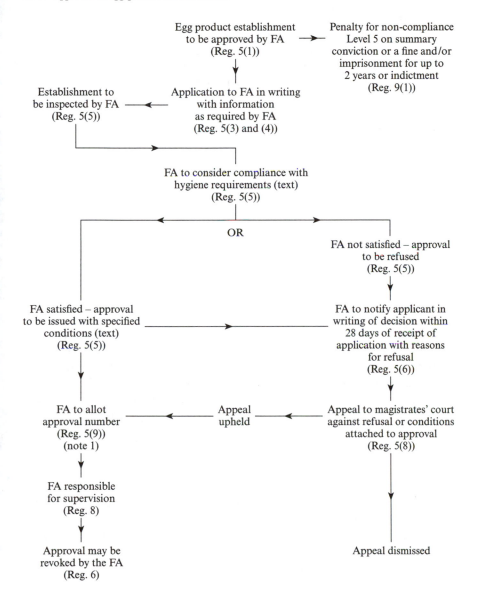

Note

1. FA to notify the FSA for listing in central register.

(c) permitted source material;
(d) microbiological specifications for finished products;
(e) other specifications including for labelling, transport and packaging.

Applications for approval

These are to be made in writing to the FA by the plant operator. No application form or particular data is specified but the applicant must provide such information as the FA may require (Reg. 5(3)).

Considerations

The FA must inspect the premises (Reg. 5(5)) and may only issue an approval if the following conditions are complied with:

(a) Schedule 5 – storage;
(b) Schedule 6 – transport;
(c) Schedule 8 – general hygiene considerations;
(d) Schedule 9 – packaging of egg products;
(e) Schedule 10 – marking of egg products (Reg. 5(5)).

Approvals

If satisfied on all of the matters listed above the FA must approve the establishment (Reg. 5(5)) and issue an approval number (Reg. 5(9)) which should be notified to the FSA for incorporation in a central register. All approvals shall have a condition requiring compliance with Schedules 1 (Requirements for the Preparation of Egg Products), Schedule 5 (Storage) and Schedule 8 (General Conditions) as appropriate to the establishment (Reg. 5(7)).

Notification of decisions

The FA must communicate its decision in writing to the applicant within 28 days of receiving it and give reasons for any refusal (Reg. 5(6)). There is an appeal to the magistrates' court against refusal or attached conditions (Reg. 5(8)).

Supervision

The supervision of the approved establishment is a duty of the FA and its particular responsibilities are specified in Schedule 7 (Reg. 8).

The AOs of the FA also have available the enforcement powers of the FSA 1990 (see FCs 75–79).

Revocation

The FA may revoke any approval where:

(a) any requirement of the regulations is being breached;
(b) any requirement has been breached and no action has been taken to ensure that a similar breach does not occur in future;
(c) any condition of approval has not been complied with.

The FA must notify the decision in writing and indicate the date on which revocation is to take place and the reasons for the decision (Reg. 6(1) and (2)). There is an appeal (Reg. 6(2)) but no provision for representation to be made before the decision is reached.

Heat treatment

Particular requirements for the pasteurization of whole eggs or yokes and the heat treatment of albumen are laid down in Schedules 2 and 3 whilst Schedule 4 identifies end product microbiological standards.

Persons supplying heat treatment processes to egg products are required to keep records of tests and processes, retain them for 2 years and make them available to the FA (Reg. 4).

Chapter 13

LICENSING

INTRODUCTORY NOTE

There are no general procedural provisions or definitions for this chapter. Each procedure contains the information relevant to the particular legislation concerned and attention is drawn to the provisions of the Local Government Act 1972 relating to notices etc. in Chapter 1.

ANIMAL WELFARE LICENCES

ANIMAL BOARDING ESTABLISHMENTS

Reference

Animal Boarding Establishments Act 1963.

Scope

A person keeping an ABE for cats and/or dogs is required to hold a licence issued by the LA (sect. 1(1)).

The Act defines the keeping of an ABE as:

'. . . shall be construed as references to the carrying on by him at premises of any nature (including a private dwelling) of a business of providing accommodation for other people's animals:
Provided that:

(a) a person shall not be deemed to keep a boarding establishment for animals by reason only of his providing accommodation for other people's animals in connection with a business of which the provision of such accommodation is not the main activity; and
(b) nothing in this Act shall apply to the keeping of an animal at any premises in pursuance of a requirement imposed under, or having effect by virtue of, the Animal Health Act 1981 (sect. 5(1)).'

FC94 Animal boarding establishments

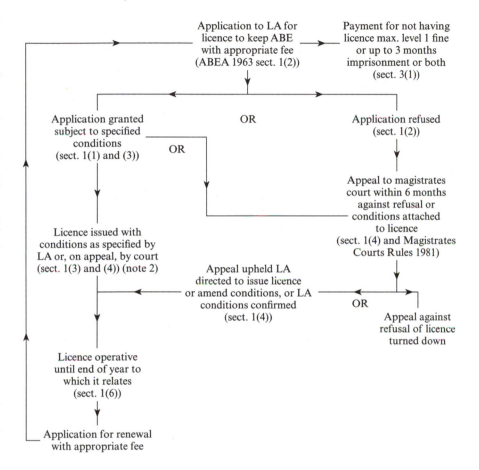

Notes

1. This procedure applies in Scotland but not in Northern Ireland (sect. 7(3)).
2. Penalty for contravening licence conditions is max. level 1 fine and up to 3 months' imprisonment or both (sect. 3(1)).
3. Conditions may only be varied on the renewal of licence.

Extent

This licensing procedure applies in England, Wales and Scotland but not in Northern Ireland (sect. 7(3)).

Applications

Applications made to the LA must specify the premises concerned and a fee as determined by the LA must be paid before any licence is granted (sect. 1(2)). The factors which may be taken into account in deciding whether or not to issue a licence are listed under 'Conditions' below but these are without prejudice to the LA's discretion to refuse a licence on any other grounds (sect. 1(3)).

Licences

Licences remain in force until the end of the year to which it relates and the latter is determined by the applicant as being either the year in which it is granted or the following year. In the first case, i.e. the year in which it is granted, the licence comes into force on the day on which it is granted and ends on the 31 December of that year, and in the second case, i.e. the year following that in which it is granted, it becomes operative on the 1 January of that following year and expires on 31 December of that year (sect. 1(5) and (6)).

On the death of a licence holder, the licence passes to his personal representative and operates for a period of 3 months from the date of death before expiring. The LA may, on application, agree to extension of that period (sect. 1(7)).

Conditions

The LA is required to specify such conditions in the licence as appears to it necessary or expedient for securing all or any of the following objectives:

(a) that animals will at all times be kept in accommodation suitable as respects construction, size of quarters, number of occupants, exercising facilities, temperature, lighting, ventilation and cleanliness;

(b) that animals will be adequately supplied with suitable food, drink and bedding material, adequately exercised, and (so far as necessary) visited at suitable intervals;

(c) that all reasonable precautions will be taken to prevent and control the spread among animals of infectious or contagious diseases, including the provision of adequate isolation facilities;

(d) that appropriate steps will be taken for the protection of the animals in case of fire or other emergency;

(e) that a register be kept containing a description of any animals received into the establishment, date of arrival and departure, and the name and address of the owner, such register to be available for inspection at all times by an officer of the local authority, veterinary surgeon or veterinary practitioner authorized under sect. 2(1) of this Act (sect. 1(3)).

Guidance on conditions relating to these licences is given in 2 booklets published by the Chartered Institute of Environmental Health in November 1995, Guide to Dog Boarding Establishments and Guide to Cat Boarding Establishments. The Institute encourages LAs to discuss with establishment owners appropriate standards and timescales based on the documents.

Power of entry

LA officers, veterinary surgeons or practitioners authorized in writing by the LA for this purpose, may, upon producing their authority if required, inspect a licensed ABE and any animals found there at all reasonable times. Persons wilfully obstructing or delaying authorized officers are subject on conviction to a maximum penalty at level 1 on the standard scale (sects. 2 and 3(2)).

There is no power of entry to unlicensed premises.

Disqualifications and cancellations

In making a conviction under the ABEA 1963 or under the Protection of Animals Act 1911, the Protection of Animals (Scotland) Act 1912 or the Pet Animals Act 1951, the court may cancel any ABE licence held by the person and may disqualify him from holding such a licence, whether or not he currently holds one, for any specified period. The cancellation or disqualification may be suspended by the court pending an appeal (sects. 3(3) and (4)).

LAs must refuse applications for ABE licences from persons disqualified under:

(a) the ABEA 1951;
(b) the Pet Animals Act 1951 from keeping a pet shop;
(c) the Protection of Animals (Amendment) Act 1954 from having the custody of animals (sect. 1(2)).

Local authority

The authorities which act as the licensing authorities are:

(a) in England:
    (i) unitary authorities and London borough councils;
    (ii) the Common Council of the City of London;
    (iii) in remaining two-tier areas, the district council;
(b) in Wales, the unitary authority;
(c) in Scotland, the unitary authority (sect. 5(2) as amended 1994).

Definitions

Animal means any cat or dog (sect. 5(2)).

DANGEROUS WILD ANIMALS

References

Dangerous Wild Animals Act 1976.
Dangerous Wild Animals Act 1976 (Modification) Order 1984.
NB. On 31/10/01 DEFRA issued a report (Review of the Dangerous Wild Animals Act 1976: Consultation Paper) that examines the effectiveness of this procedure and puts forward recommendations to address the identified shortcomings. The consultation period ended on 31/01/02 but so far no legislative changes have been made.

Scope

Any person keeping any dangerous wild animal is required to hold a licence from a LA (sect. 1(1)).

The animals covered by this provision are listed in the Schedule to the Act and this has been amended from time to time. The latest modification was in 1984 and the animals requiring their keeper to be licensed are given in Table 13.1.

This licensing procedure does not apply to animals kept in:

(a) a zoo within the meaning of the Zoo Licensing Act 1981;
(b) a circus;
(c) pet shops; and
(d) places which are designed establishments under the Animals (Scientific Procedures) Act 1986 (sect. 5).

A person is held to be the keeper of the animal if he has it in his possession and the assumption of possession continues even if the animal escapes or it is being transported etc. This removes the need for carriers or veterinary surgeons to be licensed (sect. 7(1)).

Extent

This licensing procedure applies in England, Wales and Scotland but not in Northern Ireland (sect. 10(3)).

Applications

Any application made to a LA for a licence must be made (unless in exceptional circumstances) by the person who proposes to own and possess the animal and must:

Table 13.1 Animals requiring a licensed keeper

Scientific name of kind	Common name or names

MAMMALS
Marsupials

Dasyuridae of the species *Sarcophilus harrisi*	The Tasmanian devil
Macropodidae of the species *Macropus fuliginosus*, *Macropus giganteus*, *Macropus robustus* and *Macropus rufus*	Grey kangaroos, the euro, the wallaroo and the red kangaroo

Primates

Callitrichidae of the species of the genera	Tamarins
Leontophithecus and *Saguinus Cebidae*	New-world monkeys (including capuchin, howler, saki, spider, squirrel, titi, uakari and woolly monkeys and the night monkey (otherwise known as the douroucouli))
Cercopithecidae	Old-world monkeys (including baboons, the drill, colobus monkeys, the gelada, guenons, langurs, leaf monkeys, macaques, the mandrill, mangabeys, the patas and proboscis monkeys and the talapoin)
Indriidae	Leaping lemurs (including the indri, sifakas and the woolly lemur)
Lemuriae, except the species of the genus *Hapalemur*	Large lemurs (the broad-nosed gentle lemur and the grey gentle lemur are excepted)
Pongidae	Anthropoid apes (including chimpanzees, gibbons, the gorilla and the orang-utan)

Edentates

Bradypodidae	Sloths
Dasypodidae of the species *Priodontes giganteus* (otherwise known as *Priodontes maximus*)	The giant armadillo
Myrmecophagidae of the species *Myrmecophaga tridactyla*	The giant anteater

Rodents

Erithizontidae of the species *Erithizon dorsatum*	The North American porcupine
Hydrochoeridae	The capybara
Hystricidae of the species of the genus *Hystrix*	Crested porcupines

Table 13.1 (*cont'd*)

Scientific name of kind	Common name or names
Carnivores	
Ailuropodidae (Ailuridae)	The giant panda and the red panda
Canidae, except the species of the genera *Alopex, Dusicyon, Otocyon, Nyctereutes* and *Vulpes* and the species *Canis familiaris*	Jackals, wild dogs, wolves and the coyote (foxes, the raccoon-dog and the domestic dog are excepted)
Felidae, except the species *Felis catus*	The bobcat, caracal, cheetah, jaguar, lion, lynx, ocelot, puma, serval, tiger and all other cats (the domestic cat is excepted)
Hyaenidae except the species *Proteles cristatus*	Hyaenas (except the aardwolf)
Mustelidae of the species of the genera *Arctonyx, Aonyx, Enhydra, Lutra* (except *Lutra lutra*), *Melogale, Mydaus, Pteronura* and *Taxidae* and of the species *Eira barbara, Gulo gulo, Martes pennanti* and *Mellivora capensis*	Badgers (except the Eurasian badger), otters (except the European otter), and the tayra, wolverine, fisher and ratel (otherwise known as the honey badger)
Procyonidae	Cacomistles, raccoons, coatis, olingos, the little coatimundi and the kinkajou
Ursidae	Bears
Viverridae of the species of the genus *Viverra* and of the species *Arctictis binturong* and *Cryptoprocta ferox*	The African, large-spotted, Malay and large Indian civets, the binturong and the fossa
Pinnipedes	
Odobenidae, Otariidae and Phocidae, except *Phoca vitulina* and *Halichoerus grypus*	The walrus, eared seals, sealions and earless seals (the common and grey seals are excepted)
Elephants	
Elephantidae	Elephants
Odd-toed ungulates	
Equidae, except the species *Equus asinus, Equus caballus* and *Equus asinus* x *Equus caballus*	Asses, horses and zebras (the donkey, domestic horse and domestic hybrids are excepted)
Rhinocerotidae	Rhinoceroses
Tapiridae	Tapirs
Hyraxes	
Procaviidae	Tree and rock hyraxes (otherwise known as dassies)
Aardvark	
Orycteropidae	The aardvark

Table 13.1 (*cont'd*)

Scientific name of kind	Common name or names
Even-toed ungulates	
Antilocapridae	The pronghorn
Bovidae, except any domestic form of the genera *Bos* and *Bubalus*, of the species *Capra aegagrus (hircus)* and the species *Ovis aries*	Antelopes, bison, buffalo, cattle, gazelles, goats and sheep (domestic cattle, goats and sheep are excepted)
Camelidae except the species *Lama glama* and *Lama pacos*	Camels, the guanaco and the vicugna (the domestic llama and alpaca are excepted)
Cervidae of the species *Alces alces* and *Rangifer tarandus*, except any domestic form of the species *Rangifer tarandus*	The moose or elk and the caribou or reindeer (the domestic reindeer is excepted)
Giraffidae	The giraffe and the okapi
Hippopotamidae	The hippopotamus and the pygmy hippopotamus
Suidae, except any domestic form of the species *Sus scrofa*	Old-world pigs (including the wild boar and the wart hog) (the domestic pig is excepted)
Tayassuidae	New-world pigs (otherwise known as peccaries)
Any hybrid of a kind of animal specified in the foregoing provisions of this column where one parent is, or both parents are, of a kind so specified	Mammalian hybrids with a parent or (parents) of a specified kind
BIRDS	
Cassowaries and emu	
Casuariidae	Cassowaries
Dromaiidae	The emu
Ostrich	
Struthionidae	The ostrich
REPTILES	
Crocodilians	
Alligatoridae	Alligators and caimans
Crocodylidae	Crocodiles and the false gharial
Gavialidae	The gharial (otherwise known as the gavial)
Lizards and snakes	
Colubridae of the species of the genera *Atractaspis*, *Malpolon*, *Psammophis* and *Thelatornis* and of the species *Boiga dendrophila*, *Dispholidus typus*, *Rhabdophis subminiatus* and *Rhabdophis tigrinus*	Mole vipers and certain rear-fanged venomous snakes (including the moila and montpellier snakes, sand snakes, twig snakes, the mangrove (otherwise known as the yellow-ringed catsnake), the boomslang, the red-necked keelback and the yamakagashi (otherwise known as the Japanese tiger-snake))

Table 13.1 (*cont'd*)

Scientific name of kind	Common name or names
Elapidae	Certain front-fanged venomous snakes (including cobras, coral snakes, the desert black snake, kraits, mambas, sea snakes and all Australian poisonous snakes (including the death adders)).
Helodermatidae	The gila monster and the (Mexican) beaded lizard
Viperidae	Certain front-fanged venomous snakes (including adders, the barba amarilla, the bushmaster, the copperhead, the fer-de-lance, moccasins, rattlesnakes and vipers)

INVERTEBRATES
Spiders

Ctenidae of the species of the genus *Phoneutria*	Wandering spiders
Dipluridae of the species of the genus *Atrax*	The Sydney funnel-web spider and its close relatives
Lycosidae of the species *Lycosa raptoria*	The Brazilian wolf-spider
Sicariidae of the species of the genus *Loxosceles*	Brown recluse spiders (otherwise known as violin spiders)
Theridiidae of the species of the genus *Latrodectus*	The black widow spider (otherwise known as red-back spider) and its close relatives

Scorpions

Buthidae	Buthid scorpions

(a) specify the species and number of animals to be kept;
(b) specify the premises where the animals will normally be kept;
(c) be made to the LA for those premises;
(d) be made by a person 18 years of age or over and not disqualified from holding a licence under the Act; and
(e) be accompanied by a fee stipulated by the LA at a level sufficient to meet the direct and indirect costs involved.

Applications not complying with these requirements may not be granted (sect. 1(2) and (4)).

Reports

Before granting any licence the LA is required to consider a report of an inspection of the premises by a veterinary surgeon or practitioner authorized by it (sect. 1(5)).

FC95 Dangerous wild animals

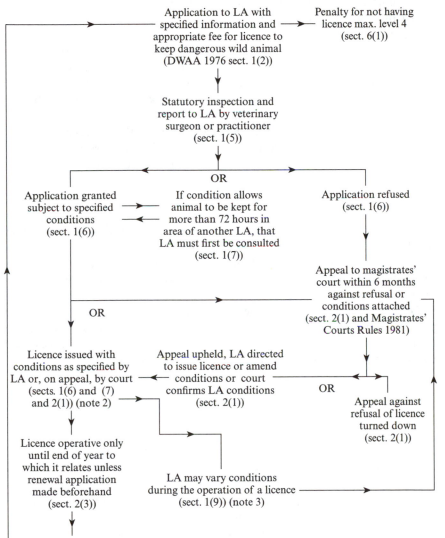

Application to LA with specified information and appropriate fee for licence to keep dangerous wild animal (DWAA 1976 sect. 1(2)) → Penalty for not having licence max. level 4 (sect. 6(1))

Statutory inspection and report to LA by veterinary surgeon or practitioner (sect. 1(5))

OR

Application granted subject to specified conditions (sect. 1(6)) → If condition allows animal to be kept for more than 72 hours in area of another LA, that LA must first be consulted (sect. 1(7))

Application refused (sect. 1(6))

Appeal to magistrates' court within 6 months against refusal or conditions attached (sect. 2(1) and Magistrates' Courts Rules 1981)

OR

Licence issued with conditions as specified by LA or, on appeal, by court (sects. 1(6) and (7) and 2(1)) (note 2)

Appeal upheld, LA directed to issue licence or amend conditions or court confirms LA conditions (sect. 2(1))

OR

Appeal against refusal of licence turned down (sect. 2(1))

Licence operative only until end of year to which it relates unless renewal application made beforehand (sect. 2(3))

LA may vary conditions during the operation of a licence (sect. 1(9)) (note 3)

Application for renewal

Notes

1. This procedure applies in England, Wales and Scotland but not in Northern Ireland (sect. 10(3)).
2. Penalty for contravening licence condition is maximum level (sect. 6(1)).
3. Unless the variation was requested by the licence holder, the LA must notify him of the variation and allow a reasonable time for compliance.

Matters for consideration

The LA may not grant a licence unless:

(a) it will not be contrary to the public interest on grounds of safety, nuisance or otherwise to issue a licence;

(b) the applicant is suitable;

(c) animals will:

    (i) be held in secure accommodation suitable in size for the animals kept and which is suitable as regards construction, temperature, lighting, ventilation, drainage and cleanliness; and

    (ii) have adequate and suitable food, drink and bedding and be visited at regular intervals;

(d) be appropriately protected in case of fire or other emergency;

(e) be subject to precautions to control infectious diseases;

(f) be provided with adequate exercise facilities (sect. 1(3)).

Licences

According to the wishes of the applicant, a licence comes into force on either the day on which it is granted, in which case it expires on the 31 December of that same year, or on the 1 January next in which case it expires on 31 December of that next year. If an application for renewal is made before the date of expiration, the licence continues until the application is determined.

On the death of a licence holder, the licence continues in the name of the personal representative for 28 days only and then expires unless application is made for a new licence within that time, in which case it continues until the new application is determined (sect. 2(2), (3) and (4)).

Conditions

The LA is required to specify conditions which:

(a) require the animals to be kept only by persons specified in the licence;

(b) require the animals to be normally held at the premises specified in the licence;

(c) require the animals not to be moved from those premises unless in circumstances allowed for in the licence;

(d) require the licence holder and person keeping the animals to be insured against liability for damage caused by the animals to the satisfaction of the LA;

(e) restrict the species and numbers of animals;

(f) require a copy of the licence to be made available by the licence holder to persons entitled to keep the animals; and

(g) any other conditions necessary or desirable to secure the objectives specified in paragraphs (c)–(f) listed under 'Matters for consideration' above (sect. 1(6) and (7)).

The LA may attach any other conditions which it thinks fit but if it is to permit the animal to be taken into another LA area for more than 72 hours, it must consult that LA (sect. 1(8)).

Conditions not required by the Act to be attached to the licence may be revoked or modified by the LA or new conditions may be added. These variations come into effect immediately if they were requested by the licence holder but otherwise the LA must notify him and allow a reasonable time for compliance (sect. 1(10)).

Disqualifications and cancellations

Where a person is convicted of an offence under the Dangerous Wild Animals Act 1976 or under:

(a) Protection of Animals Act 1911–64;
(b) Protection of Animals (Scotland) Act 1912–64;
(c) Pet Animals Act 1951;
(d) Animal Boarding Establishments Act 1963;
(e) Riding Establishments Act 1964–70;
(f) Breeding of Dogs Act 1973;

the court may cancel any licence he may hold to keep a dangerous wild animals and disqualify him, whether or not he is a current holder, from holding such a licence for such period as the court thinks fit. The cancellation or disqualification may be suspended by the court in the event of an appeal (sect. 6(2) and (3)).

Power of entry

LAs may authorize competent persons to enter premises either licensed under the Act or specified in an application for a licence, at all reasonable times, producing if required their authority, and the authorized officers may inspect these premises and an animal in them.

The penalty for wilfully obstructing or delaying an authorized officer is a maximum fine at level 4 (sects. 3 and 6(1)).

Seizure of animals

If a dangerous wild animal is being kept without the authority of a licence or in contravention of a licence condition, the LA may seize the animal and retain it, destroy it or otherwise dispose of it. The LA is not liable to compensation and may recover costs from the keeper of the animal at the time of this seizure (sect. 4).

Local authority

The licensing authorities are:

(a) in England:
  (i) unitary authorities;
  (ii) London boroughs;
  (iii) the Common Council of the City of London;
  (iv) in remaining two-tier areas, the district council;
(b) in Wales, the unitary authorities; and
(c) in Scotland, the unitary authorities (sect. 7(4) as amended 1994).

Definitions

Premises includes any place (sect. 7(4)).

DOG BREEDING ESTABLISHMENTS

Reference

Breeding of Dogs Acts 1973 and 1991.
Breeding and Sale of Dogs (Welfare) Act 1999.
The Breeding of Dogs (Licensing Records) Regulations 1999.
Home office Circular 53/1999.

Scope

A licence is required to carry out a **business** of the breeding of dogs for sale. The Act indicates that the keeping of bitches who give birth to 5 or more litters in any period of 12 months is presumed to constitute such a business and require to be licensed although LAs may conclude in the circumstances of a particular case that licensing is required for a lesser number. Where there is no sale of any puppies during the 12 months being considered licensing is not required.

Bitches count towards the qualifying total if they are kept at anytime during the 12 month period by the applicant/licence holder or by their relatives at the premises, by the applicant/licence holder elsewhere or by someone else as part of a breeding agreement. This is intended to prevent evasion of licensing by distributing bitches amongst different people and/or premises (sect. 4A).

Inspection

Upon receipt of an application for a licence for the first time, the LA must arrange for an inspection by a veterinary surgeon/practitioner **and** by an officer of the LA. Either or both may carry out inspections relating to subsequent applications (sect. 1(2A)).

Report

The LA must arrange for a report of inspection to be prepared and considered before making its' decision. (sect. 1(2B))

FC96 Dog breeding establishments

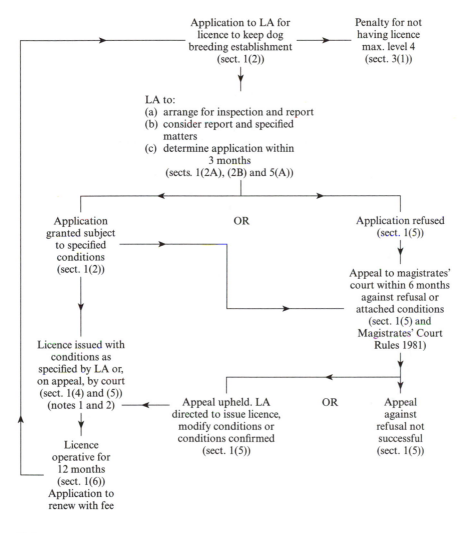

Notes
1. Penalty for non-compliance with conditions is max. fine of level 4 (sect. 3(1)).
2. Conditions may only be varied on the renewal of a licence.

Extent

This licensing procedure applies in England, Wales and, with modifications, Scotland but not in Northern Ireland (sect. 7(2)).

Matters for consideration

In deciding whether or not to grant a licence the LA must have regard to the following matters (but without prejudice to its right to refuse a licence on other grounds):

(a) that the dogs will at all times be kept in accommodation suitable as respects construction, size of quarters, number of occupants, exercising facilities, temperature, lighting, ventilation and cleanliness;
(b) that the dogs will be adequately supplied with suitable food, drink and bedding material, adequately exercised, and visited at suitable intervals;
(c) that all reasonable precautions will be taken to prevent and control the spread among dogs of infectious or contagious diseases;
(d) that appropriate steps will be taken for the protection of the dogs in case of fire or other emergency;
(e) that all appropriate steps will be taken to secure that the dogs will be provided with suitable food, drink and bedding material and adequately exercised when being transported to or from the breeding establishment;
(f) that bitches are not mated if they are less than 1 year old;
(g) that bitches do not give birth to more than 6 litters each;
(h) that bitches do not give birth to puppies before the end of the period of 12 months beginning with the day on which they last gave birth to puppies; and
(i) that accurate records are kept at the premises and made available for inspection for any authorized officer of the LA to examine. The particular records to be kept are listed in the Breeding of Dogs (Licensing Records) Regulations 1999 (sect. 1(4)).

Persons disqualified under the following provisions may not be granted a dog breeding licence:

(a) from keeping a dog breeding establishment;
(b) from keeping a pet shop under the Pet Animals Act 1951;
(c) from keeping an animal boarding establishment under the Animal Boarding Establishments Act 1963;
(d) from having the custody of animals under the Protection of Animals (Amendment) Act 1954.

(sect. 1(2) and (4))

Licences

The LA must determine the application within 3 months (sect. 1(5A)) Licences run for 12 months from its start date being either the date requested by the applicant or the date of issue, whichever is the later (sect. 1(6) and (7)). On the death of a licence holder, licences pass to his personal representative for a period of 3 months from the day of his death and then expire. The LA may extend that 3-month period at its discretion (sect. 1(8)).

Conditions

On granting a licence the LA must specify such conditions as appear necessary or expedient to achieve the objectives set out in (a)–(i) in the paragraph headed 'Matters for consideration' above (sect. 1(4)).

Disqualifications and cancellations

In making a conviction under this Act, the court may cancel any dog breeding licence held by the convicted person, disqualify him from holding a licence for such period as the court thinks fit and disqualify him from having custody of any dog for such period as specified. The cancellation or disqualification may be suspended by the court pending appeal (sect. 3(3) and (4)).

Fees

LAs may charge fees for both the applications for licences and for the related inspections. The level of fees may reflect the reasonable administrative and enforcement costs incurred and may be varied to take account of different circumstances (sect. 3A).

Appeals

Any aggrieved person may appeal to a magistrates' court (sheriff court in Scotland) against a refusal to grant a licence and against the imposition of any condition. The court may direct in either case, as it thinks proper (sect. 1(5)).

Offences/Penalties

Persons operating without a licence or breaching licence conditions are guilty of an offence and on conviction liable to a penalty of either up to 3 months imprisonment or a fine not exceeding level 4 or to both (sects. 1(9) and 3(1)).

Power of entry

LA officers, veterinary surgeons or practitioners authorized in writing by the LA may, upon producing if required their authority, enter any premises licenced as a dog breeding establishment at any reasonable time and inspect it and any animals found there (sect. 2(1)). Persons wilfully obstructing or delaying authorized officers are liable on conviction to a maximum fine of level 3 (sect. 2(2)).

AOs are also given powers of entry into premises (other than private dwellings although entry into garages, outhouses, and other structures forming part of the private dwelling is provided for) suspected to be operating without a licence. Before entry is requested a warrant must be obtained from a JP and must be effected at reasonable times. The penalty for obstruction against these powers is also a maximum fine of level 3 (1991 Act sect. 1).

Local authority

The licensing authorities are:

(a) in England:
    (i) unitary authorities;
    (ii) London boroughs;
    (iii) the Common Council of the City of London;
    (iv) in the remaining two-tier areas, the district council;
(b) in Wales, the unitary authorities; and
(c) in Scotland, the unitary authorities (sect. 5(2) as amended 1994).

PET SHOPS

Reference

Pet Animals Act 1951 as amended by the Pet Animals Act 1951 (Amendment) Act 1983.

Scope

A person keeping a pet shop requires a licence from the LA (sect. 1(1)). The Act defines 'the keeping of a pet shop' as:

'Shall be construed as references to the carrying on at premises of any nature (including a private dwelling) of a business of selling animals as pets, and as including references to the keeping of animals in any such premises as aforesaid with a view to their being sold in the course of such a business, whether by the keeper thereof or by any other person: Provided that:

(a) a person shall not be deemed to keep a pet shop by reason only of his keeping or selling pedigree animals bred by him, or the offspring of an animal kept by him as a pet;
(b) where a person carries on a business of selling animals as pets in conjunction with a business of breeding pedigree animals, and the

FC97 Pet shops

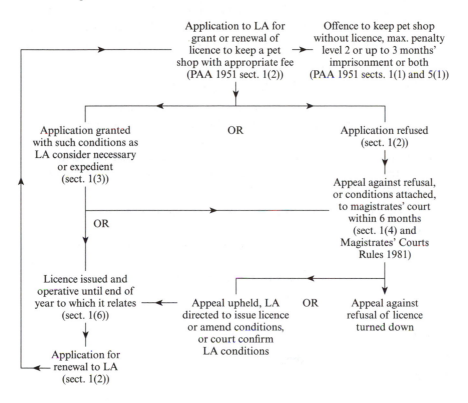

Application to LA for grant or renewal of licence to keep a pet shop with appropriate fee (PAA 1951 sect. 1(2))

Offence to keep pet shop without licence, max. penalty level 2 or up to 3 months' imprisonment or both (PAA 1951 sects. 1(1) and 5(1))

Application granted with such conditions as LA consider necessary or expedient (sect. 1(3))

OR

Application refused (sect. 1(2))

Appeal against refusal, or conditions attached, to magistrates' court within 6 months (sect. 1(4) and Magistrates' Courts Rules 1981)

OR

Licence issued and operative until end of year to which it relates (sect. 1(6))

Appeal upheld, LA directed to issue licence or amend conditions, or court confirm LA conditions

OR

Appeal against refusal of licence turned down

Application for renewal to LA (sect. 1(2))

Notes

1. Penalty for contravening licence conditions maximum level 2 or up to 3 months' imprisonment or both (sect. 5(1)).
2. This procedure applies in Scotland but not in Northern Ireland (sect. 8).
3. Conditions may only be varied on the renewal of a licence.

local authority are satisfied that the animals so sold by him (in so far as they are not pedigree animals bred by him) are animals which were acquired by him with a view to being used, if suitable, for breeding or show purposes but have subsequently been found by him not to be suitable or required for such use, the local authority may if they think fit direct that the said person shall not be deemed to keep a pet shop by reason only of his carrying on the first-mentioned business.

References in this Act to the selling or keeping of animals as pets shall be construed in accordance with the following provisions, that is to say:

(a) as respects cats and dogs, such references shall be construed as including references to selling or keeping, as the case may be, wholly or mainly for domestic purposes; and

(b) as respects any animal, such references shall be construed as including references to selling or keeping, as the case may be, for ornamental purposes. (sect. 7(1) and (2)).'

The definition of pet animal is widely drawn and includes any type of vertebrate (sect. 7(3)).

There is no definition of the word 'premises', this having been removed by the Amendment Act of 1983, but the sale of animals as pets as a business is prohibited in any part of a street or public place or at a stall or barrow in a market (sect. 2 as amended).

Extent

This licensing procedure applies in England, Wales and Scotland but not in Northern Ireland (sect. 8).

Applications

Applications made to the LA must specify the premises concerned and a fee as determined by the local authority must be paid before a licence is granted (sect. 1(2)).

The factors which may be taken into account in deciding whether or not to issue a licence are listed under 'Conditions' below but these are without prejudice to the LA's discretion to refuse a licence on any other grounds (sect. 1(3)).

Licences

Licences remain in force until the end of the year to which they relate and the latter is determined by the applicant as being either the year in which it is granted or the following year. In the first case, i.e. the year in which it is granted, the licence comes into force on the day it is granted and expires on 31 December of that year, in the second case, i.e. the year following that in which it is granted, it comes into force on the 1 January of that year and expires on 31 December of that year (sect. 1(5) and (6)).

Conditions

The LA must attach any conditions which it considers to be necessary or expedient for securing all or any of the following objectives:

(a) that animals will at all times be kept in accommodation suitable as respects size, temperature, lighting, ventilation and cleanliness;
(b) that animals will be adequately supplied with suitable food and drink and (so far as necessary) visited at suitable intervals;
(c) that animals, being mammals, will not be sold at too early an age;
(d) that all reasonable precautions will be taken to prevent the spread among animals of infectious diseases;
(e) that appropriate steps will be taken in case of fire or other emergency (sect. 1(3)).

Power of entry

LA officers, veterinary surgeons or practitioners authorized in writing by the LA for this purpose, may, upon producing his authority if required, inspect a licensed pet shop and any animals there at all reasonable times. Persons wilfully obstructing authorized officers are subject on conviction to a maximum fine of level 2 (sects. 4 and 5(2)).

There does not appear to be a power of entry for premises which do not hold a pet shop licence, e.g. which may be suspected of operating an unlicensed pet shop or in respect of which an application has been made. In the latter case refusal of the licence application would be the obvious remedy.

Disqualifications and cancellations

In making a conviction under the PAA 1951, the Protection of Animals Act 1911 or the Protection of Animals (Scotland) Act 1912, the court may cancel any pet shop licence held by the person and may disqualify him from holding such a licence, whether or not he currently holds one, for any specified period. The cancellation or disqualification may be suspended by the court pending an appeal (sect. 5(3) and (4)).

Where proceedings for cruelty or neglect are brought under the Protection of Animals Act 1911, the Protection of Animals (Amendment) Act 2000 enables courts to make orders regarding the care, disposal or slaughter of animals kept for sale in pet shops.

LAs must refuse licence applications from persons currently disqualified by a court from holding a pet shop licence (sect. 1(2)). This provision differs from that relating to Animal Boarding Establishments where disqualifications under other Acts are also relevant.

Local authority

The licensing authorities are:

(a) in England,
    (i) unitary authorities;
    (ii) London boroughs;
    (iii) the Common Council of the City of London;
    (iv) in remaining two-tier areas, the district council;
(b) in Wales, the unitary authorities; and
(c) in Scotland, the unitary authorities (sect. 7(3) as amended 1994).

Definitions

Animal includes any description of vertebrate.

Pedigree animal means an animal of any description which is by its breeding eligible for registration with a recognized club or society keeping a register of animals of that description.

Veterinary surgeon means a person who is for the time being registered in the Register of Veterinary Surgeons.

Veterinary practitioner means a person who is for the time being registered in the Supplementary Register (sect. 7(3)).

RIDING ESTABLISHMENTS

Reference

Riding Establishments Acts 1964 and 1970.

Scope

A person keeping a riding establishment is required to hold a licence issued by the LA (sect. 1(1)).

The 1964 Act defines the keeping of a riding establishment as:

'. . . the carrying on of a business of keeping horses for either or both of the following purposes, that is to say, the purpose of their being let out on hire for riding or the purpose of their being used in providing, in return for payment, instruction in riding, but as not including a reference to the carrying on of such a business:

(a) in a case where the premises where the horses employed for the purposes of the business are kept are occupied by or under the management of the Secretary of Defence; or
(b) solely for police purposes; or
(c) by the Zoological Society of London; or
(d) by the Royal Zoological Society of Scotland (sect. 6(1)).'

FC98 Riding establishments

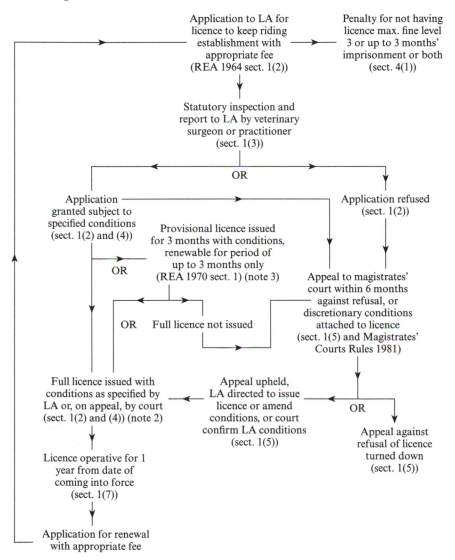

Notes
1. This procedure applies in England, Wales and Scotland but not in Northern Ireland (sect. 9(2)).
2. Penalty for contravening licence condition is maximum level 3 and up to 3 months' imprisonment or both (sect. 4(1) as amended).
3. There is no appeal against the decision to issue a provisional licence only or against the conditions attached to it.
4. Conditions may only be varied at the renewal of the licence.

Horses kept by a university providing veterinary courses are also exempt and the place at which a riding establishment is run is to be taken as the place at which the horses are kept (sect. 6(2) and (3)).

Extent

This licensing procedure applies in England, Wales and Scotland but not in Northern Ireland (sect. 9(2)).

Applications

Applications made to the LA must specify the premises concerned and a fee as determined by the LA must be paid before any licence is granted (sect. 1(2)).

Reports

The LA is required to receive a report by a listed veterinary surgeon or practitioner (see 'Power of entry' below) before making a decision and the report must be based on an inspection made not more than 12 months before the application was received (sect. 1(3)).

Matters for consideration

In deciding whether or not to grant a licence, or provisional licence, the LA must, without prejudice to its right to refuse a licence on other grounds, have regard to the following matters:

(a) whether that person appears to them to be suitable and qualified, either by experience in the management of horses or by being the holder of an approved certificate or by employing in the management of the riding establishment a person so qualified, to be the holder of such a licence; and

(b) the need for securing:

    (i) that paramount consideration will be given to the condition of the horses and that they will be maintained in good health, and in all respects physically fit and that, in the case of a horse kept for the purpose of its being let out on hire for riding or a horse kept for the purpose of its being used in providing instruction in riding, the horse will be suitable for the purpose for which it is kept;

    (ii) that the feet of all animals are properly trimmed and that, if shod, their shoes are properly fitted and in good condition;

    (iii) that there will be available at all times, accommodation for horses suitable as respects construction, size, number of occupants, lighting, ventilation, drainage and cleanliness and that these requirements be complied with not only in the case of new buildings but also in the case of buildings converted for use as stabling;

(iv) that in the case of horses maintained at grass there will be available for them at all times during which they are so maintained adequate pasture and shelter and water and that supplementary feeds will be provided as and when required;

(v) that horses will be adequately supplied with suitable food, drink and (except in the case of horses maintained at grass, so long as they are so maintained) bedding material, and will be adequately exercised, groomed and rested and visited at suitable intervals;

(vi) that all reasonable precautions will be taken to prevent and control the spread among horses of infectious or contagious disease and that veterinary first-aid equipment and medicines shall be provided and maintained in the premises;

(vii) that appropriate steps will be taken for the protection and extrication of horses in case of fire and, in particular, that the name, address and telephone number of the licence holder or some other responsible person will be kept displayed in a prominent position on the outside of the premises and that instructions as to action to be taken in the event of fire, with particular regard to the extrication of horses, will be kept displayed in a prominent position on the outside of the premises;

(viii) that adequate accommodation will be provided for forage, bedding, stable equipment and saddlery (sect. 1(4)).

Persons under 18 years old or persons or bodies corporate disqualified under the following provisions may not be given a licence:

(a) from keeping a riding establishment under the Riding Establishment Act 1964;

(b) from keeping a pet shop under the Pet Animals Act 1951;

(c) from having custody of animals under the Protection of Animals (Amendment) Act 1954;

(d) from keeping an animal boarding establishment under the Animal Boarding Establishment Act 1963 (sect. 1(2)).

Licences

Full licences continue for 1 year beginning on the day on which they came into force and then expire. The date of operation, depending upon the wishes of the applicant, is either the day on which it is granted or the 1 January next (sect. 1(6) and (7)).

Provisional licences operate for 3 months from the day on which they are granted and are used where the LA is satisfied that it would not be justified in issuing a full licence. The 3 months' period of operation may, on application before the expiration of the 3 months, be extended for a further period of not exceeding 3 months so long as this would not exceed a 6-month period in 1 year (REA 1970 sect. 1).

On the death of a licence holder, licences pass to his personal representative for a period of 3 months and then expire. The 3-month period may be extended at the LA's discretion (sect. 1(8)).

Conditions

On granting a licence the LA is required to specify conditions as appear necessary or expedient to achieve all the objectives set out in (b)(i)–(viii) in the paragraph headed 'Matters for consideration' above. In addition the following conditions are required by the Act, whether specified in the licence or not:

(a) a horse found on inspection of the premises by an authorized officer to be in need of veterinary attention shall not be returned to work until the holder of the licence has obtained at his own expense and has lodged with the LA a veterinary certificate that the horse is fit for work;

(b) no horse will be let out on hire for riding or used for providing instruction in riding without supervision by a responsible person of the age of 16 years or over unless (in the case of a horse let out for hire for riding) the holder of the licence is satisfied that the hirer of the horse is competent to ride without supervision;

(c) the carrying on of the business of a riding establishment shall at no time be left in the charge of any person under 16 years of age;

(d) the licence holder shall hold a current insurance policy which insures him against liability for any injury sustained by those who hire a horse from him for riding and those who use a horse in the course of receiving from him in return for payment, instruction in riding and arising out of the hire or use of a horse as aforesaid and which also insures such persons in respect of any liability which may be incurred by them in respect of injury to any person caused by, or arising out of, the hire or use of a horse as aforesaid;

(e) a register shall be kept by the licence holder of all horses in his possession aged 3 years and under and usually kept on the premises, which shall be available for inspection by an authorized officer at all reasonable times (sect. 1(4A)).

Disqualifications and cancellations

In making a conviction under this Act, under the Protection of Animals Act 1911, the Protection of Animals (Scotland) Act 1912, the Pet Animals Act 1951 or the Animal Boarding Establishment Act 1963, the court may cancel any riding establishment licence held by the convicted person and may disqualify him from holding such a licence, whether or not he is a current holder, for such period as the court thinks fit. The cancellation or disqualification may be suspended by the court pending an appeal (sect. 4(3) and (4)).

Power of entry

LA officers and veterinary surgeons and practitioners listed by the Royal College of Veterinary Surgeons and British Veterinary Association for this purpose, authorized in writing by the LA, may, upon producing their authority if required, enter and inspect the following premises at all reasonable times:

(a) licensed riding establishments;
(b) premises subject to an application for a licence to run a riding establishment; and
(c) unlicensed premises suspected of being used as a riding establishment.

This power extends to the inspection of any horses found on the premises (sect. 2(1)–(3)).

Persons wilfully obstructing or delaying authorized officers are subject on conviction to a maximum fine not exceeding level 2 (sects. 2(4) and 4(1)).

Local authority

The authorities which act as the licensing authority are:

(a) in England;
    (i) unitary authorities including London borough councils;
    (ii) the Common Council of the City of London;
    (iii) in remaining two-tier areas, the district council;
(b) in Wales, the unitary authority; and
(c) in Scotland, the unitary authorities (sect. 6(4) as amended 1994).

Definitions

Approved certificate means:

(a) any one of the following certificates issued by the British Horse Society, namely, Assistant Instructor's Certificate, Instructor's Certificate and Fellowship;
(b) Fellowship of the Institute of Horse; or
(c) any other Certificate for the time being prescribed by order of the SoS.

Horse includes any mare, gelding, pony, foal, colt, filly or stallion and also any ass, mule or jennet.

Premises includes land (sect. 6(4)).

PUBLIC ENTERTAINMENT

ENTERTAINMENT LICENCES

References

Local Government (Miscellaneous Provisions) Act 1982 sect. 1 and Schs. 1 and 2 (as amended).
The Entertainments (Increased Penalties) Act 1990.

Scope

The licensing requirement is applied to:
  (a) Public dancing or music or any other public entertainment of a like kind except:
      (i) to any music in a place of public religious worship or performed as an incident of a religious meeting or service;
      (ii) entertainment held in a pleasure fair;
      (iii) entertainment which takes place wholly or mainly in the open air (Sch. 1, para. 1(2) and (3)).
  (b) Sports entertainment, that is any sporting event at premises to which the public are invited as spectators.
      In this connection, premises means any permanent or temporary building and any tent or inflatable structure and includes a part of a building where the building is a sports complex but does not include part of any other building. Sports entertainment held at a pleasure fair is not covered by the licensing requirement neither is it where on the particular occasion the entertainment is not the principal purpose for the premises. This latter exclusion does not however apply within a sports complex (Sch. 1, para. 2(1)–(3) as amended by the Fire Safety and Safety of Places of Sport Act 1987).
  (c) Where the provisions have been adopted, outdoor public **musical** (i.e. a substantial musical ingredient) entertainment held:
      (i) on private land, i.e. where the public has access (whether on payment or otherwise) only by permission of the owner, occupier or lessee; and
      (ii) wholly or mainly in the open air;

but not where the music is **incidental** only at any:

      (i) garden fete, bazaar, sale of work, sporting or athletic event, exhibition, display or other function or event of a similar character whether limited to 1 day or extending over 2 or more days; or
      (ii) a religious meeting or service; or
      (iii) to any musical entertainment (incidental or not) at a pleasure fair (para. 3).

FC99a Entertainment licences: adoption of outdoor musical entertainment provisions

Outdoor musical entertainment
licensing controls may
be adopted by LA
(LG(MP)A 1982 sect. 1(2))

↓

LA pass resolution
adopting provisions
(sect. 1(2))
specifying date of operation
not less than 1 month
from date of resolution
(sect. 1(3))

↓

LA must publicize decision
by notice stating general
effect in a local newspaper on
2 successive weeks, the first
being not later than 28 days
before date the resolution
takes effect
(sect. 1(4) and (5))

↓

Licensing controls effective
on date specified in LA
resolution of adoption
(sect. 1(3))
not less than 28 days from
first public notice
(sect. 1(5))
and not less than 1 month from
resolution of adoption
(sect. 1(2))

Notes
1. These provisions are not available in Greater London, Scotland or Northern Ireland.
2. For definition of outdoor entertainment, see page 414.

FC99b Entertainment licences: applications

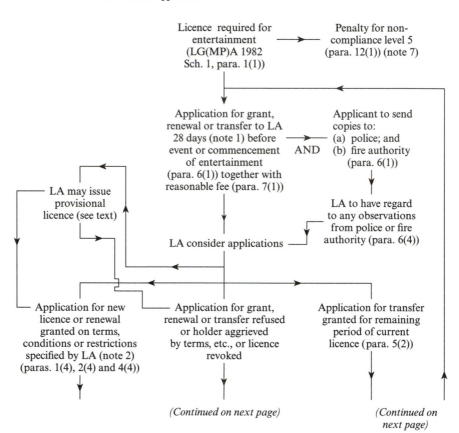

Licence required for
entertainment → Penalty for non-
(LG(MP)A 1982 compliance level 5
Sch. 1, para. 1(1)) (para. 12(1)) (note 7)

Application for grant,
renewal or transfer to LA   Applicant to send
28 days (note 1) before →   copies to:
event or commencement   AND   (a) police; and
of entertainment   (b) fire authority
(para. 6(1)) together with   (para. 6(1))
reasonable fee (para. 7(1))

LA may issue   LA to have regard
provisional   to any observations
licence (see text)   from police or fire
                                              authority (para. 6(4))

LA consider applications

Application for new   Application for grant,   Application for transfer
licence or renewal   renewal or transfer refused   granted for remaining
granted on terms,   or holder aggrieved   period of current
conditions or restrictions   by terms, etc., or licence   licence (para. 5(2))
specified by LA (note 2)   revoked
(paras. 1(4), 2(4) and 4(4))

(Continued on next page)   *(Continued on next page)*

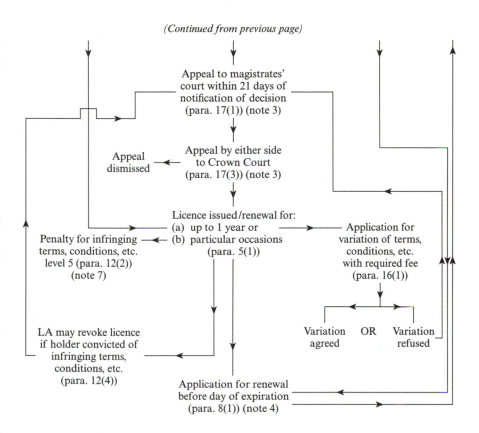

(Continued from previous page)

Appeal to magistrates' court within 21 days of notification of decision (para. 17(1)) (note 3)

Appeal dismissed ← Appeal by either side to Crown Court (para. 17(3)) (note 3)

Penalty for infringing terms, conditions, etc. level 5 (para. 12(2)) (note 7)

Licence issued/renewal for:
(a) up to 1 year or
(b) particular occasions (para. 5(1))

Application for variation of terms, conditions, etc. with required fee (para. 16(1))

LA may revoke licence if holder convicted of infringing terms, conditions, etc. (para. 12(4))

Variation agreed OR Variation refused

Application for renewal before day of expiration (para. 8(1)) (note 4)

Notes

1. The LA may grant applications where less than 28 days' notice has been given after consulting the police and fire authority.
2. The purposes for which conditions etc. on outdoor entertainment licences are specified are restricted (page 419).
3. The court may make such order as it thinks fit.
4. Provided that the application for renewal or transfer is made before the expiration of the licence, the licence continues in force beyond the original expiration date until the application is determined.
5. This procedure does not apply within Greater London, Scotland or Northern Ireland.
6. This procedure applies to all entertainment venues, i.e. public dancing or music, sports entertainment and (where adopted) outdoor, public musical entertainment.
7. The normal penalty for not holding a licence when required and for breaking conditions etc. is a fine not exceeding level 5. However, where operating without a licence (other than for sporting entertainment) and breaching a condition limiting numbers, the penalty is increased to a fine of up to £20 000 or imprisonment for up to 6 months or both (Entertainments (Increased Penalties) Act 1990).

Premises which have a justices liquor licence under the Licensing Act 1964 are not required to hold public entertainment licences where the entertainment consists of:

(a) reproduction of wireless (including television) broadcasts; or
(b) music and singing only provided by either:
   (i) the reproduction of recorded sound; or
   (ii) not more than two performers (Licensing Act 1964 sect. 182).

The word 'public' is not defined and has been the subject of consideration in many court proceedings. Entertainment which is not 'public' may be caught by the provisions of the Private Places of Entertainment (Licensing) Act 1967 which may be adopted by LAs.

Extent

This procedure applies in England and Wales other than in Greater London (for which see the London Government Act 1963 (as amended)). It does not apply in Scotland and Northern Ireland.

Applications

Applications are to be made to the LA for the area where the place at which the entertainment is to take place is situated and must specify:

(a) the place concerned;
(b) whether the application is for public entertainment or sports entertainment;
(c) whether a full licence or occasional licence is required and in the latter case the date or dates concerned (paras. 1(4) and (5), 2(4) and (5), 4(2) and (3)).

No particular form of application is specified but the LA may make regulations laying down what particulars it will require in the application and what notices (other than those required statutorily) are to be given. For example, a LA could require applications to be advertised in a local newspaper (para. 6(3)).

Copies of the applications sent to the LA must be sent to the chief officer of police and to the FA, and the application must be made at least 28 days before the entertainment is to take place although the LA may waive that requirement after consulting the police and FA (para. 6(1) and (2)).

When considering applications, the LA must have regard to any observations made to it by the police or FA (para. 6(4)).

Fees

Each applicant for grant of a new licence, renewal or transfer must pay to the LA a reasonable fee as determined by that LA and this is not repayable in the event of refusal. Fees cannot be demanded for entertainment:

(a) at a church hall, chapel hall or similar building occupied in connection with a place of public religious worship; or

(b) at a village hall, parish or community hall or other similar building.

In the case of entertainment which is:

(a) of an educational or other like character; or

(b) given for charitable or other like purposes;

the LA may decide to forego the whole or part of any fee that would otherwise be payable (para. 7).

Home Office circular 13/2000 gives LAs guidance on the fee levels that may be set with a particular aim of ensuring that these are reasonable.

Licences

There is no specified form of licence but the place subject to the licence and, in respect of occasional licences, the particular dates concerned, must be stated.

Except for occasional licences, licences may be issued for any period of up to 1 year (para. 5).

Conditions etc.

Other than for outdoor public musical entertainment, the LA may attach to the licence such terms, conditions or restrictions as it thinks fit and the provisions draw no boundaries to this discretion (subject to appeal). The LA may make standard conditions applying to all or any class of entertainment. For outdoor public musical entertainment, however, conditions etc. (including standard conditions) may **only** be attached where they are designed to:

(a) secure the safety of performers or other persons present;

(b) secure adequate access for emergency vehicles;

(c) secure adequate sanitary appliances;

(d) prevent unreasonable disturbance by noise to persons in the neighbourhood (paras. 1(4), 2(4) and 4(4)).

Home Office circular 13/2000 gives LAs guidance on the imposition of conditions. The circular aims to ensure that these are relevant to the aims of public entertainment licensing, are not adding unreasonable burdens and are not duplicating or overlapping other legislation.

Hypnotism

The LA is able to attach conditions to public entertainment licences regulating or prohibiting exhibitions, demonstrations or performances of hypnotism on persons at the premises concerned.

At places not licensed for public entertainment, hypnotism on persons in connection with entertainment where the public are admitted must have the authorization of the LA (Hypnotism Act 1952 sects. 1 and 2).

Refusals

No particular grounds for the refusal of applications are contained within these provisions and refusals are, therefore, at the discretion of the LA but subject to appeal.

Transfers

The LA may, on application, transfer the licence to any person it thinks fit (para. 5(2)) and where such an application is made, the licence will remain in force beyond its date of expiration until the application is determined (para. 8(2)). In the event of the death of the licence holder, the person carrying on the entertainment is deemed to be the licence holder until such time as legal representatives of the deceased have been constituted or the licence is transferred (para. 9).

Variations

The holder of the licence may apply to the LA at any time during its operation for variation to any of the terms, conditions or restrictions on which it may have been issued. The LA may deal with such applications by either:

(a) refusal;
(b) approval; or
(c) making variations, including or not as it thinks fit modifications requested by the applicant (para. 16).

Applications for variations are subject to such reasonable fee as the LA may determine (para. 16A).

Appeals

The licensing procedure provides for appeals to the magistrates' courts against:

(a) refusal of grant, renewal or transfer;
(b) refusal of application for variation of terms, etc.;
(c) revocation of the licence; or
(d) by the holder who is aggrieved by any term, condition, etc. attached to the licence (para. 17(1)).

Appeals are to be made within 21 days from the date on which the person is notified of the decision concerned (para. 17(1) and (2)).

Either side may appeal to the Crown Court against the decision of the magistrates' court (para. 17(3)).

Revocation

If the holder of the licence is convicted of breaching the terms, conditions or restrictions attached to the licence, the LA may at its discretion revoke the licence (para. 12(4)).

Power of entry

The provisions provide a power of entry in the following circumstances:

(a) Where a public entertainment licence is operative, the following officers may enter the place either during or just before the entertainment concerned is taking place or is about to take place:
  (i) police constable;
  (ii) authorized officer of the LA;
  (iii) authorized officer of the FA.
  No prior warning is required.
(b) Authorized officers of the FA may enter a place where a licence is operative after giving not less than 24 hours' notice in order to inspect fire precautions including terms, conditions and restrictions relating to fire precautions.
(c) Authorized officers of the LA and police constables may enter **any** place if authorized by a warrant granted by a JP in order to check:
  (i) if entertainment is taking place which would require a licence;
  (ii) if entertainment is taking place which is not authorized by the licence;
  (iii) if the terms, conditions and restrictions of any licence are being complied with.
  In each case, authorized officers must, if requested, produce their authority to enter. The penalty for refusal to permit entry is level 3 (para. 14).

Provisional licences

If the LA is satisfied that, if the premises are completed in accordance with the submitted plans, it would grant a licence, it may issue a provisional licence. This licence will have no effect until confirmed but is an indication to the applicant of the LA's intentions. When the LA is satisfied that the premises have been satisfactorily completed in accordance with the plans, modified if necessary with the approval of the LA, **and** it considers that the licence is held by a fit and proper person, it may confirm the provisional licence (para. 15).

The LA can only consider the issue of a provisional licence when it has received an application for a full licence and, therefore, the provisions relating to appeals and the necessity to pass copies to the police and FA still apply.

Liquor licensing

There are two ways in which public entetainment licences are affected by or are relevant to liquor licences issued by the magistrates' court:

(a) where a special order of exemption has been granted under sect. 74(4) of the Licensing Act 1964, the hours of any public entertainment licence are automatically extended to those specified in the special order;

(b) in granting a special hours certificate, the permitted hours are extended on weekdays from 12.30 pm to 3.00 pm and from 6.30 pm to 2.00 am, except on Good Friday. On Maundy Thursday and Easter Eve, the final hour is midnight. The certificate is conditional upon there being in force a public entertainment licence and upon the premises being structurally adapted and bona fide used for the provision of music and dancing and substantial refreshment to which the provision of intoxicating liquor is ancillary.

Local authority

The licensing authorities are:

(a) unitary authorities;

(b) in remaining two-tier areas, district councils; and

(c) the Council of the Isles of Scilly (sect. 1(11) as amended 1994).

Definitions

Place of public religious worship means a place of public religious worship which belongs to the Church of England or to the Church in Wales (within the meaning of the Welsh Church Act 1914), or which is for the time being certified as required by law as a place of religious worship (para. 22).

Pleasure fair means any place:

(a) which is for the time being used wholly or mainly for providing, whether or not in combination with any other entertainment, any entertainment to which this section applies; and

(b) for admission to which, or for use of contrivances in which, a charge is made (PHA 1961 sect. 75(2)(a)).

The entertainments to which the above applies are the following:

(a) circuses;

(b) exhibitions of human beings or of performing animals;

(c) merry-go-rounds, roundabouts, swings, switchback railways;

(d) coconut shies, hoop-las, shooting galleries, bowling alleys;

(e) dodgems or other mechanical riding or driving contrivances;

(f) automatic or other machines intended for entertainment or amusement;

(g) anything similar to any of the foregoing (PHA 1961 sect. 75(3)).

Sport includes any game in which physical skill is the predominant factor and any form of physical recreation which is also engaged in for the purpose of competition or display, except dancing in any form.

Sports complex means a building:

(a) which provides accommodation and facilities for both those engaging in sport and spectation; and
(b) the parts of which are so arranged that one or more sports can be engaged in simultaneously in different parts of the building.

Sporting event means any contest, exhibition or display of sport. (para. 2.6 as substituted by the Fire Safety and Safety of Places of Sport Act 1987).

CINEMA LICENCES

Reference

Cinemas Act 1985.

Scope

A licence is required for film exhibitions at any premises other than:

(a) private houses to which the public are not admitted;
(b) other premises where either the public are not admitted or are admitted without charge; and
(c) in any case where the society or organization concerned has been granted a certificate of exemption (sects. 1(1), 5 and 6).

Film exhibition means any exhibition of moving pictures, which is produced otherwise than by the simultaneous reception and exhibition of programmes included in a programme service within the meaning of the Broadcasting Act 1990 (sect. 21).

This definition will include video exhibitions in commercial premises, e.g. public houses, but does not extend to video games or to amusement arcades or similar premises.

Extent

This procedure applies in England and Wales including Greater London, but with some variations, and in Scotland.

Licensing authority

The LAs charged with the licensing responsibilities are:

(a) London borough councils;
(b) the Common Council of the City of London;

FC100 Cinema licences

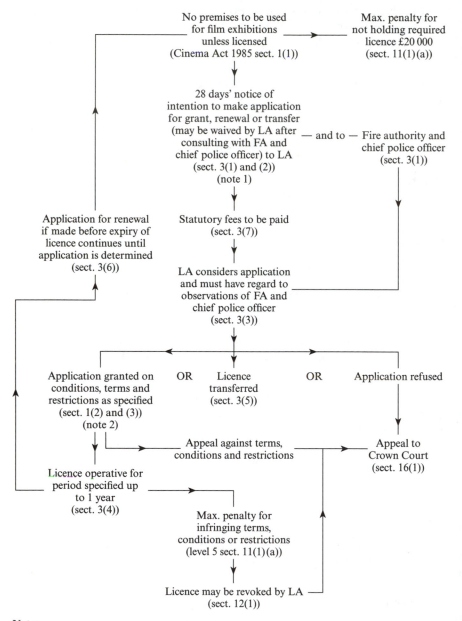

Notes
1. In the Greater London area, applications may also be considered for a provisional grant of a licence based upon plans and specification (sect. 17).
2. Variations to conditions etc. before renewal are only possible in Greater London (sect. 18).

(c) unitary authorities in England and the district councils in the remaining two-tier areas;

(d) unitary authorities in Wales;

(e) unitary authorities in Scotland (sect. 21(1)).

Application

No particular details are specified by the Act. Fees are determined by the SoS by Fees for Cinema Licences (Variation) Orders (sect. 3(7)).

Conditions etc.

Licences may be granted by the LA on such terms, conditions and restrictions as they may determine (sect. 1(2)). However, the authority is under a duty to prohibit the admission of children to exhibitions which the LA considers unsuitable and to consider what other conditions etc. should be specified where children are to be admitted (sect. 1(3)).

All such premises are also subject to the Cinematograph (Safety) Regulations 1955, and conditions etc. of licences may not alter those requirements.

Power of entry

Authorized officers of the LA may enter to check licensed premises being used or about to be used for film exhibition at any time without notice. Officers of the fire authority are required to give 24 hours' notice. Warrants are obtainable by authorized officers to check compliance with the Act where there is reason to suspect an offence. The maximum penalty for obstruction is level 3 on the standard scale (sect. 13).

THEATRE LICENCES

Reference

Theatres Act 1968.

Scope

Any premises used for the public performance of a play is required to be licensed (sect. 12(1)).

Licensing authority

The authorities charged with the licensing responsibility are district councils (unitary and continuing two-tier districts), London borough councils, the Common Council of the City of London and in Wales county or county borough councils (unitary authorities).

FC101 Theatre licences

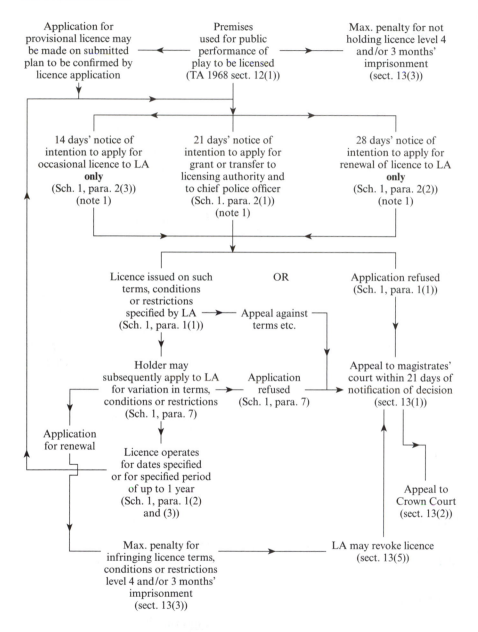

Note

1. Applications to be accompanied by such particulars, notices, etc., which the LA may require in its regulations and by a fee determined by the LA (Sch. 1, para. 2(1) and (3)).

Extent

The Act applies in England, Wales and Scotland (sect. 18(1)).

Applications

Whilst no particular form of application is specified, each authority can make its own regulations detailing what particulars are to be submitted or notices, etc. given in relation to application (Sch. 1, para. 2(1)).

Consultation

The procedure requires notice to be given by the applicant to the chief police officer for other than renewals and occasional licences. The regulations to be made by each LA could widen this and could include the chief fire officer for example. There is no statutory requirement for the LA to consider representations from the chief police officer as with cinema licences but they would no doubt wish to do so.

Fees

Each LA may set its own level of fee so long as it is reasonable (see Home Office Circular 13/2000) but no fee is required for plays which are of an educational or like character or are to be performed for a charitable or like purpose (Sch. 1, para. 3).

Conditions etc.

Licences may be granted subject to such terms, conditions or restrictions as may be specified by the LA (see Home Office Circular 13/2000) (Sch. 1, para. 1(1)).

Provisional licences

Applications, accompanied by plans etc., may be made for a provisional licence for premises which are to be provided, altered or extended. Such licences are of no effect until subsequently confirmed by a full licence but this must be granted if the premises have been constructed in accordance with the plans submitted for a provisional licence and the licence is to be held by a fit and proper person (Sch. 1, para. 6).

Power of entry

Authorized officers of the LA may enter licensed premises at any reasonable time to check compliance with licence conditions etc. Access to premises without licences is only possible with a warrant issued by a JP. The maximum penalty for obstruction is level 1 on the standard scale (sect. 15).

Definitions

Play means:

(a) any dramatic piece, whether involving improvisation or not, which is given wholly or in part by one or more persons actually present and performing and in which the whole or a major proportion of what is done by the person or persons performing, whether by way of speech, singing or action, involves the playing of a role; and

(b) any ballet given wholly or partly by one or more persons actually present and performing, whether or not it falls within paragraph (a) of this definition.

Premises includes any place.

Public performance includes any performance in a public place within the meaning of the Public Order Act 1936, and any performance which the public or any section thereof are permitted to attend, whether on payment or otherwise.

LATE NIGHT REFRESHMENT HOUSES

Reference

Late Night Refreshment Houses Act 1969 (as amended by Local Government Act 1974 and Local Government (Miscellaneous Provisions) Act 1982).

Scope

The procedure requires any person keeping a late night refreshment house to hold an annual licence (sect. 2(1)).

A late night refreshment house is a house, room, shop or building kept open for public refreshment, resort and entertainment at any time between 10 pm and 5 am on the following morning other than exempt licensed premises.

An exempt licensed premises is a house, room, shop or building which is:

(a) licensed for the sale of beer, cider, wine or spirits; and

(b) is not kept open for public refreshment and entertainment between normal evening closure (30 minutes after the evening closing hour permitted by Licensing Act 1964 sect. 60 or 10 pm where there are no permitted hours in the evening) and 5 am the following morning.

This exemption for licensed premises does not, however, apply in Greater London (sect. 1).

Licences

A licence must be issued unless the applicant has been disqualified by a court. For licences granted between 31 March and 1 May, the operative

FC102 Late night refreshment houses

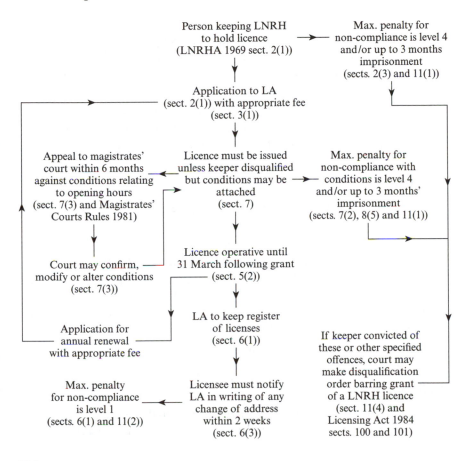

Notes

1. The LA does not have power to refuse a licence unless the applicant is disqualified. Control lies in the conditions which may be applied.
2. This procedure does not apply to Scotland or Northern Ireland (sect. 15(3)).
3. For closing orders for take-aways not subject to licensing under this procedure, see FC114a and b.

date is 1 April and, for licences granted at other times, the date of grant. All licences expire on 31 March of the year following grant and are then subject to annual renewal (sect. 5).

Fees

The LA may charge a fee of such an amount as appears to be appropriate (sect. 3). Where the first licence is granted on or after 1 July and a licence has not been held for the previous 2 years, the fee for the period up to 31 March may be reduced:

(a) by ³/₄ if licence taken out July, August or September;
(b) by ½ if licence taken out October, November, December; or
(c) by ¼ if licence taken out January, February or March (sect. 4).

Conditions

At the discretion of the LA, conditions may be attached to the licence to cover the following:

(a) in order to avoid unreasonable disturbances to residents of any neighbourhood, prohibiting the opening of the LNRH at any time between 11 pm and 5 am (the applicant may appeal against this condition to a magistrates' court (sect. 7); and
(b) requiring the display of a tariff of charges between 10 pm and 5 am so as to be visible before entering the premises (sect. 8).

There is no appeal against this condition.

Death of a licensee

In this situation, the LA may authorize the personal representative, widow or child to operate the LNRH until the 31 March following (sect. 5(3)).

Register

The LA is required to keep a register of all licences issued showing the home address of each licensee and the name and description of the licensed premises. If required by the Clerk of the Justices, a copy of the register or an extract from it must be sent to that office (sect. 6).

Disorderly conduct etc.

A licensee knowingly permitting unlawful gaming or permitting prostitutes, thieves or drunken or disorderly persons to assemble at the premises is guilty of an offence. Persons who are drunk, quarrelsome or disorderly in these premises and who refuse to leave when requested by the occupier or manager are also guilty of an offence (sect. 9).

Disqualification

Persons convicted of offences under this procedure relating to not having a licence, breach of licence conditions, allowing disorderly conduct, etc., or obstructing a constable, or convicted of offences of selling intoxicating liquor without a licence or relating to the supply of liquor at parties, may be disqualified by the court on conviction from holding a LNRH licence (sect. 11).

Power of entry

Officers of the LA are not given any powers of entry or inspection. Police constables, however, may enter **licensed** premises at any time (sect. 10).

Licensing authority

The authorities responsible for issuing licences under this procedure are London borough councils and the Common Council of the City of London, in England unitary authorities and remaining two-tier districts and in Wales, unitary authorities (sect. 2(1)).

OTHER LICENCES

SKIN PIERCING

References

Local Government (Miscellaneous Provisions) Act 1982 sects. 13–17.
Body Art – Skin piercing, cosmetic therapies and other special treatments. CIEH Good Practice Guidelines 2001.

Adoption

These provisions do not operate unless they are adopted by a LA in respect of its own area by the procedure shown in FC103. The adoption may relate to all or any of the skin piercing activities mentioned below and different dates of operation may be applied to different activities (sect. 13).

Scope

The skin piercing activities which may be controlled by registration and subsequent application of bye-laws are:

(a) acupuncture,
(b) tattooing,

FC103 Skin piercing

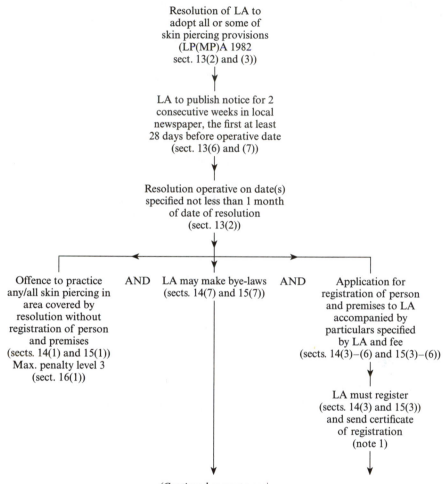

Resolution of LA to
adopt all or some of
skin piercing provisions
(LP(MP)A 1982
sect. 13(2) and (3))

LA to publish notice for 2
consecutive weeks in local
newspaper, the first at least
28 days before operative date
(sect. 13(6) and (7))

Resolution operative on date(s)
specified not less than 1 month
of date of resolution
(sect. 13(2))

Offence to practice
any/all skin piercing in
area covered by
resolution without
registration of person
and premises
(sects. 14(1) and 15(1))
Max. penalty level 3
(sect. 16(1))

AND

LA may make bye-laws
(sects. 14(7) and 15(7))

AND

Application for
registration of person
and premises to LA
accompanied by
particulars specified
by LA and fee
(sects. 14(3)–(6) and 15(3)–(6))

LA must register
(sects. 14(3) and 15(3))
and send certificate
of registration
(note 1)

(Continued on next page)

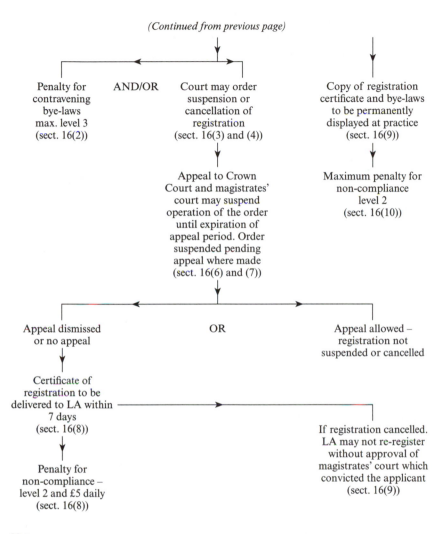

(Continued from previous page)

| Penalty for contravening bye-laws max. level 3 (sect. 16(2)) | AND/OR | Court may order suspension or cancellation of registration (sect. 16(3) and (4)) | Copy of registration certificate and bye-laws to be permanently displayed at practice (sect. 16(9)) |

Appeal to Crown Court and magistrates' court may suspend operation of the order until expiration of appeal period. Order suspended pending appeal where made (sect. 16(6) and (7))

Maximum penalty for non-compliance level 2 (sect. 16(10))

Appeal dismissed or no appeal

OR

Appeal allowed – registration not suspended or cancelled

Certificate of registration to be delivered to LA within 7 days (sect. 16(8))

Penalty for non-compliance – level 2 and £5 daily (sect. 16(8))

If registration cancelled. LA may not re-register without approval of magistrates' court which convicted the applicant (sect. 16(9))

Note

1. Where a previous registration has been cancelled by that LA, future registrations require the consent of the magistrates' court which convicted the applicant (sect. 16(8)(b)).

(c) ear piercing,
(d) electrolysis.

Businesses carried on under the supervision of a registered medical practitioner do not require registration under these provisions (sects. 14(8) and 15(8)) and they do not apply to skin piercing on animals (sect. 16(12)).

Extent

The provisions apply only in England and Wales.

In London powers also exist through the London Government Act 1991 to license body piercing of any form and not just the restricted provisions of the procedure dealt with here.

Registration

Both the practitioner and the premises at which the skin piercing is to be carried out are required to be registered (sects. 14(1) and (2) and 15(1) and (2)). Where a registered practitioner sometimes visits people to give treatment, the premises where the treatment takes place is not required to be registered (sects. 14(2) and 15(2)).

Applications are to be accompanied by such particulars as the LA may reasonably require and these include details of:

(a) the premises where the applicant desires to practise; and
(b) any convictions for non-compliance with skin piercing bye-laws;

but the LA cannot require information about people to whom treatment services have been given (sects. 14(5) and 15(5)).

Registration is required to be effected by the LA unless the applicant has had a previous registration cancelled by a court, in which case the consent of that court is required (sects. 14(3), 15(3) and 16(8)(b)).

Fees

The LA may charge reasonable fees for registration at their discretion (sects. 14(6) and 15(6)).

Bye-laws

The LA operating these procedures may (but does not need to) make bye-laws which may cover:

(a) cleanliness of premises and fittings;
(b) cleanliness of persons;
(c) cleansing and sterilization of instruments, materials and equipment (sects. 14(7) and 15(7)).

Power of entry

This is only available through a warrant by a JP and may be granted if the JP is satisfied that there is reasonable ground for entry on suspicion that an offence is being committed and that:

(a) admission has been refused; or
(b) admission is apprehended; or
(c) the case is one of urgency; or
(d) application for admission would defeat the object of entry.

Unless the situation falls under (c) or (d) above, notice of intention to apply for the warrant must be given to the occupier.

The warrant is operative for 7 days or until entry is secured, whichever is the shorter period. There is no mention in these provisions of entry by force and authorized officers effecting entry may be required to show their authority.

The maximum penalty for refusing to permit entry to an authorized officer acting under warrant is level 3 (sect. 17).

Local authority

The procedure applies only in England and Wales and the authorities which may adopt and operate it are:

(a) London boroughs;
(b) the Common Council of the City of London;
(c) in England outside of London, unitary authorities or in continuing two-tier areas, district councils; and
(d) in Wales, unitary authorities (sect. 13(11) as amended 1994).

STREET TRADING

Reference

Local Government (Miscellaneous Provisions) Act 1982 sect. 3 and Sch. 4.

Adoption

The procedures only operate through a specific resolution of adoption by the LA (sect. 3). There are no requirements about the advertisement etc. of any intention to adopt these provisions.

Extent

This procedure applies in England and Wales but does not apply in Greater London (for which see the London Local Authorities Act 1990) or in Scotland and Northern Ireland.

FC104a Street trading: designation of

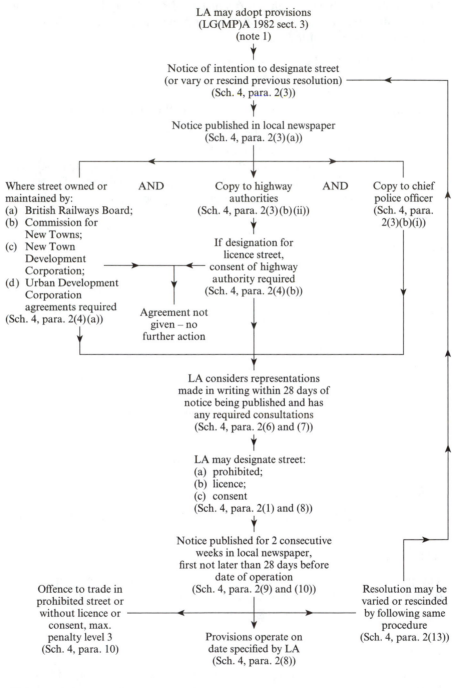

LA may adopt provisions
(LG(MP)A 1982 sect. 3)
(note 1)

Notice of intention to designate street
(or vary or rescind previous resolution)
(Sch. 4, para. 2(3))

Notice published in local newspaper
(Sch. 4, para. 2(3)(a))

Where street owned or maintained by:
(a) British Railways Board;
(b) Commission for New Towns;
(c) New Town Development Corporation;
(d) Urban Development Corporation
agreements required
(Sch. 4, para. 2(4)(a))

AND

Copy to highway authorities
(Sch. 4, para. 2(3)(b)(ii))

AND

Copy to chief police officer
(Sch. 4, para. 2(3)(b)(i))

If designation for licence street, consent of highway authority required
(Sch. 4, para. 2(4)(b))

Agreement not given – no further action

LA considers representations made in writing within 28 days of notice being published and has any required consultations
(Sch. 4, para. 2(6) and (7))

LA may designate street:
(a) prohibited;
(b) licence;
(c) consent
(Sch. 4, para. 2(1) and (8))

Notice published for 2 consecutive weeks in local newspaper, first not later than 28 days before date of operation
(Sch. 4, para. 2(9) and (10))

Offence to trade in prohibited street or without licence or consent, max. penalty level 3
(Sch. 4, para. 10)

Resolution may be varied or rescinded by following same procedure
(Sch. 4, para. 2(13))

Provisions operate on date specified by LA
(Sch. 4, para. 2(8))

Note

1. No advertisement, publicity, etc. required at this stage.

FC104b Street trading: licences

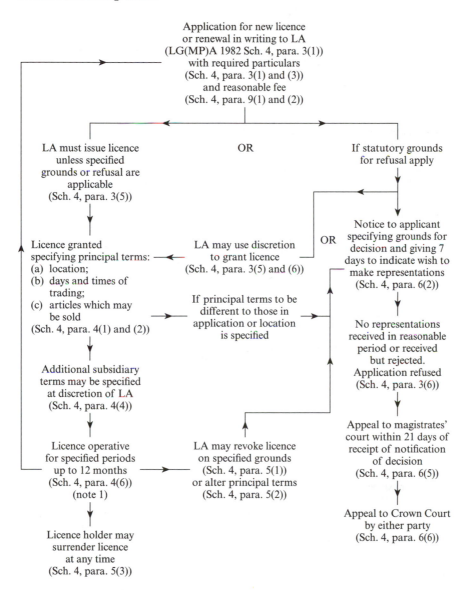

Application for new licence
or renewal in writing to LA
(LG(MP)A 1982 Sch. 4, para. 3(1))
with required particulars
(Sch. 4, para. 3(1) and (3))
and reasonable fee
(Sch. 4, para. 9(1) and (2))

LA must issue licence
unless specified
grounds or refusal are
applicable
(Sch. 4, para. 3(5))

OR

If statutory grounds
for refusal apply

Licence granted
specifying principal terms:
(a) location;
(b) days and times of
trading;
(c) articles which may
be sold
(Sch. 4, para. 4(1) and (2))

LA may use discretion
to grant licence
(Sch. 4, para. 3(5) and (6))

OR

Notice to applicant
specifying grounds for
decision and giving 7
days to indicate wish to
make representations
(Sch. 4, para. 6(2))

Additional subsidiary
terms may be specified
at discretion of LA
(Sch. 4, para. 4(4))

If principal terms to be
different to those in
application or location
is specified

No representations
received in reasonable
period or received
but rejected.
Application refused
(Sch. 4, para. 3(6))

Licence operative
for specified periods
up to 12 months
(Sch. 4, para. 4(6))
(note 1)

LA may revoke licence
on specified grounds
(Sch. 4, para. 5(1))
or alter principal terms
(Sch. 4, para. 5(2))

Appeal to magistrates'
court within 21 days of
receipt of notification
of decision
(Sch. 4, para. 6(5))

Licence holder may
surrender licence
at any time
(Sch. 4, para. 5(3))

Appeal to Crown Court
by either party
(Sch. 4, para. 6(6))

Note

1. The maximum penalty for breaching the principal terms of the licence is level 3. There is no offence committed in breaching the subsidiary terms but this could be taken into account in any consideration of renewal or revocation (Sch. 4, para. 10(4)).

FC104c Street trading: consents

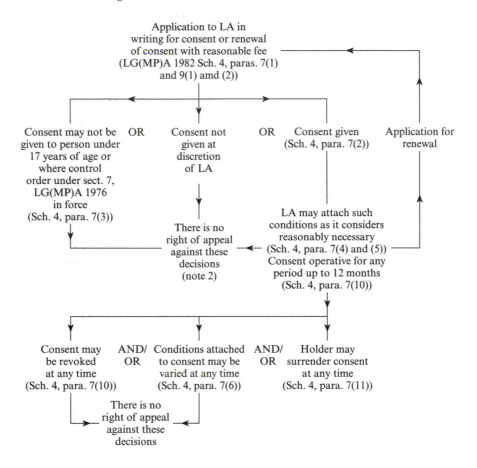

Application to LA in writing for consent or renewal of consent with reasonable fee (LG(MP)A 1982 Sch. 4, paras. 7(1) and 9(1) amd (2))

Consent may not be given to person under 17 years of age or where control order under sect. 7, LG(MP)A 1976 in force (Sch. 4, para. 7(3))

OR

Consent not given at discretion of LA

OR

Consent given (Sch. 4, para. 7(2))

Application for renewal

There is no right of appeal against these decisions (note 2)

LA may attach such conditions as it considers reasonably necessary (Sch. 4, para. 7(4) and (5)) Consent operative for any period up to 12 months (Sch. 4, para. 7(10))

Consent may be revoked at any time (Sch. 4, para. 7(10))

AND/ OR

Conditions attached to consent may be varied at any time (Sch. 4, para. 7(6))

AND/ OR

Holder may surrender consent at any time (Sch. 4, para. 7(11))

There is no right of appeal against these decisions

Notes

1. The maximum penalty for contravening a condition of consent is level 3 (Sch. 4, para. 10(4)).

2. Although no appeal procedure exists, in the case of *R* v *Bristol City Council, ex parte Pearce and Another* (1984) the judge commented that the LA should tell the applicant of the content of their objectives and give him an opportunity to comment.

Local authority

These procedures are operated by unitary and two-tier district councils in England, unitary authorities in Wales and in the Isles of Scilly.

Designation

Following adoption, the LA may designate streets or parts of streets to be either:

(a) prohibited streets in which street trading is not allowed;
(b) licence streets in which licences to trade are required; and
(c) consent streets where the consent requirements operate.

These designations may be rescinded or changed from one type to another at any time using the full procedure (Sch. 4, para. 2).

Notice of intention to designate is required with copies to the chief police officer and to the highways authority. If the designation is to be for licence streets, the consent of the highways authority is required.

The notice must contain a draft of the resolution and say that written representations may be made during a specified period not less than 28 days from publication of the notice (Sch. 4, para. 2(6)).

Scope

Street trading is defined as being the selling or exposing or offering for sale of any article (or living thing) in a street but the following are not considered to be street trading:

(a) trading by a person acting as a pedlar under the authority of a pedlar's certificate granted under the Pedlars Act 1871;
(b) anything done in a market or fair the right to hold which was acquired by virtue of a grant (including a presumed grant) or acquired or established by virtue of an enactment or order;
(c) trading in a trunk road picnic area provided by the SoS under sect. 112 of the Highways Act 1980;
(d) trading as a news vendor;
(e) trading which:
  (i) is carried on at premises used as a petrol filling station; or
  (ii) is carried on at premises used as a shop or in a street adjoining premises so used and as part of the business of the shop;
(f) selling things, or offering, or exposing them for sale, as a roundsman;
(g) the use for trading under Part VIIA of the Highways Act 1980 of an object or structure placed on, in or over a highway;
(h) the operation of facilities for recreation or refreshment under Part VIIA of the Highways Act 1980;
(i) the doing of anything authorized by regulations made under sect. 5 of the Police, Factories, etc. (Miscellaneous Provisions) Act 1916.

The reference to trading as a news vendor in (d) above is a reference to trading where:

(a) the only articles sold or exposed or offered for sale are newspapers or periodicals; and

(b) they are sold or exposed or offered for sale without a stall or receptacle for them or with a stall or receptacle for them which does not:
    (i) exceed 1 m in length or width or 2 m in height;
    (ii) occupy a ground area exceeding 0.25 m²; or
    (iii) stand on the carriageway of a street (Sch. 4, para. 1(2) and (3)).

Licences

Applications must be in writing and must give:

(a) full name and address of applicant;

(b) street, day and times of proposed trading;

(c) description of articles, stalls or containers;

(d) any other particulars required by the Council which can include two photographs of the applicant (Sch. 4, para. 3(2) and (3)).

Unless one or more of the following grounds of refusal are applicable, the LA must grant the licence and may even grant it if the grounds of refusal are available:

(a) that there is not enough space in the street for the applicant to engage in the trading in which he desires to engage without causing undue interference or inconvenience to persons using the street;

(b) that there are already enough traders trading in the street from shops or otherwise in the goods in which the applicant desires to trade;

(c) that the applicant desires to trade on fewer days than the minimum number specified in a resolution under para. 2(11);

(d) that the applicant is unsuitable to hold the licence by reason of having been convicted of an offence or for any other reason;

(e) that the applicant has at any time been granted a street trading licence by the council and has persistently refused or neglected to pay fees due to them for it or charges due to them under para. 9(6) of Sch. 4, LG(MP)A 1982, for services rendered by them to him in his capacity as licence-holder;

(f) that the applicant has at any time been granted a street trading consent by the council and has persistently refused or neglected to pay fees due to them for it;

(g) that the applicant has without reasonable excuse failed to avail himself to a reasonable extent of a previous street trading licence (para. 3(6)).

Also licences must be refused if the applicant is under 17 years of age or the location is covered by a control order (road-side sales) under sect. 7 LG(MP)A 1976.

The licence issued must state:

(a) the street in which and days and times between which the holder is able to trade; and
(b) the description of articles in which he may trade and may also state a particular location for trading (para. 4(1)).

These are known as the principal terms of the licence and a breach may result in prosecution (Sch. 4, para. 4(1)–(3)).

In addition, subsidiary terms may be applied at the discretion of the LA and these may include:

(a) specifying the size and type of any stall or container which the licence-holder may use for trading;
(b) requiring that any stall or container so used shall carry the name of the licence-holder or the number of his licence or both; and
(c) prohibiting the leaving of refuse by the licence-holder or restricting the amount of refuse which he may leave or the places in which he may leave it (Sch. 4, para. 5).

Licences may be revoked if:

(a) owing to circumstances which have arisen since the grant or renewal of the licence, there is not enough space in the street for the licence-holder to engage in the trading permitted by the licence without causing undue interference or inconvenience to persons using the street;
(b) the licence-holder is unsuitable to hold the licence by reason of having been convicted of an offence or for any other reason;
(c) since the grant or renewal of the licence, the licence-holder has persistently refused or neglected to pay fees due to the council for it or charges due to them under para. 9(6), Sch. 4, LG(MP)A 1982, for services rendered by them to him in his capacity as licence-holder; or
(d) since the grant or renewal of the licence, the licence-holder has without reasonable excuse failed to avail himself of the licence to a reasonable extent (para. 5(1));

and the LA may also vary the principal terms by altering the days or times of trading or restricting the type of goods sold, subject to notice and appeal (Sch. 4, para. 5(2)).

The licence street provisions are most appropriate for the formalized street market situations and imply a positive will to promote trading on behalf of the LA to the extent that refusal powers are limited.

Consents

There are no specified particulars for consent applications although they must be in written form (Sch. 4, para. 7(1)). Unless the applicant is under 17 years of age or the location is covered by a control order under sect. 7 of the

LG(MP)A 1976 in which situations refusal is mandatory, consents are entirely at the discretion of the LA as it sees fit. If consent is given the LA may attach conditions at its discretion including those to prevent obstruction and nuisance or annoyance.

Specific consent to trade from a vehicle or portable stall is required.

Consent procedures are most applicable where trading is to be itinerant or infrequent and there is no appeal against refusals or conditions to be applied. There is no requirement in these provisions to hear applicants but in the case of *R* v *Bristol City Council, ex parte Pearce and Another* (1984), it was indicated that the LA should tell applicants of the contents of any objections and give them an opportunity to comment (Sch. 4, para. 7).

Fees

The LA may charge such fees as they consider reasonable for both licences and consents with different fee levels being possible for different types of licence/consents, periods, streets and articles. Fees may be paid in instalments and the initial application need only be accompanied by a deposit at the discretion of the LA.

Fees are returnable on surrender, revocation and refusal.

Separate and additional charges may be levied on licence holders for the collection of refuse, street sweeping and any other services rendered by the LA (Sch. 4, para. 9).

Definition

Street includes:

(a) any road, footway, beach or other area to which the public have access without payment; and
(b) a service area as defined in sect. 329 of the Highways Act 1980;

and also includes any part of a street (Sch. 4, para. 1(1)).

LICENSING OF CARAVAN SITES

Reference

Caravan Sites and Control of Development Act 1960 Part I (as amended by Local Government (Miscellaneous Provisions) Act 1982).

Scope

The occupier of any land must not cause or permit the land to be used as a caravan site unless he holds a site licence (sect. 1(1)).

FC105 Licensing of caravan sites

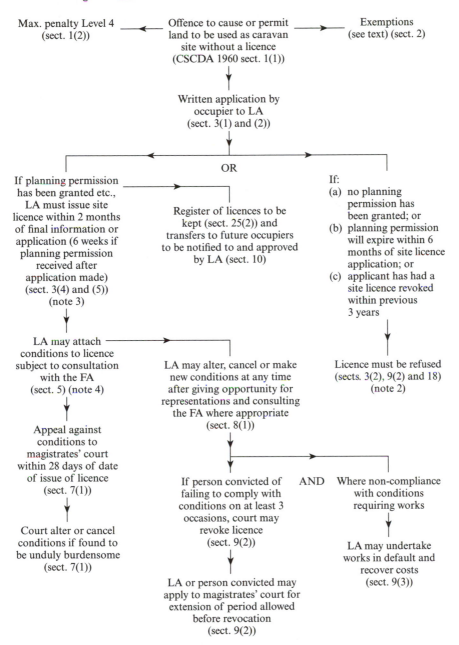

Max. penalty Level 4 ← Offence to cause or permit → Exemptions
(sect. 1(2)) land to be used as caravan (see text) (sect. 2)
 site without a licence
 (CSCDA 1960 sect. 1(1))

Written application by
occupier to LA
(sect. 3(1) and (2))

OR

If planning permission →
has been granted etc., Register of licences to be
LA must issue site kept (sect. 25(2)) and
licence within 2 months transfers to future occupiers
of final information or to be notified to and approved
application (6 weeks if by LA (sect. 10)
planning permission
received after
application made)
(sect. 3(4) and (5))
(note 3)

If:
(a) no planning
 permission has
 been granted; or
(b) planning permission
 will expire within 6
 months of site licence
 application; or
(c) applicant has had a
 site licence revoked
 within previous
 3 years

LA may attach →
conditions to licence LA may alter, cancel or make Licence must be refused
subject to consultation new conditions at any time (sects. 3(2), 9(2) and 18)
with the FA after giving opportunity for (note 2)
(sect. 5) (note 4) representations and consulting
 the FA where appropriate
 (sect. 8(1))

Appeal against
conditions to
magistrates' court
within 28 days of date
of issue of licence
(sect. 7(1))

If person convicted of AND Where non-compliance
failing to comply with with conditions
conditions on at least 3 requiring works
occasions, court may
revoke licence
(sect. 9(2))

Court alter or cancel
conditions if found to
be unduly burdensome
(sect. 7(1))

LA may undertake
works in default and
recover costs
(sect. 9(3))

LA or person convicted may
apply to magistrates' court for
extension of period allowed
before revocation
(sect. 9(2))

Notes
1. For licensing of camping sites (other than for caravans), see FC7.
2. There is no power for the LA to refuse the licence for reasons other than these.
3. The licence is not to be issued for a limited period unless this is the case with the planning consent, in which case the two must be brought together (sect. 4).
4. Maximum penalty for failure to comply with conditions level 4 (sect. 9(1)).

Extent

The procedure applies in England, Wales and with modifications in Scotland. It does not apply in Northern Ireland (sect. 50).

Exemptions

The long list of exemptions from licensing requirements is set out in Sch. 1 of the Act in the following categories:

(a) incidental use within the curtilage of a dwelling-house;
(b) a single caravan for not more than two nights and 28 days in 12 months;
(c) holdings of 5 acres or more if not more than 28 days in 12 months and maximum three caravans at a time;
(d) sites occupied and supervised by organizations exempted by the Minister;
(e) sites approved by exempted organizations for up to five caravans;
(f) meetings organized by exempted organizations;
(g) agriculture and forestry workers;
(h) building and engineering sites;
(i) travelling showmen;
(j) sites occupied by a local authority; and
(k) gipsy sites occupied by county councils or regional councils (sect. 2 and Sch. 1).

Applications

The occupier of the land must apply to the LA in writing and must specify the land in question and, either at the time of making the application or later, give the LA such information as it may reasonably require (sect. 3). No fee is payable.

Issue of licence

The LA cannot issue the licence if:

(a) there has not been a formal grant of planning permission (sect. 3(3); or
(b) the applicant has had a site licence revoked within the previous 3 years (sect. 3(6));

Licences must be issued within 2 months of the giving to the LA of any information which it requires, or such longer period as may be agreed in writing between them (sect. 3(4)). If it does not occur, no offence is committed in relation to not having a site licence (sect. 6).

Licences may not be time limited unless this is the case with the planning permission in which case the licence must expire on the same date (sect. 4).

Conditions

The LA may attach to the licence any conditions which it sees necessary or desirable to impose in the interest of persons living in the caravans, or of other classes of persons, or of the public at large. No conditions may however control the material to be used in the construction of the caravans. In forming conditions, the LA must have regard to model standards issued by the Minister from time to time.

In respect of conditions which require the carrying out of works, the works may be required within a specified period and caravans may be prohibited or restricted until the works have been completed. The LA has power to undertake these works in default at the end of the period allowed and recover costs (sects. 5 and 9(3)).

The LA is required to consult with the FA before granting a site licence or making or varying conditions relating to fire precautions (sect. 5(3A) and (3B)).

Power of entry

An authorized officer of the LA upon:

(a) producing if required a duly authenticated document giving his authority; and
(b) giving 24 hours' notice to the occupier;

may demand entry to any land used as a caravan site or in respect of which a site licence application has been made for the purpose of:

(a) determining the conditions to be attached to licences;
(b) ascertaining if there have been or are any contraventions;
(c) ascertaining if the LA should take any action or carry out any work; or
(d) to take action or undertake any work authorized by the Act.

Where a JP is satisfied that:

(a) admission has been refused or is apprehended; or
(b) the occupier is temporarily absent and the case is one of urgency; or
(c) that an application for admission would defeat the object of entry; and
(d) there is reasonable ground for entering the land;

he may authorize the LA by warrant to enter, if need be by force. The officer exercising the warrant may take with him any other person as may be necessary. The maximum penalty for obstruction is level 1 (sect. 26).

Local authority

In England and Wales the licensing authorities are:

(a) in England:
   (i) London boroughs;
   (ii) the Common Council of the City of London;
   (iii) outside of London, unitary authorities and two-tier districts;
   (iv) the Council of the Isles of Scilly; and
(b) in Wales, the unitary authorities (sect. 29(1) as amended 1994).

Definitions

Caravan means any structure designed or adapted for human habitation which is capable of being moved from one place to another (whether by being towed, or by being transported on a motor vehicle or trailer) and any motor vehicle so designed or adapted but does not include:

(a) any railway rolling stock which is for the time being on rails forming part of a railway system; or
(b) any tent (sect. 29(1));

but a structure designed or adapted for human habitation which:

(a) is composed of not more than two sections separately constructed and designed to be assembled on a site by means of bolts, clamps or other devices; and
(b) is, when assembled, physically capable of being moved by road from one place to another (whether by being towed, or by being transported on a motor vehicle or trailer);

shall not be treated as not being (or as not having been) a caravan within the meaning of Part 1 of the Caravan Sites and Control of Development Act 1960, by reason only that it cannot be so moved on a highway when assembled.

For the purposes of Part 1 of the Caravan Sites and Control of Development Act 1960, the expression 'caravan' shall not include a structure designed or adapted for human habitation which falls within paragraphs (a) and (b) of the foregoing subsection if its dimensions when assembled exceed any of the following limits, namely:

(a) length (exclusive of any drawbar): 60 ft (18 m);
(b) width: 20 ft (6 m);
(c) overall height of living accommodation (measured internally from the floor at the lowest level to the ceiling at the highest level): 10 ft (3 m) (CSA 1968 sect. 13).

Caravan site means land upon which a caravan is stationed for the purposes of human habitation and land which is used in conjunction with land on which a caravan is so stationed (sect. 1(4)).

Occupier means, in relation to any land, the person who, by virtue of an estate or interest therein held by him, is entitled to possession thereof or would be so entitled but for the rights of any other person under any licence granted in respect of the land (sect. 1(3)).

Works. Any reference . . . to the carrying out of works shall include a reference to the planting of trees and shrubs and the carrying out of other operations for preserving or enhancing the amenity of the land (sect. 29(2)).

SCRAP METAL DEALERS

Reference

Scrap Metal Dealers Act 1964.

Scope

Scrap metal dealers (definition on page 449) are required to register with the LA (see definitions page 450) in whose area they are carrying on the business (sect. 1(1)). The business is said to be carried on in the area of a particular LA if either:

(a) he occupies a place as a scrap metal store in that area; or
(b) he resides in that area **but** does not occupy a scrap metal store in that area or elsewhere; or
(c) a place is occupied by him wholly or partly for his scrap metal business in that area but he does not occupy a scrap metal store in that area or elsewhere (sect. 1(2)).

Extent

The provisions apply in England and Wales but not to either Northern Ireland or to Scotland (sect. 11(3)).

'Appropriate particulars'

These are the details which must be included in the application for registration and which form the information contained within the register. The details are:

(a) full name of the dealer;
(b) the address:
  (i) for an individual, the usual place of residence;
  (ii) for a body corporate, its registered or principal office;
(c) the address of each place occupied as a scrap metal store;
(d) if no scrap metal stores are occupied, a statement to that effect;
(e) if a place other than a scrap metal store is occupied by him the address of that place (sect. 1(4)).

No registration fee is payable.

FC106 Scrap metal dealers

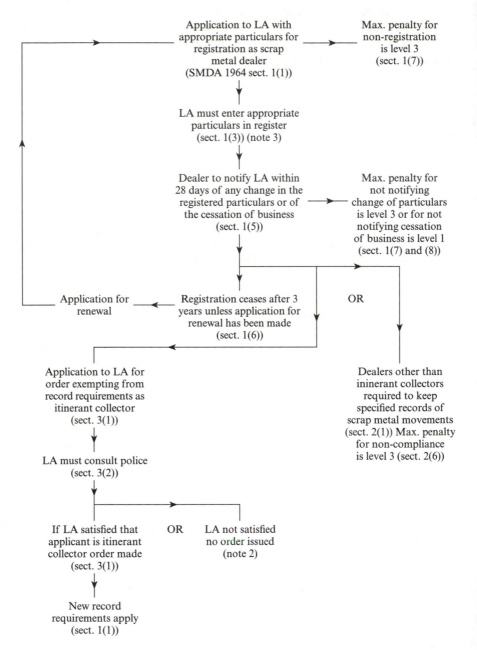

Application to LA with appropriate particulars for registration as scrap metal dealer (SMDA 1964 sect. 1(1))

Max. penalty for non-registration is level 3 (sect. 1(7))

LA must enter appropriate particulars in register (sect. 1(3)) (note 3)

Dealer to notify LA within 28 days of any change in the registered particulars or of the cessation of business (sect. 1(5))

Max. penalty for not notifying change of particulars is level 3 or for not notifying cessation of business is level 1 (sect. 1(7) and (8))

Application for renewal

Registration ceases after 3 years unless application for renewal has been made (sect. 1(6))

OR

Application to LA for order exempting from record requirements as itinerant collector (sect. 3(1))

Dealers other than ininerant collectors required to keep specified records of scrap metal movements (sect. 2(1)) Max. penalty for non-compliance is level 3 (sect. 2(6))

LA must consult police (sect. 3(2))

If LA satisfied that applicant is itinerant collector order made (sect. 3(1))

OR

LA not satisfied no order issued (note 2)

New record requirements apply (sect. 1(1))

Notes
1. If the LA is itself acting as a scrap metal dealer, although this procedure as a whole does not apply, particulars must be kept in the register (sect. 1(10)).
2. There is no appeal against this decision.
3. There is no discretion here provided the appropriate particulars have been submitted.

Records

Each dealer must keep a book containing records of all scrap metal received, processed and dispatched and these are modified in relation to itinerant collectors certified as such by order of the LA (sects. 2 and 3).

A police constable has the power to enter a place at all reasonable times, examine these records and take copies or extracts (sect. 6(1)). These powers are not available to LA officers.

Restrictions by courts

If a person is convicted of not being registered, of not keeping the required records or involves an offence of dishonesty, the court may by order, effective for up to 2 years, impose restrictions on him:

(a) no scrap metal to be received between 6 pm and 8 am; and/or
(b) scrap metal received, unless processed, to be kept for at least 72 hours.

The maximum penalty for non-compliance with an order is level 3 (sect. 4).

Power of entry

In addition to the powers given to a police constable ('Records' above) an authorized officer of a LA may enter any place at any reasonable time if he has reasonable grounds for believing that the place is being used as a scrap metal store and that particulars are not entered in the register. On application, a magistrates' court may issue a warrant to enter, by force if necessary. The maximum penalty for obstruction is level 1 (sect. 6).

LA officers have no right to enter registered places.

Definitions

Scrap metal dealer. For the purposes of this Act a person carries on business as a scrap metal dealer if he carries on a business which consists wholly or partly of buying and selling scrap metal, whether the scrap metal sold is in the form in which it was bought or otherwise, other than a business in the course of which scrap metal is not bought except as materials for the manufacture of other articles and is not sold except as a by-product of such manufacture or as surplus materials bought but not required for such manufacture; and 'scrap metal dealer' (where that expression is used in this Act otherwise than in a reference to carrying on business as a scrap metal dealer) means a person who (in accordance with the preceding provisions of this subsection) carries on business as a scrap metal dealer.

Article includes any part of an article.

Itinerant collector means a person regularly engaged in collecting waste materials, and old, broken, worn out or defaced articles, by means of visits from house to house.

Local authority means the council of a district (unitary authority or remaining two-tier district), the Common Council of the City of London or the council of a London borough but in Wales, the council of a county or county borough (unitary authorities).

Place includes any land, whether consisting of enclosed premises or not.

Scrap metal includes any old metal and any broken, worn out, defaced or partly manufactured articles made wholly or partly of metal, and any metallic wastes, and also includes old, broken, worn out or defaced tooltips or dies made of any of the materials commonly known as hard metal or of cemented or sintered metallic carbides.

Scrap metal store means a place where scrap metal is received or kept in the course of the business of a scrap metal dealer.

Any reference in the preceding provisions of this Act to *metal*, except in the phrases 'hard metal' and 'metallic carbides', shall be taken as a reference to any of the following metals, that is to say, aluminium, copper, iron, lead, magnesium, nickel, tin and zinc, or, subject to the next following subsection, to brass, bronze, gunmetal, steel, white metal or any other alloy of any of the said metals.

For the purpose of this Act, a substance being an alloy referred to in the last preceding subsection shall not be treated as being such an alloy if, of its weight, 2% or more is attributable to gold or silver or any one or more of the following metals, that is to say, platinum, iridium, osmium, palladium, rhodium and ruthenium (sect. 9).

Chapter 14

MISCELLANEOUS

INTRODUCTORY NOTE

There are no general procedural provisions or definitions for this chapter. Each procedure contains the information relevant to the particular legislation concerned but attention is drawn to the general provisions of the Local Government Act 1972 relating to notices etc. in Chapter 1.

REMOVAL OF PERSONS IN NEED OF CARE

References

National Assistance Act 1948 sect. 47.
National Assistance (Amendment) Act 1951.

Scope

The procedure is aimed at securing the necessary care and attention of persons who:

(a) are suffering from grave chronic illness or, being aged, infirm or physically incapacitated, are living in insanitary conditions; **and**

(b) are unable to devote themselves, and are not receiving from other persons, proper care and attention (NA 1948 sect. 47(1)).

Certificate of proper officer

If the officer designated by the LA (definition page 454) as its proper officer for this procedure, after thorough enquiry and consideration of:

(a) the interests of the person; or

(b) the need to prevent injury to the health of, or serious nuisance to, other persons;

is satisfied that the situation as in (a) and (b) in 'Scope' above exists, he certifies such in writing to the LA (NAA 1948 sect. 47(2)).

FC107a Removal of persons in need of care: normal procedure

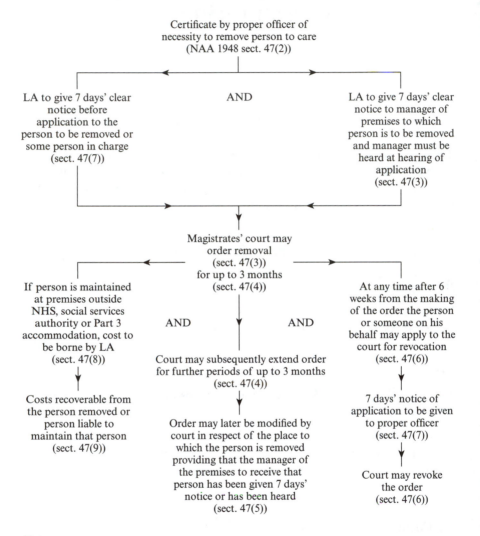

Certificate by proper officer of
necessity to remove person to care
(NAA 1948 sect. 47(2))

LA to give 7 days' clear
notice before
application to the
person to be removed or
some person in charge
(sect. 47(7))

AND

LA to give 7 days' clear
notice to manager of
premises to which
person is to be removed
and manager must be
heard at hearing of
application
(sect. 47(3))

Magistrates' court may
order removal
(sect. 47(3))
for up to 3 months
(sect. 47(4))

If person is maintained
at premises outside
NHS, social services
authority or Part 3
accommodation, cost to
be borne by LA
(sect. 47(8))

AND

AND

At any time after 6
weeks from the making
of the order the person
or someone on his
behalf may apply to the
court for revocation
(sect. 47(6))

Court may subsequently extend order
for further periods of up to 3 months
(sect. 47(4))

Costs recoverable from
the person removed or
person liable to
maintain that person
(sect. 47(9))

7 days' notice of
application to be given
to proper officer
(sect. 47(7))

Order may later be modified by
court in respect of the place to
which the person is removed
providing that the manager of
the premises to receive that
person has been given 7 days'
notice or has been heard
(sect. 47(5))

Court may revoke
the order
(sect. 47(6))

Notes
1. For accelerated procedure, see FC107b.
2. Maximum penalty for wilfully disobeying or obstructing the execution of an order is £50 (sect. 47(11)).
3. All notices may be served by post (sect. 47(14)).

FC107b Removal of persons in need of care: accelerated procedure

Certificate by proper officer and
another registered medical
practitioner of necessity
to remove person to care
without delay
(NAA 1948 sect. 47(2) and
NA(A)A 1951 sect. 1(1))

↓

LA or proper officer
(if authorized) may apply
to magistrates court or single
JP for removal order
(NA(A)A 1951 sect. 1)

↓

Court or JP may order removal
but manager of premises to
which person is to be removed
must be heard
(NAA 1948 sect. 47(3) and
NA(A)A 1951 sect. 1(4))

If person is maintained
at premises outside
NHS, social services
department or Part 3
accommodation, cost to
be borne by LA
(NAA 1948 sect. 47(8))

↓

Order may be made for a
period not exceeding 3 weeks
(NA(A)A 1951 sect. 1(4))

↓

Costs recoverable from
the person removed or
person liable to
maintain that person
(sect. 47(a))

Orders may be extended by
adoption of the normal
procedure in FC107a

Notes
1. For normal procedure, see FC107a.
2. There is no provision to allow application for the revocation of an order made under this accelerated procedure.
3. The maximum penalty for wilfully disobeying or obstructing the execution of a removal order is £50 (sect. 47(11)).

If the proper officer, joined in the certificate by another registered medical practitioner, considers that it is necessary in the interests of the person to remove him without delay the accelerated procedure may be invoked (NA(A)A 1951 sect. 1(1)).

Removal orders

The application for an order is made by the LA (or its proper officer if authorized to do so) to a magistrates' court or, if the accelerated procedure is being used, a single JP. If the court or JP is satisfied that it is expedient to do so they may order the removal of the person by a specified officer of the LA to a suitable hospital or other place and order his detention and maintenance there (NAA 1948 sect. 47(3)).

Period of order

An order may be made for not exceeding 3 months (3 weeks for accelerated procedure) but may be extended by further periods of up to 3 months. If the first order was made by use of the accelerated procedure, subsequent applications need to be made by the normal procedure with appropriate notice of application etc. being made (NAA 1948 sect. 47(4)).

Applications for revocation of order (other than one made under the accelerated procedure) may be made to the magistrates' court at any time after 6 weeks from the making of the order (NAA 1948 sect. 47(6)).

Local authority

The LAs charged with duties under this procedure are unitary authorities, district councils, London borough councils, the Sub-Treasurer of the Inner Temple and the Under-Treasurer of the Middle Temple (NAA 1948 sect. 47(12)).

LA ROLE IN CONTROLS OVER RADIOACTIVE SUBSTANCES

Reference

Radioactive Substances Act 1993 (as amended by the Environment Act 1995).

Scope

The main controls are by way of the registration of users of radioactive material and the authorization of the accumulation and disposal of radioactive waste. These are exercised by the Environment Agency or, in Scotland, the Environmental Protection Agency but, in so doing, there are

FC108a LA role in controls over radioactive substances: registration of users

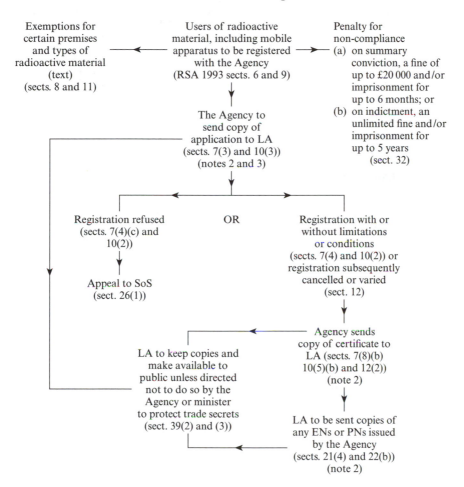

Notes
1. This procedure demonstrates the LA role in the registration process but does not show the full detail of that process.
2. The minister may direct the Agency not to send copies to the LA in the interests of national security (sect. 25).
3. The LA has no specified powers to make representations in respect of the application.

FC108b LA role in controls over radioactive substances: authorizations for the accumulation and disposal of waste

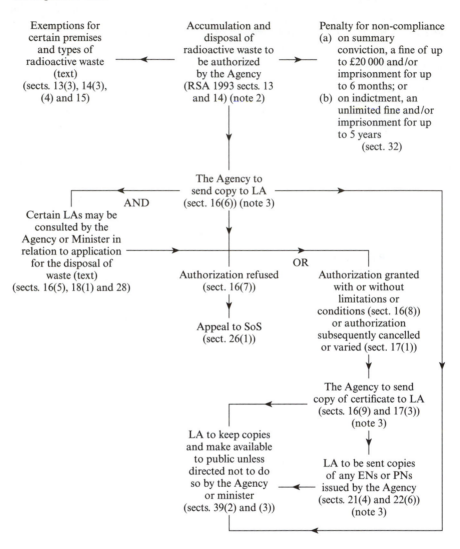

Exemptions for certain premises and types of radioactive waste (text) (sects. 13(3), 14(3), (4) and 15)

Accumulation and disposal of radioactive waste to be authorized by the Agency (RSA 1993 sects. 13 and 14) (note 2)

Penalty for non-compliance
(a) on summary conviction, a fine of up to £20 000 and/or imprisonment for up to 6 months; or
(b) on indictment, an unlimited fine and/or imprisonment for up to 5 years (sect. 32)

The Agency to send copy to LA (sect. 16(6)) (note 3)

AND

Certain LAs may be consulted by the Agency or Minister in relation to application for the disposal of waste (text) (sects. 16(5), 18(1) and 28)

OR

Authorization refused (sect. 16(7))

Appeal to SoS (sect. 26(1))

Authorization granted with or without limitations or conditions (sect. 16(8)) or authorization subsequently cancelled or varied (sect. 17(1))

The Agency to send copy of certificate to LA (sects. 16(9) and 17(3)) (note 3)

LA to keep copies and make available to public unless directed not to do so by the Agency or minister (sects. 39(2) and (3))

LA to be sent copies of any ENs or PNs issued by the Agency (sects. 21(4) and 22(6)) (note 3)

Notes
1. This procedure demonstrates the LA role in the authorization process but does not show the full detail of that process.
2. Authorization for the disposal of radioactive waste from nuclear sites also requires the approval of the appropriate minister (sect. 16(3)).
3. The minister may direct the Agency not to send copies to the LA in the interests of national security (sect. 25).

requirements to involve LAs by way of information and, in some cases, consultation. The procedures here illustrate that LA involvement but, whilst setting out the overall pattern of the control processes, they do not indicate the full content of the processes undertaken by the Agencies.

The licensing of nuclear sites is provided for outside of this procedure by the Nuclear Installations Act 1965 in which the appropriate minister may direct any applicant for a nuclear site licence to copy it to any LA he so specifies. Those LAs are then able to make representations to the minister within 3 months (Nuclear Installations Act 1965 sect. 3).

Extent

These provisions apply to England, Wales, Scotland and to Northern Ireland (sect. 51).

Registration of users of radioactive material

Registration is required for any person keeping or using radioactive material on any premises used for an undertaking carried on by him (sect. 6).

Exemptions from this requirement are:

(a) where the premises is the subject of a nuclear site licence under the Nuclear Installations Act 1965;
(b) premises used for the keeping and use of clocks and watches which are of radioactive material except where there are processes which involve the use of luminous material in which case registration is required (sect. 8(4) and (5));
(c) where the the Agency by order grants exemptions for specified classes of premises and undertakings and descriptions of radioactive material (sect. 8(6)).

Registration is also required for the use of mobile radioactive apparatus used:

(a) for testing, measuring or other investigation of the characteristics of substances or articles; or
(b) which release quantities of radioactive material into the environment or introduce such material to organisms (sect. 9).

The Agency may grant exemption by order for mobile apparatus used by specified classes of persons, or specified description of mobile apparatus (sect. 11).

Authorization of the disposal or accumulation of radioactive waste

Authorization is required for:

(a) the disposal of any radioactive waste from any premises used as an undertaking or for causing or permitting such activity (sect. 13(1));

(b) disposal of radioactive waste arising from any mobile radioactive apparatus (sect. 13(2));
(c) releasing or disposal of radioactive waste (sect. 13(3)); and
(d) the accumulation of radioactive waste with a view to its subsequent disposal (sect. 14(1)).

Exemptions from these requirements are:

(a) accumulation and disposal of radioactive waste on nuclear sites (sects. 13(4) and 14(3));
(b) waste accumulated for less than 3 months (sect. 14(4));
(c) disposal and accumulation of radioactive waste arising from clocks and watches other than where the processes involve the use of luminous material (sect. 15(1); and
(d) exemptions specified by order by the SoS in relation to particular descriptions of radioactive waste (sect. 15(2)).

Information to LAs

The information which is required to be sent to them by the Agency is:

(a) copies of application for registration as users (sect. 7(3)) including those relating to mobile apparatus (sect. 10(3));
(b) copies of applications for authorization to accumulate or dispose of radioactive waste (sect. 16(6)).
(c) copies of certificates of registration and authorization (sects. 7(3), 10(3) and 16(9));
(d) copies of revocation or variation to authorization and regulations (sects. 12(2) and 17(3)); and
(e) copies of any ENs or PNs served by the Agency (sects. 21(4) and 22(6)).

The SoS for the Environment, Food and Rural Affairs has the power to direct the Agency that, on the grounds of national security, knowledge about particular applications should be restricted (sect. 25(1)). In these cases the Agency is prevented from sending copies of the documents in (a)–(e) above to LAs (sect. 25(3)).

Having received the specified documents each LA must make them available to the public (not necessarily in documentary form). The public have the right of inspection at all reasonable times and, on payment of a reasonable fee, may be supplied with copies (sect. 39(2)–(4)).

The Agency may indicate to an LA that specified documents should not be made available to the public in order to protect trade secrets (sect. 39(2) and (3)).

The documents supplied to LAs by the Agency under these provisions are for information purposes only.

Consultation with LAs

The provision to consult with certain LAs applies only in respect of the disposal of radioactive waste and, in these cases, the Agency and the appropriate minister may, when they consider appropriate in the circumstances, give LAs the opportunity of appearing before an inquiry into a proposal to:

(a) refuse an application;
(b) attach limitations or conditions to an authorization;
(c) vary an existing authorization; and
(d) revoke an authorization (sect. 28(1)).

In relation to disposal of waste from nuclear sites or premises situated on nuclear sites, the Agency must consult such LAs as appear proper (sect. 16(5)).

One further LA involvement occurs where the Agency or appropriate minister feels that some special precautions need to be taken by a LA in respect of a particular application for authorization to dispose of radioactive waste. In these cases those LAs must first be consulted before the granting of any authorization (sect. 18(1)). There are similar provisions relating to sewerage and water undertakers (sect. 18(1)).

LAs involved in the provision of special precautions have a power to charge the person to whom the authorization is granted (sect. 18(2)).

Definitions

Radioactive material means anything which, not being waste, is either a substance to which this situation applies or an article made wholly or partly from, or incorporating, such a substance (sect. 1(1)).

These substances are:

(a) radioactive elements specified in Sch. 1 of the Act;
(b) substances possessing radioactivity wholly or partly due to a process of nuclear fisson.

Radioactive waste means waste which consists wholly or partly of:

(a) a substance or article which, if it were not waste, would be radioactive material; or
(b) a substance or article which has been contaminated in the course of the production, keeping or use of radioactive material, or by contact with or proximity to other waste falling within paragraph (a) or this paragraph (sect. 2).

Mobile radioactive apparatus means any apparatus, equipment, appliance or other thing which is radioactive material and:

(a) is constructed or adapted for being transported from place to place; or
(b) is portable and designed or intended to be used for releasing radioactive material into the environment or introducing it into organisms (sect. 3).

Disposal in relation to waste, includes its removal, deposit, destruction, discharge (whether into water or into the air or into a sewer or drain or otherwise) or burial (whether underground or otherwise) and 'dispose of' shall be construed accordingly (sect. 47(1)).

Local authority means:

(a) in England and Wales, the council of a county, district or London borough or the Common Council of the City of London or an authority established by the Waste Regulation and Disposal (Authorities) Order 1985 (this means outside of London, the unitary authorities in Wales and England and in England concurrent involvement of counties and districts in continuing two-tier areas).

(b) in Scotland, a regional, islands or district council (unitary authority); and

(c) in Northern Ireland, a district council (sect. 47(1)).

Nuclear site means:

(a) any site in respect of which a nuclear site licence is for the time being in force; or

(b) any site in respect of which, after the revocation or surrender of a nuclear site licence, the period of responsibility of the licensee has not yet come to an end (sect. 47(1)).

Waste includes any substance which constitutes scrap material or an effluent or other unwanted surplus substance arising from the application of any process, and also includes any substance or article which requires to be disposed of as being broken, worn out, contaminated or otherwise spoilt (sect. 47(1)).

CONTROL OF RATS AND MICE

Reference

Prevention of Damage by Pests Act 1949 sects. 2–7.

Extent

This procedure applies in England, Wales and, with amendments, to Scotland but not in Northern Ireland (sects. 1 and 29).

Duty of LA

Each LA is under a statutory duty to take such steps as may be necessary to secure so far as practicable that its district is free from rats and mice including:

FC109 Control of rats and mice

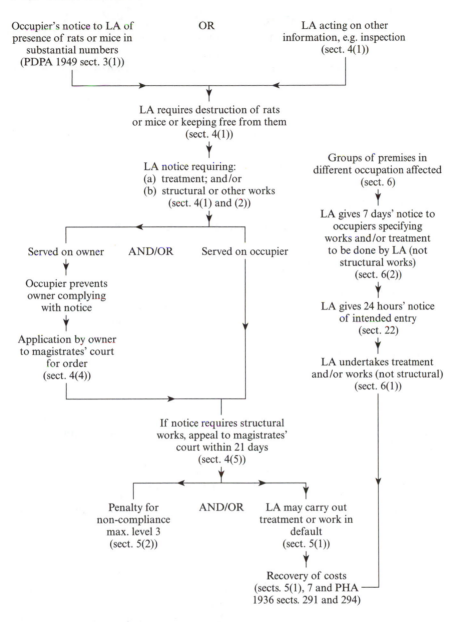

Occupier's notice to LA of presence of rats or mice in substantial numbers (PDPA 1949 sect. 3(1))

OR

LA acting on other information, e.g. inspection (sect. 4(1))

LA requires destruction of rats or mice or keeping free from them (sect. 4(1))

LA notice requiring:
(a) treatment; and/or
(b) structural or other works (sect. 4(1) and (2))

Groups of premises in different occupation affected (sect. 6)

LA gives 7 days' notice to occupiers specifying works and/or treatment to be done by LA (not structural works) (sect. 6(2))

Served on owner — AND/OR — Served on occupier

Occupier prevents owner complying with notice

LA gives 24 hours' notice of intended entry (sect. 22)

Application by owner to magistrates' court for order (sect. 4(4))

LA undertakes treatment and/or works (not structural) (sect. 6(1))

If notice requires structural works, appeal to magistrates' court within 21 days (sect. 4(5))

Penalty for non-compliance max. level 3 (sect. 5(2))

AND/OR

LA may carry out treatment or work in default (sect. 5(1))

Recovery of costs (sects. 5(1), 7 and PHA 1936 sects. 291 and 294)

Notes

1. The following general provisions of the PHA 1936 are applicable to this procedure:
   (a) Authentication and service of notices (sects. 284–6).
   (b) Appeals against notices (only where structural works are required) (sect. 290(3)–(5) inclusive and 300–302 inclusive).
   (c) Recovery of expenses (sects. 291 and 294).
   These provisions are set out on pages 12–15.
2. This procedure applies with certain modifications to non-seagoing ships and to aircraft (sect. 23).

(a) carrying out inspections;
(b) destroying rats and mice on premises which the LA occupies and keeping the premises free from infestation;
(c) enforcing duties on owners and occupiers under PDPA 1949 (sect. 2).

Local authorities

The authorities charged with the implementation of these procedures in England are:

(a) The Common Council of the City of London
(b) Unitary authorities and
(c) District councils.

In Wales and Scotland the authorities are the unitary councils (sect. 1).

Occupier's duty to notify LA

When an occupier of any land becomes aware of an infestation by rats or mice in substantial numbers he must notify the LA in writing but this does not apply to agricultural land, or where food in certain categories of premises is infested when notification is then given to the Minister of the Environment, Food and Rural Affairs or, in Scotland, the Secretary of State (sect. 3).

Notices

Notices, which must be in writing, may be served on the owner or the occupier or, where the owner is not the occupier, on both. The notice may require either or both of the following:

(a) treatment at times which may be prescribed;
(b) the carrying out of structural or other works.

The time allowed for compliance must be reasonable and should be not less than the period allowed for appeal, i.e. 21 days (sect. 4).

Block treatment

If rats or mice are found in substantial numbers in a group of premises in different occupation, the LA may take such steps as it considers necessary (other than carrying out structural works), after giving at least 7 days' notice to each occupier specifying the steps which it intends to take, and may subsequently recover its costs (sect. 6).

Power of entry

Authorized officers of LAs, producing if required evidence of authority, may enter any land after, in the case of occupied land, giving at least 24 hours notice, for the following purposes:

(a) inspection;
(b) compliance with notices or other requirements placed on owners or occupiers;
(c) enforcing notices served under sects. 5 and 6 including the carrying out of treatment of works (sect. 22).

Definitions

Land includes land covered with water and any building or part of a building (sect. 28).

SANITARY CONVENIENCES AT PLACES OF ENTERTAINMENT ETC.

References

Local Government (Miscellaneous Provisions) Act 1976 sects. 20 and 21.
Chronically Sick and Disabled Persons Act 1970 sect. 6.
Disabled Persons Act 1981 sect. 4.

Extent

This procedure does not apply in Scotland or Northern Ireland (sect. 83).

Scope

The premises covered by these provisions are termed 'relevant places' and are premises used, or proposed to be used, occasionally or permanently for the following purposes:

(a) the holding of any entertainment, exhibition or sporting event to which the public is admitted as spectators or otherwise;
(b) the sale of food and drink to the public for consumption on the premises;
(c) a betting office (sect. 20(9)).

In respect of (a), a licence for the holding of entertainment may also be required – see FC99a and b.

Notices

Notices must be in writing on either the owner or the occupier of the premises and can be served in anticipation of the use of a premises as a 'relevant place'. Notices may require:

(a) the provision of a specified number and type of sanitary appliances at specified positions within a defined period, which should be not less than 6 weeks;

FC110 Sanitary conveniences at places of entertainment etc.

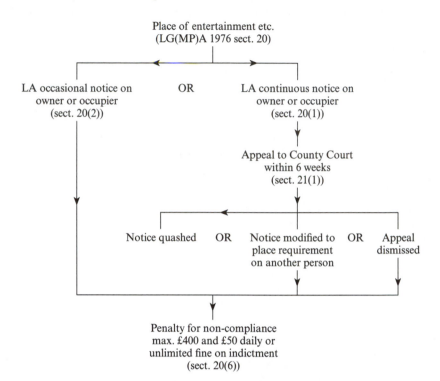

Place of entertainment etc.
(LG(MP)A 1976 sect. 20)

LA occasional notice on owner or occupier (sect. 20(2))

OR

LA continuous notice on owner or occupier (sect. 20(1))

Appeal to County Court within 6 weeks (sect. 21(1))

Notice quashed

OR

Notice modified to place requirement on another person

OR

Appeal dismissed

Penalty for non-compliance max. £400 and £50 daily or unlimited fine on indictment (sect. 20(6))

Notes

1. Notices under this section may be served by a district council, London borough council, the Common Council and the Council of the Isles of Scilly (LG(MP)A 1976 sects. 20(1) and 44(1)).
2. There is no provision for appeal against 'occasional' notices.
3. The notice must be in writing (LG(MP)A 1976 sect. 33(1)).
4. For the licensing of public entertainment see FC99a and b.

(b) the maintenance and cleaning of the appliances;
(c) the provision and maintenance of washing facilities including hot and cold water and other facilities for use in connection with the sanitary appliances, e.g. hand drying facilities etc.;
(d) that the appliance and associated facilities should be made available to the public resorting to the premises, if required by the notice, free of charge.

The notice may require that existing sanitary appliances must be made available but cannot require either moveable sanitary appliances at a betting office or, in the case of new buildings, provisions in excess of requirements of the building regulations (sect. 20(1), (2) and (3)).

Occasional notices

As an alternative to permanent provision, the notice may require provision of the specified appliances and associated facilities only on occasions which are specified in the notice (sect. 20(2)).

Disabled persons

A person upon whom a sect. 20 notice is served must make provision for the needs of the disabled so far as is practicable and reasonable (CS and DPA 1970 sect. 6). The notice served under sect. 20 must draw attention to the requirements of the CS and DPA 1970 and to the Code of Practice for Access for the Disabled to Buildings (DPA 1981 sect. 4).

Appeals

So far as notices requiring permanent provision of facilities are concerned, appeal may be made within 6 weeks of service to the county court on the following grounds:

(a) a requirement is unreasonable;
(b) it would have been fairer to have served the notice on the owner or occupier as the case may be (sect. 21).

There is no appeal against an occasional notice but in any subsequent proceedings it will be a defence to prove situations as in (a) and (b) above.

Defences

In proceedings for non-compliance with notices served under sect. 20 it will be a defence to prove:

(a) that at the time of the failure the person on whom the notice was served was neither the owner nor occupier and that he did not cease to be the owner or occupier with a view to avoiding compliance with the notice; or

(b) where the contravention relates to a particular day, that the relevant place was closed to members of the public or not used as a relevant place; or

(c) in respect of occasional notices only that:
    (i) the requirement was unreasonable; or
    (ii) that it would have been fairer to have served the notice on another person (sect. 20(7) and (8)).

Power of entry

Authorized officers producing evidence of such may enter a 'relevant place' at any reasonable time to determine if a notice should be served or is being complied with. Penalty for obstruction is max. level 3 (sect. 20(5)).

Local authority

The LAs who may serve notices under sect. 20 are:

(a) unitary authorities and district councils;
(b) London borough councils;
(c) the Common Council of the City of London; and
(d) the Council of the Isles of Scilly (sects. 20(1) and 44(1)).

Definitions

Betting office means a place for which a betting office licence within the meaning of the Betting, Gaming and Lotteries Act 1963 is in force (LG(MP)A 1976 sect. 20(9)).

Owner in relation to any land, place or premises, means a person who, either on his own account or as agent or trustee for another person, is receiving the rackrent of the land, place or premises or would be entitled to receive it if the land, place or premises were let at a rackrent and 'owned' shall be construed accordingly (LG(MP)A 1976 sect. 44(1)).

Sanitary appliances means water closets, other closets, urinals and wash basins (LG(MP)A 1976 sect. 20(9)).

PROTECTION OF BUILDINGS

Reference

Local Government (Miscellaneous Provisions) Act 1982 sects. 29–32 inclusive.

FC111 Protection of buildings

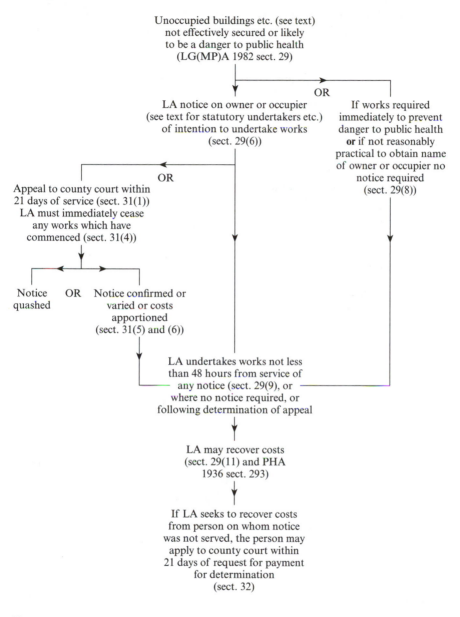

Unoccupied buildings etc. (see text)
not effectively secured or likely
to be a danger to public health
(LG(MP)A 1982 sect. 29)

OR

LA notice on owner or occupier
(see text for statutory undertakers etc.)
of intention to undertake works
(sect. 29(6))

If works required
immediately to prevent
danger to public health
or if not reasonably
practical to obtain name
of owner or occupier no
notice required
(sect. 29(8))

OR

Appeal to county court within
21 days of service (sect. 31(1))
LA must immediately cease
any works which have
commenced (sect. 31(4))

Notice   OR   Notice confirmed or
quashed   varied or costs
apportioned
(sect. 31(5) and (6))

LA undertakes works not less
than 48 hours from service of
any notice (sect. 29(9), or
where no notice required, or
following determination of appeal

LA may recover costs
(sect. 29(11) and PHA
1936 sect. 293)

If LA seeks to recover costs
from person on whom notice
was not served, the person may
apply to county court within
21 days of request for payment
for determination
(sect. 32)

Notes
1. No offences are committed under this procedure.
2. For operational land of British Railways Board or any statutory undertaker, in carrying out works the LA must comply with any reasonable request which British Railways Board or the statutory undertakers may impose for the protection or safety of their undertaking (sect. 30).
3. The procedure can be applied to any building in the specified condition including ones subject to orders under the HAs.

Extent

This procedure does not apply in Scotland or Northern Ireland.

Scope

This procedure applies to any building:

  (a) which is unoccupied; or
  (b) where the occupier is temporarily absent;

and which is:

  (a) not effectively secured against unauthorized entry; **or**
  (b) likely to become a danger to public health (sect. 29(1)).

Works

The LA may undertake any works necessary to prevent unauthorized entry or to prevent the building becoming a danger to public health. If either:

  (a) works are immediately necessary to prevent it becoming a danger to public health; or
  (b) the name and address of the owner, or the whereabouts of the occupier are not reasonably practical to obtain;

a notice of intention is not required. In all other cases at least 48 hours' written notice must be given (sect. 29(6) and (8)).

LA notices

  (a) **Person.** The notice is to be served on either the owner or the occupier (sect. 29(6)) or on British Railways Board or any statutory undertaker in respect of their operational land (sect. 30(2)).
  (b) **Content.** The notice must specify the works which the LA intends to carry out (sects. 29(7) and 30(3)).
  (c) **Service.** This procedure is not incorporated in the provisions of either the PHAs or the HAs and therefore service should be by one of the methods specified in sect. 233 of the LGA 1972 (page 8).

Appeals

A person in receipt of the LA notice of intention may appeal to the County Court within 21 days of service on any of the following grounds:

  (a) the works specified are not authorized by the sect. 29 procedure;
  (b) the works are unnecessary;
  (c) it was unreasonable for the LA to undertake the works (sect. 31(1) and (2)).

In the event of an appeal the LA must stop any works which it has commenced and await the determination of the court. If the works had been completed, the court may still determine the position regarding recovering of costs (sect. 31(4)).

The court may confirm, vary or quash the notice and make provision in its order relating to recovery of costs by the LA (sect. 31(5) and (6)).

Persons not in receipt of a LA notice but from whom costs are being recovered may appeal to the County Court within 21 days of the request for payment and ask the court for a declaration that either:

(a) the works were unnecessary; or
(b) it was unreasonable for the LA to undertake them (sect. 32).

Power of entry

A person authorized by the LA may enter:

(a) the building on which works are to be undertaken;
(b) any land appurtenant to that building;
(c) any other land if it appears to be unoccupied and where it would be impossible to undertake works without entering it (sect. 29(10)).

There does not appear to be a power of entry in respect of the assessment of the building before a decision about the carrying out of works is made.

Definitions

Building includes structure (sect. 29(3)).

Local authority means a district council (i.e. unitary authorities and continuing district councils), a London borough council and the Common Council of the City of London (sect. 29(4)).

ABANDONED VEHICLES

References

Refuse Disposal (Amenity) Act 1978 sects. 3–5.
Removal and Disposal of Vehicles Regulations 1986 (as amended).
Removal and Disposal of Vehicles (England) (Amendment) Regulations 2002.
Removal, Storage and Disposal of Vehicles (Prescribed Sums and Charges etc.) Regulations 1989 (as amended 1993).

Extent

This procedure applies in England, Wales and Scotland but not in Northern Ireland (sect. 13(5)).

FC112a Abandoned vehicles: removal

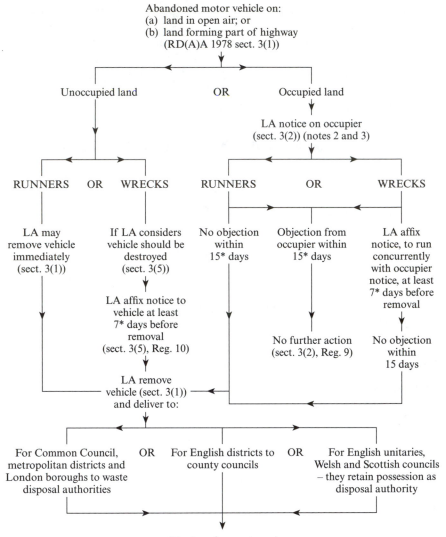

Abandoned motor vehicle on:
(a) land in open air; or
(b) land forming part of highway
(RD(A)A 1978 sect. 3(1))

Unoccupied land   OR   Occupied land

LA notice on occupier
(sect. 3(2)) (notes 2 and 3)

RUNNERS   OR   WRECKS   RUNNERS   OR   WRECKS

LA may remove vehicle immediately (sect. 3(1))

If LA considers vehicle should be destroyed (sect. 3(5))

No objection within 15* days

Objection from occupier within 15* days

LA affix notice, to run concurrently with occupier notice, at least 7* days before removal

LA affix notice to vehicle at least 7* days before removal (sect. 3(5), Reg. 10)

No further action (sect. 3(2), Reg. 9)

No objection within 15 days

LA remove vehicle (sect. 3(1)) and deliver to:

For Common Council, metropolitan districts and London boroughs to waste disposal authorities

OR   For English districts to county councils

OR   For English unitaries, Welsh and Scottish councils – they retain possession as disposal authority

(Continued on next page)

Notes
1. All regulation numbers refer to the Removal of Vehicles Regulations 1986.
2. If the land is occupied by the LA itself, no 'occupier' notice is required.
3. For powers to remove vehicles illegally, obstructively or dangerously parked, see Road Traffic Regulations Act 1984 and Removal of Vehicles Regulations 1986 (as amended 1993).
4. For power of entry, see sect. 8.
* As at March 2002 there are DEFRA proposals to reduce these notice periods from 7 days to 24 hrs and from 15 days to a shorter period.

FC112b Abandoned vehicles: disposal

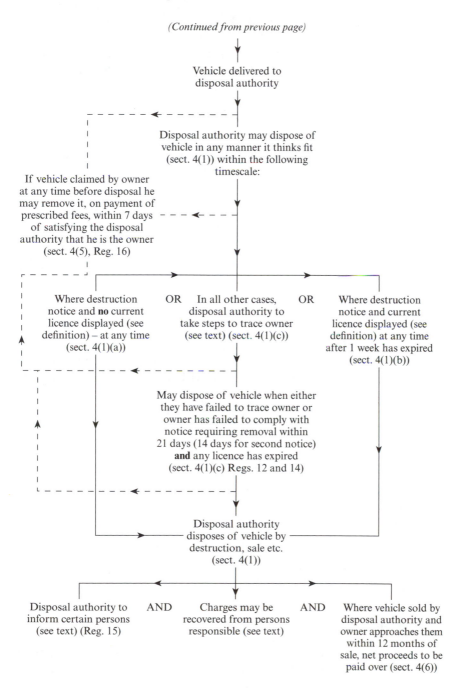

(Continued from previous page)

Vehicle delivered to disposal authority

Disposal authority may dispose of vehicle in any manner it thinks fit (sect. 4(1)) within the following timescale:

If vehicle claimed by owner at any time before disposal he may remove it, on payment of prescribed fees, within 7 days of satisfying the disposal authority that he is the owner (sect. 4(5), Reg. 16)

Where destruction notice and **no** current licence displayed (see definition) – at any time (sect. 4(1)(a))

OR   In all other cases, disposal authority to take steps to trace owner (see text) (sect. 4(1)(c))

OR   Where destruction notice and current licence displayed (see definition) at any time after 1 week has expired (sect. 4(1)(b))

May dispose of vehicle when either they have failed to trace owner or owner has failed to comply with notice requiring removal within 21 days (14 days for second notice) **and** any licence has expired (sect. 4(1)(c) Regs. 12 and 14)

Disposal authority disposes of vehicle by destruction, sale etc. (sect. 4(1))

Disposal authority to inform certain persons (see text) (Reg. 15)

AND   Charges may be recovered from persons responsible (see text)

AND   Where vehicle sold by disposal authority and owner approaches them within 12 months of sale, net proceeds to be paid over (sect. 4(6))

Scope

LAs are placed under a statutory duty to remove motor vehicles unlawfully abandoned on any land in the open air or on any other land forming part of a highway except that, where the cost of removal to the nearest convenient carriageway would be unreasonably high, the procedure is discretionary (sect. 3(1) and (3)).

The removal of vehicles illegally, obstructively or dangerously parked are matters for the police and traffic wardens under the Removal and Disposal of Vehicles Regulations 1986 (as amended 1993).

Runners

Unless on occupied land, vehicles which are **not** to be taken for destruction may be removed immediately without notice (sect. 3(1)).

Wrecks

If, in the opinion of the LA, the vehicle is in such a condition that it ought to be destroyed, notice needs to be fixed to the vehicle that removal for destruction is intended in not less than 24 hours' time. There is no specified form for this purpose (sect. 3(5)).

Occupied land

If the vehicle (either runner or wreck) is situated on occupied land, a further notice procedure is required. In this case, the LA must give at least 7 days written notice to the occupier (sect. 3(2)).

Delivery of vehicle

Having removed the vehicle, the LA must deliver it by arrangement:

(a) in London and English metropolitan areas to the waste disposal authorities; or
(b) in the rest of England to the county council.

In England, Wales and Scotland, the unitary authority retains possession and continue to operate the procedure from that point as refuse disposal authority (sect. 3(6) and (7)).

Arrangements

Where two authorities are involved, delivery to the disposal authority is required to be in accordance with arrangements agreed between the LA and the disposal authority, and these arrangements may include sharing of expenses incurred or income received. In default of agreement, there is provision for arbitration or, in the case of London, reference to the SoS (sect. 3(6) and (7)).

Penalties

No penalties are specified for not recovering vehicles after notices have been served but sect. 2(1) of the Act creates an offence of abandoning the vehicle in the open air or on land forming part of a highway, and penalties are not exceeding level 4 on the standard scale and/or imprisonment for up to 3 months (sect. 2(1)).

Safe custody

For vehicles other than those intended for destruction with the appropriate notice attached, the LA or disposal authority having custody of the vehicle is responsible for taking reasonable steps to ensure its safe custody (sect. 3(8)). In the event of this being in question approaches may be made by the person concerned to the SoS who must hold a public inquiry (sects. 1(5) and 3(7)).

Disposal of vehicles

The disposal authority may dispose of vehicles delivered to them in any manner it thinks fit in the timescales shown FC112b (sect. 4(1)).

The steps to be taken to trace the owner (i.e. other than where destruction notice given) are specified in Reg. 12 of the 1986 Regulations.

Notice to owners requiring removal of the vehicle are to be served by either:

(a) delivering it to the owner;
(b) leaving it at the usual or last known place of abode;
(c) sending it by a prepaid registered letter or recorded delivery addressed to him at his usual or last known abode;
(d) where the owner is an incorporated company or body, delivering it to the secretary or clerk or sending it addressed to that person by either prepaid registered letter or recorded delivery at their registered or principal office (Reg. 13).

The persons to whom information must be given following the disposal of the vehicle are detailed in Reg. 15 of the 1986 Regulations.

Charges for storage and removal

These are statutory charges prescribed from time to time by the Removal and Disposal of Vehicles (Amendment) Regulations. Charges may be recovered from:

(a) the owner at the time the vehicle was abandoned unless he shows that he was not aware of or concerned in the abandonment; or
(b) any person who put it in the place from which it was removed; or
(c) persons convicted of abandoning the vehicle under sect. 2(1).

Definitions

Current licence includes a reference to a licence which was current during any part of the period of 14 days ending with the day preceding that on which the removal of the vehicle in question took place; and . . . reference to the expiration of a licence shall be construed as a reference to that expiration of the period of 14 days beginning with the day following that on which the licence expired (sect. 4(2)).

Local authority means:

(a) in relation to England, a district council (unitary or continuing two-tier district), London borough council or the Common Council;
(b) in relation to Scotland, the unitary authority; and
(c) in relation to Wales, the unitary authority.

Motor vehicle means a mechanically propelled vehicle intended or adapted for use on roads, whether or not it is in a fit state for such use, and includes any trailer intended or adapted for such use as an attachment to such a vehicle, any chassis or body, with or without wheels, appearing to have formed part of such a vehicle or trailer and anything attached to such a vehicle or trailer (sect. 11(1)).

REMOVAL AND DISPOSAL OF ABANDONED REFUSE

References

Refuse Disposal (Amenity) Act 1978 sect. 6.
Removal of Refuse Regulations 1967.

Extent

This procedure applies in England, Wales and Scotland but not in Northern Ireland (sect. 13(5)).

Scope

These provisions deal with anything (other than motor vehicles) abandoned without lawful authority on any land in the open air or on any other land forming part of a highway (sect. 6(1)).

Unlike the provisions relating to abandoned cars etc., there is no duty placed here on the LA to act, the powers are discretionary.

FC113 Removal and disposal of abandoned refuse

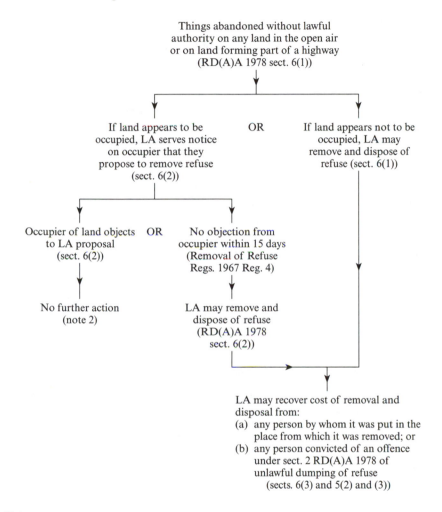

Things abandoned without lawful
authority on any land in the open air
or on land forming part of a highway
(RD(A)A 1978 sect. 6(1))

If land appears to be
occupied, LA serves notice
on occupier that they
propose to remove refuse
(sect. 6(2))

OR

If land appears not to be
occupied, LA may
remove and dispose of
refuse (sect. 6(1))

Occupier of land objects
to LA proposal
(sect. 6(2))

OR

No objection from
occupier within 15 days
(Removal of Refuse
Regs. 1967 Reg. 4)

No further action
(note 2)

LA may remove and
dispose of refuse
(RD(A)A 1978
sect. 6(2))

LA may recover cost of removal and
disposal from:
(a) any person by whom it was put in the
place from which it was removed; or
(b) any person convicted of an offence
under sect. 2 RD(A)A 1978 of
unlawful dumping of refuse
(sects. 6(3) and 5(2) and (3))

Notes
1. This procedure does not apply to abandoned cars, see FC112a and b.
2. The provisions do not provide for any action beyond this point but see following notes for alternative procedures.
3. For accumulations constituting a statutory nuisance, see FC33a and b.
4. For removal of rubbish resulting from demolition, see FC20.
5. For removal of rubbish seriously detrimental to the amenities of a neighbourhood, see FC9.
6. This procedure applies in Scotland but not in Northern Ireland.
7. The TCPA 1990 sect. 215 contains a power for the LA to require the tidying up of a site which is in such a state to adversely affect the amenities of a neighbourhood.

Unoccupied land

Where the land in question appears to be unoccupied, the LA may immediately remove the articles without notice (sect. 6(1)).

Occupied land

In this situation, the LA is first required to serve notice on the occupier stating its intention to remove the abandoned articles. If the occupier objects in writing to the proposal within 15 days of service of the notice, the LA cannot proceed further unless other legislative procedures are available for the particular circumstances (notes to FC113). If no objection within the 15-day period is received, the LA may remove the articles (sect. 6(2)).

Notice

The form of the notice is specified in the Removal of Refuse Regulations 1967 which, although made under the Civic Amenities Act 1967 by a provision now repealed, has been continued in force by sect. 12(3) of the RD(A)A 1978. With appropriate rewording, this form should, therefore be used.

For the authentication and service of notices, see LGA 1972, pages 8–10.

Recovery of costs

The LA is entitled to recover its costs of removal from either:

(a) the person who left the articles on the land; or
(b) any person convicted under sect. 2(1) of the 1978 Act for illegally abandoning the articles (sect. 6(4)).

Sums are recoverable as a simple contract debt (sect. 5(3)) and courts making a conviction under sect. 2(1) may, on application by the LA, order the person convicted to reimburse to the LA their costs of removal (sect. 6(6)).

Penalties

This procedure itself carries no penalties but prosecution for abandoning the articles may be taken under sect. 2(1) of the 1978 Act with penalties not exceeding level 4 on the standard scale or imprisonment for up to 3 months or both.

Definitions

Local authority means:

(a) in relation to England, a district council (or continuing two-tier district), London borough council or the Common Council;
(b) in relation to Scotland, the unitary authority;
(c) in relation to Wales, the unitary authority.

Motor vehicle means a mechanically propelled vehicle intended or adapted for use on the roads, whether or not it is in a fit state for such use, and includes any trailer intended or adapted for use as an attachment to such a vehicle, any chassis or body, with or without wheels, appearing to have formed part of such a vehicle or trailer and anything attached to such a vehicle or trailer (sect. 11(1)).

CLOSING ORDERS FOR TAKE-AWAY FOOD SHOPS

Reference

Local Government (Miscellaneous Provisions) Act 1982 sects. 4–6 inclusive.

Extent

This procedure applies only in England and Wales.

Scope

The provisions apply to any premises where meals or refreshments are supplied for consumption off the premises except:

(a) late night refreshment houses;
(b) premises licensed to sell liquor (sect. 4(1)).

Late night refreshment houses are any house, room, shop or building kept open for public refreshment, resort and entertainment at any time between 10 pm and 5 am on the following morning (FC101). The licensed premises excluded from the scope of these provisions are houses, rooms, shops or buildings which are:

(i) licensed for the sale of beer, cider, wine or spirits; and
(ii) not kept open for public refreshment and entertainment between normal evening closure (30 minutes after the evening closing hour permitted by the Licensing Act 1964 sect. 60, or 10 pm where there are no permitted hours in the evening) and 5 am the following morning. This licensed premises exemption does not apply in Greater London (LNRHA 1969 sect. 1).

This procedure, therefore, may be used on take-away food shops open between midnight and 5 am, subject to the exemptions.

Complaints

The LA cannot make a closing order or a variation order with more restrictive hours unless it has received complaints from residents in the neighbourhood about disturbance from or related to the premises (sect. 5(1)).

FC114a Closing orders for take-away food shops: closing orders

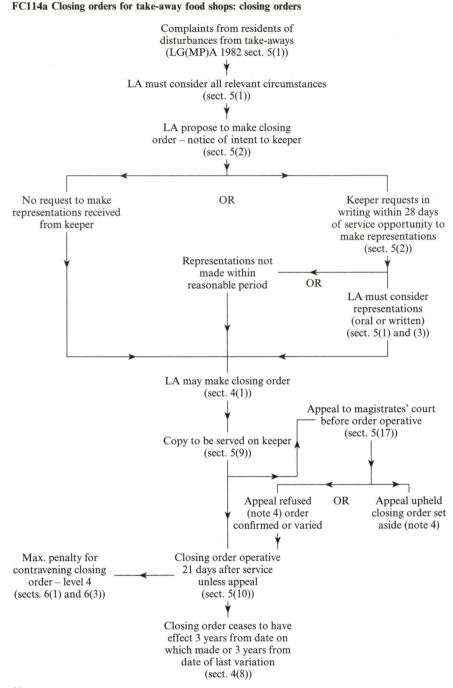

Complaints from residents of
disturbances from take-aways
(LG(MP)A 1982 sect. 5(1))

LA must consider all relevant circumstances
(sect. 5(1))

LA propose to make closing
order – notice of intent to keeper
(sect. 5(2))

No request to make
representations received
from keeper

OR

Keeper requests in
writing within 28 days
of service opportunity to
make representations
(sect. 5(2))

Representations not
made within
reasonable period

OR

LA must consider
representations
(oral or written)
(sect. 5(1) and (3))

LA may make closing order
(sect. 4(1))

Appeal to magistrates' court
before order operative
(sect. 5(17))

Copy to be served on keeper
(sect. 5(9))

Appeal refused
(note 4) order
confirmed or varied

OR

Appeal upheld
closing order set
aside (note 4)

Max. penalty for
contravening closing
order – level 4
(sects. 6(1) and 6(3))

Closing order operative
21 days after service
unless appeal
(sect. 5(10))

Closing order ceases to have
effect 3 years from date on
which made or 3 years from
date of last variation
(sect. 4(8))

Notes
1. For revocation and variation, see FC114b.
2. This procedure applies in England and Wales only.
3. For licensing of late night refreshment houses, see FC101.
4. There is provision for an appeal to the Crown Court against a decision of the magistrates court (sect. 5(19)).

FC114b Closing orders for take-away food shops: variation and revocation of closing orders

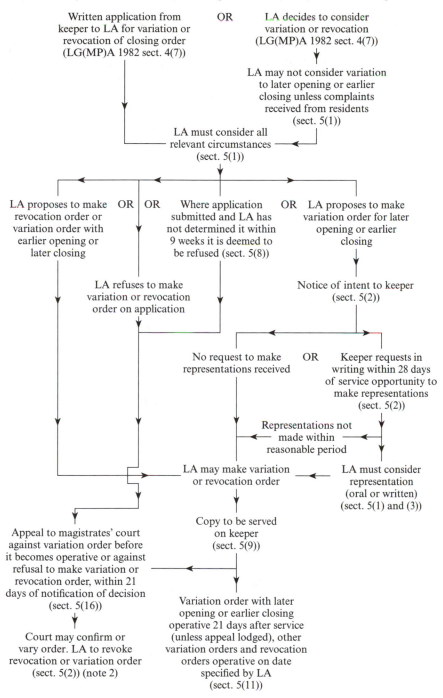

Written application from keeper to LA for variation or revocation of closing order (LG(MP)A 1982 sect. 4(7))

OR

LA decides to consider variation or revocation (LG(MP)A 1982 sect. 4(7))

LA may not consider variation to later opening or earlier closing unless complaints received from residents (sect. 5(1))

LA must consider all relevant circumstances (sect. 5(1))

LA proposes to make revocation order or variation order with earlier opening or later closing

OR OR

Where application submitted and LA has not determined it within 9 weeks it is deemed to be refused (sect. 5(8))

OR LA proposes to make variation order for later opening or earlier closing

LA refuses to make variation or revocation order on application

Notice of intent to keeper (sect. 5(2))

No request to make representations received

OR

Keeper requests in writing within 28 days of service opportunity to make representations (sect. 5(2))

Representations not made within reasonable period

LA may make variation or revocation order

LA must consider representation (oral or written) (sect. 5(1) and (3))

Appeal to magistrates' court against variation order before it becomes operative or against refusal to make variation or revocation order, within 21 days of notification of decision (sect. 5(16))

Copy to be served on keeper (sect. 5(9))

Court may confirm or vary order. LA to revoke revocation or variation order (sect. 5(2)) (note 2)

Variation order with later opening or earlier closing operative 21 days after service (unless appeal lodged), other variation orders and revocation orders operative on date specified by LA (sect. 5(11))

Notes
1. For making of closing orders, see FC114a.
2. There is provision for appeal to the Crown Court against a decision of the magistrates court (sect. 5(19)).

Closing orders

Having taken all the relevant circumstances into consideration and being satisfied that an order is desirable to prevent neighbouring residents being unreasonably disturbed either by persons using the premises or by the use of the premises, the LA may make a closing order. Noise etc. arising outside the premises, e.g. cars and motor cycles stopping and starting, groups of people staying to eat the food purchased etc., can be taken into account (sect. 4(1)).

The closing order must relate to an individual premises and prohibits opening between such hours as are stated between midnight and 5 am. Different hours can be specified for different days (sect. 4(2) and (3)). Orders are effective for 3 years but further orders may be made but only by the LA operating the full procedure for the making of an order (sect. 4(8)).

Notices

Where the LA intends to make a closing order or vary an existing order by stipulating more restrictive opening hours, it must serve notice of its intention on the keeper of the premises who may ask within 28 days to be allowed to make written or oral representations. Oral representations must be made to either a committee or subcommittee. Having indicated a wish to make representations but having not done so within what the LA considers to be a reasonable time, the LA may proceed (sect. 5(2)–(6)).

Where revocation orders or less restrictive variation orders are intended, no prior notification to the keeper is required.

Variation orders

Orders varying the closure hours may be made either following representations from the keeper or on the initiative of the LA. When variation orders are made, the closing order to which it relates continues for a 3-year period from the last variation (sect. 4(6) and (9)).

Where the keeper has applied for variation and revocation and the LA have not determined the matter within 8 weeks, the application is deemed to be refused and the appeal provisions are available (sect. 5(8)).

Revocation orders

As with variation orders, revocation orders may be made either on the initiative of the LA or on application by the keeper (sect. 4(7)).

Service of notices, orders

Orders and notices may be served on the keeper by hand or by post. The document may be addressed to the keeper by name or to 'the keeper' of the premises (describing them). If delivery is to be by post, prepaid registered letter or the recorded delivery services must be used (sect. 5(12)–(15)).

Operation of orders

Revocation orders and variations with less restrictive hours become operative as soon as made. All other orders become effective 21 days after service of the copy on the keeper except where an appeal has been made in which case the order is not effective until the appeal has been determined or withdrawn (sect. 5(10) and (17)),

Appeals

The keeper may appeal to a magistrates' court against:

 (a) closing orders or variation orders; and
 (b) refusal of the LA to make a revocation or variation order;

at any time up to 21 days after the notification of refusal or before the order becomes effective (sect. 5(16)–(18)).

Both the keeper and the LA may appeal to the Crown Court against a decision of the magistrates' court (sect. 5(19)).

Definitions

Keeper means the person having the conduct or management of the premises (sect. 4(10)).

Local authority means district councils in England and Wales (sects. 4 and 49), i.e. unitary authorities and continuing two-tier districts.

CONTROLS OVER TRADING ON SUNDAYS AT LARGE SHOPS

Reference

Sunday Trading Act 1994.

Extent

These procedures apply only in England and Wales (sect. 9(4)).

Scope

By virtue of this Act Sunday Trading was deregulated except in relation to large shops where generally trading is prohibited except in accordance with the '6-hour trading' procedures in FC115.

A LA may designate its area to be a Loading Control Area in order to control loading and unloading before 9 am on Sundays in accordance with procedures in FC116.

FC115 Control over trading on Sundays at large shops

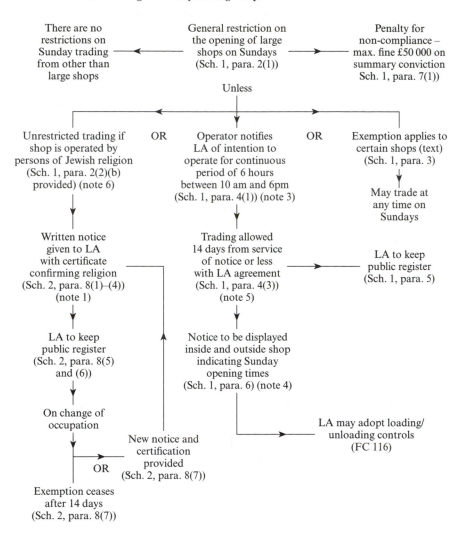

Notes
1. Penalty not exceeding level 5 for making false statements (Sch. 2, para. 8(10)) and the LA may cancel notice (Sch. 2, para. 8(11)).
2. A fresh certificate is not required if previously provided for that occupier – this is at the LA discretion (Sch. 2, para. 8(8)).
3. This notice may be cancelled or the hours varied at any time by further notice to the LA (Sch. 1, para. 4(2)).
4. Penalty not exceeding level 2 for non-compliance (Sch. 1, para. 7(2)).
5. Maximum penalty for trading outside of notified hours £50 000 (Sch. 1, para. 7(1)).
6. These shops must be closed on the Jewish Sabbath – sundown Friday to sundown on Saturday.

FC116 Controls over loading and unloading at large shops on Sundays

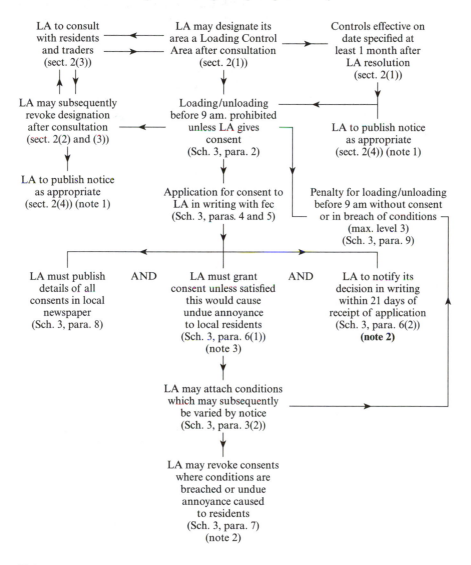

LA to consult
with residents
and traders
(sect. 2(3))

LA may designate its
area a Loading Control
Area after consultation
(sect. 2(1))

Controls effective on
date specified at
least 1 month after
LA resolution
(sect. 2(1))

LA may subsequently
revoke designation
after consultation
(sect. 2(2) and (3))

Loading/unloading
before 9 am. prohibited
unless LA gives
consent
(Sch. 3, para. 2)

LA to publish notice
as appropriate
(sect. 2(4)) (note 1)

LA to publish notice
as appropriate
(sect. 2(4)) (note 1)

Application for consent to
LA in writing with fec
(Sch. 3, paras. 4 and 5)

Penalty for loading/unloading
before 9 am without consent
or in breach of conditions
(max. level 3)
(Sch. 3, para. 9)

LA must publish
details of all
consents in local
newspaper
(Sch. 3, para. 8)

AND

LA must grant
consent unless satisfied
this would cause
undue annoyance
to local residents
(Sch. 3, para. 6(1))
(note 3)

AND

LA to notify its
decision in writing
within 21 days of
receipt of application
(Sch. 3, para. 6(2))
(note 2)

LA may attach conditions
which may subsequently
be varied by notice
(Sch. 3, para. 3(2))

LA may revoke consents
where conditions are
breached or undue
annoyance caused
to residents
(Sch. 3, para. 7)
(note 2)

Notes
1. No particular means of publication is specified, this is for the LA to decide.
2. There is no provision for appeal against a refusal of consent, conditions attached or revocation.
3. There is no requirement on a LA to keep a register of consents.

Large shops

A 'large shop' is one which has a relevant floor area exceeding 280 square metres.

The 'relevant floor area' is defined as being the internal floor area of so much of the shop as consists of or comprises in a building excluding any part of the shop which throughout the week ending with the Sunday in question is used neither for the serving of customers in connection with the sale of goods nor for the display of goods.

Shop is defined as any premises where there is carried on a trade or business consisting wholly or mainly of the sale of goods. Outlets which offer a service e.g. hairdressers, shoe repair shops, restaurants, public houses, etc. are therefore excluded from Sunday trading restrictions (Sch. 1, para. 1).

Exempted large shops

Some specified types of large shops are exempted from the 6-hour restriction. These are:

(a) Farm shops.
(b) Motor and cycle supply shops.
(c) Stands at exhibitions.
(d) Shops at railway stations and airports.
(e) Petrol filling stations.
(f) Pharmacists for the sale of medicines and medicinal and surgical appliances.
(g) Shops servicing ocean-going ships (Sch. 1, para. 3).

Shops occupied by persons observing the Jewish Sabbath are also exempt form the 6-hour limit on Sunday trading provided the procedure identified in FC115 has been followed and the shop is closed on the Jewish Sabbath.

The certificate required to be sent to the LA with the notification is one to be signed by an authorized person (a Minister or Secretary of a Synagogue or a person nominated for the purpose by the Board of Deputies of British Jews) and indicates that the person giving the notice is of the Jewish religion. Such exemption is also available to persons of any religious body who regularly observes the Jewish Sabbath but certification is required from the religious body concerned (Sch. 2, paras. 8 and 9).

The 6-hour provision

Unless exempt or occupied by Jews, large shops may trade for 6 continuous hours on Sundays between 10 am and 6 pm provided notification has been given to the LA. The choice of the 6 hour-trading period is a matter for each trader and this may be altered by subsequent notification to the LA (Sch. 1, para. 4).

The LA must keep a register of such shops which gives details of:

(a) the name and address of the shop; and
(b) the permitted hours.

This register must be kept available for public inspection and may be kept in computerized form (Sch. 1, para. 5).

Shops using the 6-hour trading provision must display conspicuous notices both inside and outside of the building specifying the permitted Sunday opening times (Sch. 1, para. 6).

Controls over loading and unloading

FC116 shows the procedure through which LAs may at their discretion operate a scheme of control over loading/unloading of goods from vehicles before 9 am on Sundays at large shops. There are no such restrictions on small shops.

The purpose of the procedure is to allow LAs to prevent local residents from being caused undue annoyance and requires the LA, once a Loading Control Area has been declared, to sanction loading and unloading at each premises before 9 am and specifying appropriate conditions. A scheme can only apply to the whole of the LA area.

Before adopting or subsequently revoking a scheme the LA must consult people likely to be affected, residents and traders, and must publicize its decision to adopt or revoke (sect. 2, and Sch. 3).

Enforcement

LAs (definition below) have a duty to enforce these provisions and must appoint inspectors who have powers of entry to large shops to check on possible contraventions, inspect and take copies of records (including computerized records) and to take measurements and photographs (Sch. 2).

Definitions

Local authority means any unitary authority or any district council so far as they are not a unitary authority.
   Unitary authority means:

(a) the council of any county so far as they are the council for an area in which there are no district councils;
(b) the council of any district comprising an area for which there is no county council;
(c) a county borough council;
(d) a London borough council;
(e) the Common Council of the City of London; or
(f) the Council of the Isles of Scilly (sect. 8).

Retail sale means any sale other than a sale for use or resale in the course of a trade or business, and references to retail purchase shall be construed accordingly.

Sale of goods does not include:

(a) the sale of meals, or refreshments or intoxicating liquor for consumption on the premises on which they are sold; or

(b) the sale of meals or refreshments prepared to order for immediate consumption off those premises.

Shop means any premises where there is carried on a trade or business consisting wholly or mainly of the sale of goods.

Stand in relation to an exhibition means any platform, structure, space or other area provided for exhibition purposes (Sch. 1, para. 1).

REMOVAL OF UNAUTHORIZED CAMPERS

References

Criminal Justice and Public Order Act 1994 Sects. 77–80.
DoE Circular 18/94 Gypsy Sites Policy and Unauthorized Camping (amended 2000).
DETR 1998 Managing Unauthorised Camping: A Good Practice Guide (revised 2000).

Extent

This procedure is applicable in England and Wales but not in Scotland or Northern Ireland.

Scope

Under this procedure LAs are able to direct persons residing unlawfully in vehicles:

(a) on highway land;

(b) on unoccupied land; or

(c) on occupied land without the consent of the occupier;

to leave the land and remove their vehicles and any other property they have with them (sect. 77(1)).

The procedure may be applied to all persons residing in vehicles (as defined below) including New Age Travellers; however, in DoE Circular 18/94 LAs are asked not to use these powers needlessly in relation to gypsy encampments, particularly where they are causing no nuisance (paras. 6–9). Complementary provisions in sect. 61 give powers to the police to deal with collective trespass or nuisance on land in circumstances where the trespassers have refused requests from the occupier to leave and either:

FC117 Removal of unauthorized campers

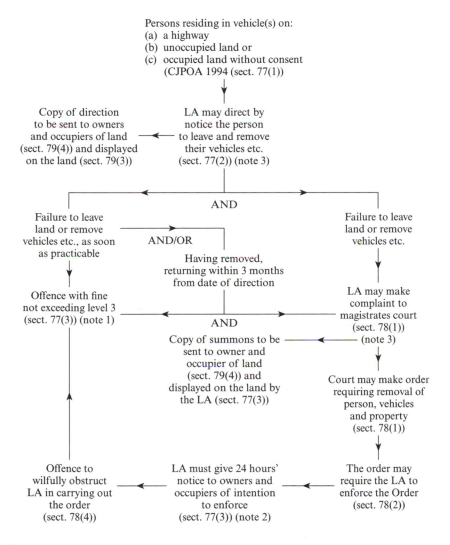

Persons residing in vehicle(s) on:
(a) a highway
(b) unoccupied land or
(c) occupied land without consent
(CJPOA 1994 (sect. 77(1))

Copy of direction to be sent to owners and occupiers of land (sect. 79(4)) and displayed on the land (sect. 79(3))

LA may direct by notice the person to leave and remove their vehicles etc. (sect. 77(2)) (note 3)

AND

Failure to leave land or remove vehicles etc., as soon as practicable

AND/OR

Failure to leave land or remove vehicles etc.

Having removed, returning within 3 months from date of direction

Offence with fine not exceeding level 3 (sect. 77(3)) (note 1)

AND

LA may make complaint to magistrates court (sect. 78(1)) (note 3)

Copy of summons to be sent to owner and occupier of land (sect. 79(4)) and displayed on the land by the LA (sect. 77(3))

Court may make order requiring removal of person, vehicles and property (sect. 78(1))

Offence to wilfully obstruct LA in carrying out the order (sect. 78(4))

LA must give 24 hours' notice to owners and occupiers of intention to enforce (sect. 77(3)) (note 2)

The order may require the LA to enforce the Order (sect. 78(2))

Notes
1. It is a defence to show that failure to leave or remove vehicles etc. was due to illness, mechanical breakdown or other immediate emergencies (sect. 77(5)).
2. Notice is not required where after reasonable enquiry, the LA is unable to obtain the names and addresses of the person(s) subject to the court direction (sect. 78(3)).
3. For service of notice of direction and summons of complaint see text.
4. For police powers to remove trespassers on land, including those with at least 6 vehicles (including caravans) see CJPOA 1994 sect. 61.

(a) they have caused damage to land/property; or
(b) threatening, abusive or insulting behaviour has been used; or
(c) there are at least 6 vehicles (sect. 61).

The operation of both provisions therefore requiries close liaison between the Police and the LA.

Vehicles

Both provisions deal with the presence of vehicles, defined as including:

(a) any vehicle, whether or not it is in a fit state for use on roads, and includes any body, with or without wheels, appearing to have formed part of such a vehicle, and any load carried by, and anything attached to, such a vehicle; and
(b) a caravan as defined in sect. 29(1) of the Caravan Sites and Control of Development Act 1960;

and a person may be regarded for the purposes of this section as residing on any land notwithstanding that he has a home elsewhere (sects 61(a) and 77(6)).

Gypsies

In Circular 18/94 LAs are asked in enforcing this procedure to take account of the definition of 'gypsies' given by the Court of Appeal in *R* v *South Hams DC, ex parte Gibb* (*The Times* 8 June 1994; i.e. 'gipsies' meant any persons who wandered or travelled for the purpose of making or seeking their livelihood, and did not include persons who moved from place to place without any connection between their movement and their means of livelihood).

Directions

Directions to be given by LAs must be served on the persons to whom the direction is to be applied but where there is more than one person the direction can merely specify the land and be addressed to all occupants of the vehicles without naming them (sect. 77(2)).

Where it is impracticable to serve on the persons named in the direction, a copy may be fixed prominently on the vehicle concerned and if directed to, unnamed persons may be served in the same manner on every vehicle (sect. 79(2)).

Copies of the direction must also be displayed on the land (other than on a vehicle) and copies sent to the owner and any occupier of the land, unless after reasonable enquiry they cannot be found (sect. 79(3) and (4)).

The direction requires the persons concerned to leave the land and remove the vehicles etc., as soon as practicable. It is an offence:

(a) not to comply; and
(b) having removed the vehicles etc., to again enter the land with a vehicle within 3 months of the day on which the direction was given (sect. 77(3)).

Defences

In any proceedings taken by the LA for non-compliance with a direction, it will be a defence to show that non-compliance was due to:

(a) illness; or
(b) mechanical breakdown; or
(c) other immediate emergency (sect. 77(5)).

Complaint to magistrates' court

In addition to seeking a fine for non-compliance with the directive, the LA may apply to the magistrates' court for an order requiring the removal of the vehicles and other property (sect. 77(1)).

Any order made may authorize the LA to enforce it by entering the land and removing the vehicles etc., but, in relation to occupied land, at least 24 hours' notice must be given to the owner and lawful occupiers (sect. 78(3)).

There is no provision for the recovery of the LA costs in enforcing these orders.

Provisions for the service of summons in relation to an LA complaint are similar to the service of directions above.

Definitions

Land means land in the open air.
Local authority means:

(a) in Greater London, a London borough or the Common Council of the City of London;
(b) in England outside Greater London, a county council, a district council or the Council of the Isles of Scilly (unitary authorities or concurrent powers for counties and districts in continuing two-tier areas);
(c) in Wales, a county council or a county borough council (unitary authorities).

ANTI-SOCIAL BEHAVIOUR ORDERS (ASBOs)

References

The Crime and Disorder Act 1998 sects. 1 and 4.
The Magistrates' Courts (Sex Offenders and Anti-social Behaviour Orders) Rules 1998.

FC118 Anti-social behaviour orders

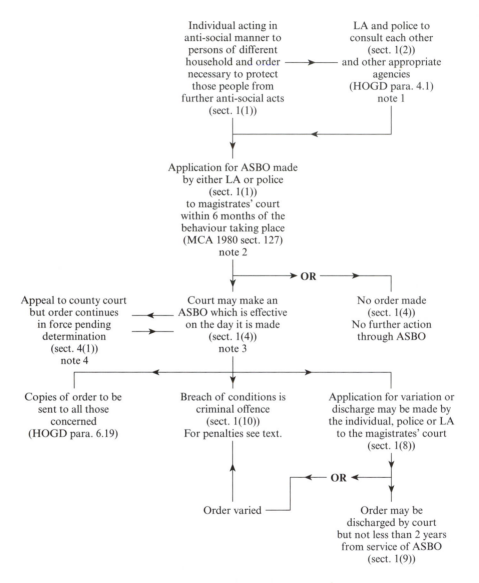

Individual acting in anti-social manner to persons of different household and order necessary to protect those people from further anti-social acts (sect. 1(1)) → LA and police to consult each other (sect. 1(2)) and other appropriate agencies (HOGD para. 4.1) note 1

Application for ASBO made by either LA or police (sect. 1(1)) to magistrates' court within 6 months of the behaviour taking place (MCA 1980 sect. 127) note 2

OR

Appeal to county court but order continues in force pending determination (sect. 4(1)) note 4 ← Court may make an ASBO which is effective on the day it is made (sect. 1(4)) note 3 → No order made (sect. 1(4)) No further action through ASBO

Copies of order to be sent to all those concerned (HOGD para. 6.19)

Breach of conditions is criminal offence (sect. 1(10)) For penalties see text.

Application for variation or discharge may be made by the individual, police or LA to the magistrates' court (sect. 1(8))

OR

Order varied

Order may be discharged by court but not less than 2 years from service of ASBO (sect. 1(9))

Notes

1. The sharing of information in the consultation process is covered by sect. 115 CDA 1998.
2. The form of application is set out in schedule 4 of the 1998 Rules.
3. The form of the ASBO is specified in schedule 6 of the 1998 Rules.
4. There may be an appeal to the High Court against the decision of the crown court or an application for judicial review.

Home Office Guidance Document (HOGD) 1999 – The Crime and Disorder Act 1998 – Anti-Social Behaviour Orders – Guidance on drawing up local ASBO protocols. (Home Office and Local Government Association) June 2000 (G on P).

Extent

This procedure applies in England and Wales but the Act makes a similar provision for Scotland.

Scope

An ASBO may be issued by a magistrates' court on application by either the police or a LA (see below) against an individual aged 10 or over:

(a) who has acted in an anti-social manner – that is in a manner that caused or was likely to cause harassment, alarm or distress to one or more persons not of the same household and

(b) the order is necessary to protect persons against whom the harassment etc. was directed from further anti-social acts by that person.
(sect. 1(1))

The purpose of an ASBO is to restrain serious anti-social behaviour caused to individuals who cannot be expected to tolerate the behaviour unchecked. This could include situations, e.g. noise, that cannot be dealt with effectively or appropriately through other remedies e.g. as statutory nuisances under sect. 80 of the EPA 1990.

The LAs that may apply for orders are district councils and unitary authorities including the London boroughs (sect.1 (12)).

Usage

Home Office guidance (HOGD 1999) suggests that orders should be used as a last resort to control serious anti-social behaviour where other controls that do not involve court proceedings have failed or are inappropriate. Types of anti-social behaviour where an ABSO may be appropriate include:

(i) where individuals intimidate neighbours and others through threats or violence or a mixture of unpleasant actions;

(ii) persistently unruly behaviour by a small group of individuals on a housing estate or other local area, who may dominate others and use minor damage to property and fear of retaliation, possibly at unsociable hours, as a means of intimidating other people;

(iii) where there are families whose anti-social behaviour, when challenged, leads to verbal abuse, vandalism, threats and graffiti, sometimes using children as the vehicle for action against neighbouring families;

(iv) persistent abusive behaviour towards elderly people or towards mentally ill or disabled people causing them fear and distress;

(v) serious and persistent bullying of children on an organized basis in public recreation grounds or on the way to school or within the school grounds if normal disciplinary procedures do not stop the behaviour;

(vi) persistent racial harassment or homophobic behaviour;

(vii) persistent anti-social behaviour as a result of drugs or alcohol abuse. (HOGD para. 3.9)

However, there is no requirement to demonstrate that every other remedy has been exhausted before applying for an ASBO. The key is that one should be used where it is the most appropriate remedy, is necessary to protect an individual, a group of people or a community and is the proportionate step to take (G on P paras. 6.15 and 6.16).

Partnership

Action on ASBOs should be taken within the framework of partnership arrangements under sect. 5 of the C and DA 1998 between the police, LA, health authority and the probation service and be subject to a local protocol for ASBOs which also involves social landlords, organizations representing local residents and agencies such as social services and the youth offending teams (G on P paras. 1.1–1.4).

The protocol should set out agreed procedures for considering and making applications for ASBOs, for their monitoring and applying for the variation and discharge of them. As part of this protocol an ASBO coordinator should be identified for each LA district (G on P para. 2.1).

Consultation

Before making application for an ASBO the authority (police or LA) must consult with each other (sect. 1(2)). This consultation should involve all the relevant departments and also include other agencies as appropriate e.g. social services, health service, landlord etc. and be conducted through a consultative group established as part of the local protocol (G on P para. 6.2).

In some cases where the offending individual lives outside of the local government area in which the harassment etc. is being caused, additional authorities will need to be involved.

Agreement between the police and LA is not a requirement but in the absence of it the magistrates' court must be informed when the application is made.

Applications

Applications for orders are made by way of complaint to the magistrates' court for the area where the harassment etc. is taking place. (sect. 1(3)).

Such complaint must be made within 6 months of the offending behavior occurring (Magistrates' Courts Act 1980 sect. 127).

Because the court acts in a civil capacity in relation to these applications the civil standard of proof will apply i.e. 'on the balance of probabilities'.

A separate application is to be made against each individual even though that person may be part of a family or larger group causing the problems. The form of application is specified in schedule 4 of the Magistrates' Courts (Sex Offenders and Anti-social Behaviour Orders) Rules 1998.

Guidance on the preparation of evidence to be presented at the hearing of an application is given in the G on P section 11.

Orders

If it is shown that the conditions in 'Scope' above exist the magistrates' court may make an ASBO against a named individual which prohibits that person from doing anything described in the order (sect. 1 (4)). Such prohibitions are to be those necessary to prevent further anti-social behaviour by that individual and may relate to more than one local government area, providing that the required consultation has taken place – see above (sect. 1(5) and (6)).

Prohibitions should be specific in time and place so that it is readily apparent both to the defendant and to those enforcing the order what does or does not constitute a breach. Orders should contain a prohibition on inciting/encouraging the commission of specified anti-social acts within the meaning of the Act, including where appropriate by minors in the household, and by others over whom the person subject to the order has control. Any requirements in the order must be negative. There is no power to compel an individual to do anything, only not to take particular actions (HOGD para. 6.10).

Orders are effective for the period specified in it, with a minimum of 2 years, unless subsequently varied or discharged – see below (sect. 1(7)). The form of the order is set out in schedule 6 to the 1998 Rules.

Copies of the order should be sent to all those concerned including the police, LA and other relevant people e.g. the landlord (HOGD para. 6.19).

Appeals

Appeals against the making of an ASBO are made to the Crown Court which may make such orders as necessary to give effect to its decision including consequential or incidental orders (sect. 4).

Challenge to a decision of the Crown Court is to the High Court by way of case stated or by an application for judicial review.

Effect of an ASBO

Breach of the prohibitions in the order is a criminal offence (sect. 1(10)) and the standard of proof required is therefore that of 'beyond reasonable doubt'. All prosecutions are to be taken by the Crown Prosecution Service (HOGD para. 8.3).

Cases are triable either way with penalties on summary conviction of imprisonment for up to 6 months and a fine not exceeding the statutory maximum, or both, or on indictment imprisonment for up to 5 years or an unlimited fine, or both. There is a defence of 'reasonable excuse' (sect. 1(10)).

Variation and discharge of orders

The police, LA or the individual against whom the order was made may apply for these by way of complaint to the magistrates' court. The order cannot be discharged less than 2 years after its service (not making) without the agreement of both the police and the LA (sect. 1(8) and (9)).

If successful the variation could for example change the prohibitions. Copies of varied orders and of orders for discharge should be sent to all those who received copies of the original order (HOGD para. 6.21).

Acceptable Behaviour Contracts (ABCs)

These are informal arrangements that may be used as an alternative to ASBOs where the latter are not considered to be justified. An ABC is an individual written agreement by a young person with a partner agency and the police not to carry on with certain identifiable acts that could be construed as anti-social behaviour. Whilst not a legally binding document, a breach of it could be cited in any application for an ASBO. Guidance is given in appendix G of the G on P.

Chapter 15

HOUSING ACTS

The Housing Acts which legislate for the procedures in this chapter are:

1. *Housing Act 1985*
   - part 6 Repair notices (as amended by the Housing Grants, Construction and Regeneration Act 1996)
   - part 9 Slum Clearance (as amended by the Housing Grants, Construction and Regeneration Act 1996)
   - part 10 Overcrowding
   - part 11 Houses in Multiple Occupation (as amended by the Housing Act 1996)
2. *Local Government and Housing Act 1989*
   - part 7 Renewal Areas
3. *Housing Act 1996*
   - part 2 Houses in Multiple Occupation (amends part 11 of the Housing Act 1985)
4. *Housing Grants, Construction and Regeneration Act 1996*
   - part 1 Grants etc. for Renewal of Private Sector Housing as amended by the Regulatory Reform (Housing Assistance) (England and Wales) Order 2002.
   - part 4 Grants etc. for Regeneration, Development and Relocation (to be repealed from a date yet to be determined).

GENERAL PROCEDURAL PROVISIONS

Unless otherwise indicated, the following general provisions are applicable to the procedures in this chapter.

Local housing authority

The local authorities charged with responsibilities under the above named Housing Acts are known as local housing authorities and are:

(a) district councils;
(b) London borough councils;
(c) the Common Council of the City of London;

(d) a Welsh county council or county borough council;

(e) the Council of the Isles of Scilly (sect. 1 HA 1985).

Throughout this chapter these are referred to as the local authority (LA). The unitary authorities in England and Wales are local housing authorities.

Extent

The procedures in this chapter apply only in England and Wales and not in Scotland or Northern Ireland (HA 1985 sect. 625(5) and HGCRA 1996 sect. 148).

Notices

(a) **Form.** Notices must be in writing (LGA 1972 sect. 233). The SoS has power to prescribe the particular form of any notice or other document (HA 1985 sect. 614 and HGCRA 1996 sect. 89). This power has been widely used and prescribed forms are noted, where appropriate, in each flow chart.

(b) **Authentication.** Notices, orders and other documents may be signed on behalf of the authority by the proper officer and 'signature' includes a facsimile (LGA 1972 sect. 234).

(c) **Service.** Where the LA has to serve a document on a person who is either:

  (i) a person having control of the premises; or

  (ii) a person managing the premises; or

  (iii) any person having an estate or interest in the premises, the authority must take reasonable steps to identify these persons (HA 1985 sect. 617(1)).

A person having an estate or interest in a premises may give notice to the LA of that interest and the LA must record it (HA 1985 sect. 617(2)).

Notices may be served by:

(a) delivering it;

(b) leaving it at the proper address;

(c) sending it by post to the proper address.

The proper address referred to is:

(a) the last known address; or

(b) in the case of a body corporate, the registered or principal office; or

(c) in the case of a partnership, the principal office (LGA 1972 sect. 233).

If it is not possible after reasonable enquiry to ascertain the name and address of the person having control of the premises, the notice may be served by addressing it to the person by the description of 'person having control of' the premises (naming them) and delivering it to some person on the premises or, if there is no person to whom it can be delivered, affixing it, or a copy of it to some conspicuous part of the premises (HA 1985 sect. 617(3)).

The notice may be served on more than one person if necessary in the case where more than one person comes within the description of those upon whom the notice shall be served, e.g. person in control, owner, etc. (sect. 617(4)).

There is a similar provision in the LGA 1972 sect. 233 for service on other persons where the name and address cannot be ascertained.

For the general power to require information about interest in any land/premises, see LG(MP)A 1976 sect. 16, on page 10.

Powers of entry

Specific powers of entry are provided within each Part of the Housing Act 1985. In relation to repair notices (sect. 197), slum clearance (sect. 319) and compulsory purchase (sect. 600) the period of notice required before entry can be demanded is 7 days while for overcrowding it is 24 hours (sects. 337 and 340).

In relation to houses in multiple occupation, normally 24 hours' notice is required (sect. 395) except in the following cases where entry is authorized without prior notice at any reasonable time:

(a) failure to comply with a registration scheme (sect. 348G);
(b) failure to comply with a direction order (sect. 355(2));
(c) contravention of an overcrowding notice (sect. 358(4));
(d) use of a house without adequate means of escape from fire (sect. 368(3));
(e) contravention of management regulations (sect. 369(5));
(f) failure to comply with notice requiring works (sect. 376(1) or (2)).

If required, the AO must produce his authorization. The penalty for obstruction is not exceeding level 3.

There is provision for a LA to apply to a JP for a warrant to enter HIMOs (sect. 397). Such warrants allow entry by force where necessary.

DEFINITIONS

Definitions for this chapter are shown separately for each section or in an individual procedure.

UNFIT PREMISES

STANDARD OF FITNESS*

Any dwelling-house is deemed to be fit for human habitation unless it fails to meet one or other of the requirements below and by reason of that failure is not reasonably suitable for occupation:

* Also see 'Proposal to replace fitness standard etc.' page 500.

(a) it is structurally stable;
(b) it is free from serious disrepair;
(c) it is free from dampness prejudicial to the health of the occupants (if any);
(d) it has adequate provision for lighting, heating and ventilation;
(e) it has an adequate piped supply of wholesome water;
(f) there are satisfactory facilities in the dwelling for the preparation and cooking of food, including a sink with a satisfactory supply of hot and cold water;
(g) it has a suitably located water closet for the exclusive use of the occupants (if any);
(h) it has, for the exclusive use of the occupants (if any), a suitably located fixed bath or shower and a wash hand basin each of which is provided with a satisfactory supply of hot and cold water; and
(i) it has an effective system for the draining of foul water and surface water.

Whether or not a dwelling-house which is a flat satisfies the above requirements, it is unfit for human habitation if the building or part of the building outside the flat fails to meet one or more of the requirements in paragraphs (a) to (e) below and by reason of that failure the flat is not reasonably suitable for occupation:

(a) the building or part is structurally stable;
(b) it is free from serious disrepair;
(c) it is free from dampness;
(d) it has adequate provision for ventilation; and
(e) it has an effective system for the draining of foul, waste and surface water.

(sect. 604 HA1985)

Dwelling-houses and flats

The standard of fitness applies to:

(a) Dwelling-houses – these are defined as including any yard, garden, outhouses or appurtenances belonging to it or usually enjoyed with it (HA 1985 sect. 623(2));
(b) Flats – a flat is defined as a dwelling-house which is not a house, and, in this context, a dwelling-house is a house if, and only if, it is a structure reasonably so called so that:
　(i) where a building is divided horizontally, the flats or other units into which it is divided are not houses;
　(ii) where a building is divided vertically, the units into which it is divided may be houses;
　(iii) where a building is not structurally detached, it is not a house if a material part of it lies above or below the remainder of the structure.

A dwelling-house which is not a house is a flat.
(sects. 183 and 623 HA 1985)

Houses in multiple occupation

The fitness standard also applies to HIMOs in the same way as it does to ordinary dwelling-houses occupied by single people and family units (sect. 604(3) HA 1985). Where the unsuitability of a HIMO for the number of its occupants is to be dealt with, the procedure in FC129 is a available.

Guidance on fitness standard

Detailed guidance on the interpretation of the housing fitness standard is given in Annex A to DoE Circular 17/96. This guidance is advisory. LAs are asked to have regard to it when applying the standard but must form their opinion in the light of all relevant circumstances.

FITNESS ENFORCEMENT

Code of Guidance

In reaching a decision on 'the most satisfactory course of action' in respect of premises which have been identified as unfit for human habitation LAs are required to have regard to the Code of Guidance which is contained in Annex B of DoE circular 17/96 (sect. 604A HA1985).

Pre-formal enforcement action

The Code indicates that formal enforcement action should generally be viewed as a last resort unless immediate enforcement action is considered by the LA to be necessary, e.g. there is an imminent risk to the health and safety of occupiers.

Normally, LAs should discuss the condition of the premises and the options available with owners and agree the way forward.

Formal options

The fitness enforcement options available to LAs are:

(a) Repair notices (HA1985 sect. 189(1) or 1A-FC120).
(b) Deferred action notices (sects. 81 and 84 HGCRA 1996-FC119).
(c) Closing orders (HA 1985 sect. 264-FC121c).
(d) Demolition orders (HA1985 sect. 265-FC121b).
(e) Clearance Areas (HA 1985 sect. 289-FC122).
(f) Renovation grants (HGCRA 1996 part 1-FC132).
(g) Renewal Areas (LGHA 1989 sect. 89-FC131).

Assessment

The LA in considering the options available is required to determine 'the most satisfactory course of action'. The code recommends that the LA should use the Neighbourhood Renewal Assessment (NRA) technique which includes a series of sequential steps designed to enable the LA to explore costs and socio-environmental implications of each course of action. The process is detailed in Annex C3 (Appendix 4) of circular 17/96.

Relevant factors

In addition to the NRA assessment the LA is required by the Code to give consideration to the following factors in arriving at a decision on 'the most satisfactory course of action':

1. the wider context of the local private sector housing renewal strategy and the resources available;
2. the practicality of the options;
3. the life expectancy if repaired;
4. the relationship with neighbouring properties and their condition;
5. proposals for the future of the area and its status, e.g. conservation area;
6. the circumstances and wishes of owners and occupiers;
7. the management record of any landlord;
8. the effect on the community in the area;
9. the effect of each option on the local environment and the overall appearance of the locality.

There are additional relevant factors which relate to each enforcement option and these are noted in the text to the individual procedures which follow.

Proposal to replace fitness standard etc.

In March 2001 the then DETR issued a consultation paper that included proposals to replace the Housing Fitness Standard by a Housing Health and Safety Rating System (HHSRS). There are also other proposals to change many of the enforcement procedures included in this chapter. These changes will require new legislation for which there is no current timetable. In the meantime LAs have been asked to familiarize themselves with HHSRS and to adopt the principles of the system informally alongside their existing enforcement work, and as part of local housing stock condition surveys. The HHSRS and other proposals are set out in *Health and Safety in Housing: Replacement of the Housing Fitness Standard by the Housing Health and Safety Rating System. A Consultation Paper.* DETR March 2001.

DEFERRED ACTION NOTICES (DANs)

References

Housing Grants, Construction and Regeneration Act 1996 sects. 81–85.
The Housing (Fitness Enforcement Procedures) Order 1996.
The Housing (Deferred Action and Charge for Enforcement Action) (Forms) Regulations 1996.
The Housing (Maximum Charge for Enforcement Action) Order 1996.
DOE Circular 17/96 Annex B Fitness Enforcement-Code of Guidance for Dealing with Unfit Premises.

Scope

DANs are to be served where the LA considers in respect of a dwelling-house or HIMO that this is the most satisfactory course of action (see below and page 500) (sect. 81(1)).

These notices were introduced to provide LAs with additional flexibility in dealing with unfit properties and in particular to relate to the discretionary renovation grant system (see FC132). The previous obligation on LAs to take action on an unfit properly can be deferred by the use of these notices and in particular their service does not create any grant entitlement.

Decision to serve DAN

In addition to the general guidance (see pages 499–500), in deciding whether to serve a DAN the Code of Guidance requires a LA to consider:

  (a) the circumstances and wishes of the owner and occupants;
  (b) the health and needs of the owner and occupants;
  (c) the physical condition of the property, e.g. does it constitute an immediate health and safety risk to the occupants;
  (d) the cost and nature of the works; and
  (e) whether the LA is willing to provide grant aid (CoP para. 34).

Minded to take action notice (MTN)*

Except in the case where the LA considers that immediate action is necessary (e.g. imminent risk to health and safety) the LA must first serve a MTN which gives the reasons why it is considering the service of a DAN (or renewing one) and seeks any written representations within not less than 14 days. Oral representations may also be made if requested within 7 days of the service of the MTN. The LA must properly consider any representations made. If such a procedure is not followed this is a ground of appeal against any subsequent DAN (The Housing (Fitness Enforcement Procedures) Order 1996 Regs. 3 and 4).

---

* The DLTR intend to repeal this provision from a date yet to be determined.

FC119a Deferred action notices (DANs)

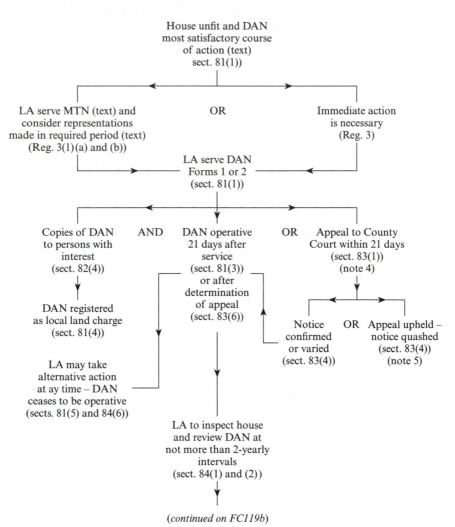

House unfit and DAN
most satisfactory course
of action (text)
sect. 81(1))

LA serve MTN (text) and   OR   Immediate action
consider representations   is necessary
made in required period (text)   (Reg. 3)
(Reg. 3(1)(a) and (b))

LA serve DAN
Forms 1 or 2
(sect. 81(1))

Copies of DAN   AND   DAN operative   OR   Appeal to County
to persons with   21 days after   Court within 21 days
interest   service   (sect. 83(1))
(sect. 82(4))   (sect. 81(3))   (note 4)
                        or after
                        determination
DAN registered   of appeal   Notice   OR Appeal upheld –
as local land charge   (sect. 83(6))   confirmed   notice quashed
(sect. 81(4))   or varied   (sect. 83(4))
                                             (sect. 83(4))   (note 5)

LA may take
alternative action
at ay time – DAN
ceases to be operative
(sects. 81(5) and 84(6))

LA to inspect house
and review DAN at
not more than 2-yearly
intervals
(sect. 84(1) and (2))

(continued on FC119b)

FC119b Deferred action notices (DANs) – review and renewal

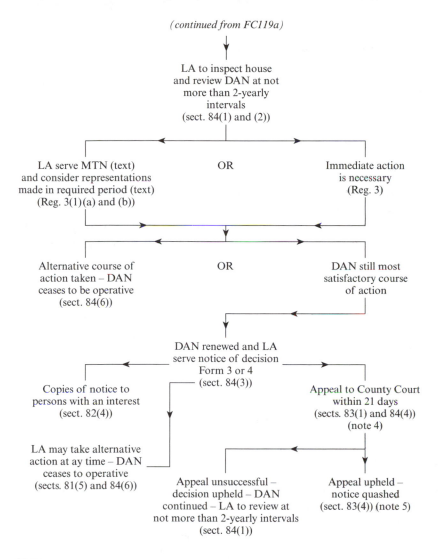

(continued from FC119a)

LA to inspect house
and review DAN at not
more than 2-yearly
intervals
(sect. 84(1) and (2))

LA serve MTN (text)
and consider representations
made in required period (text)
(Reg. 3(1)(a) and (b))

OR

Immediate action
is necessary
(Reg. 3)

Alternative course of
action taken – DAN
ceases to be operative
(sect. 84(6))

OR

DAN still most
satisfactory course
of action

DAN renewed and LA
serve notice of decision
Form 3 or 4
(sect. 84(3))

Copies of notice to
persons with an interest
(sect. 82(4))

Appeal to County Court
within 21 days
(sects. 83(1) and 84(4))
(note 4)

LA may take alternative
action at ay time – DAN
ceases to operative
(sects. 81(5) and 84(6))

Appeal unsuccessful –
decision upheld – DAN
continued – LA to review at
not more than 2-yearly intervals
(sect. 84(1))

Appeal upheld –
notice quashed
(sect. 83(4)) (note 5)

Notes

1. Section numbers refer to the HGCRA 1996.
2. Regulation numbers refer to the Housing (Fitness Enforcement Procedures) Order 1996.
3. Form numbers refer to the Housing (Deferred Action and Charges for Enforcement Action) (Forms) Regulations 1996.
4. In the event of an appeal against a DAN the notice is suspended pending appeal (sect. 83(6)) but a decision to renew a DAN continues until the appeal is determined (sect. 84(5)).
5. Where one of the reasons for allowing the appeal is that there is a more satisfactory course of action, if requested by either party the judge shall include this in his judgement (sect. 83(5)).
6. The LA may recover charges for operating this procedure – see text.

Content of DANs

The DAN must:

  (a) state that the premises are unfit;
  (b) specify the works required to make it fit; and
  (c) identify the other courses of action available to the LA (e.g. repairs notice etc.) (sect. 81(2)).

Service of DANs

DANs are to be served:

  (a) in the case of a dwelling-house, on the person having control as defined in sect. 207 of the HA 1985 (see page 510);
  (b) in the case of a HIMO, on the person having control as defined in sect. 398 of the HA 1985 or on the person managing the house; and
  (c) in the case of a flat, on the person having control as defined in sect. 207 of the HA 1985.

In addition, copies are to be served on any other person having an interest and copies may be served on any licensees to occupy (sect. 82, see also page 496).

Operation of DANs

Subject to appeal, notices become operative 21 days after service (sect. 81(3)).

Review of DANs

LAs may review the situation at any time and must do so at least every two years. If the LA is satisfied that the DAN still remains the most satisfactory course of action the notice is renewed but only following the service of a MTN and consideration of any representations.

If not so satisfied, the LA will proceed by an alternative course of action and the DAN ceases to be operative (sect. 84).

Appeals

There is an appeal to the county court against a DAN or its renewal within 21 days of service. DANs are suspended until the appeal is finally determined but the decision to renew a DAN remains operative until determination. A ground of appeal is that there is a more satisfactory course of action and court must take account of the Code of Guidance – see pages 499–500 (sects. 81(3), 83 and 84(5)).

Alternative courses of action

Irrespective of the existence of a DAN and independently of any formal review of a DAN, the LA may take an alternative course of action at any time provided the appropriate procedural steps for that alternative are taken, e.g. the service of a MTN etc. (sect. 81(5)).

Charges

The LA may make a reasonable charge for its costs in taking action through a DAN or its renewal. The costs may relate to:

(a) determining whether to serve the notice;
(b) identifying the works;
(c) serving the notice;
(d) deciding whether to review the notice; and
(e) serving notice of the LAs decisions.

The maximum charge is set at £300. A court in allowing an appeal may also reduce, quash or require payment of these charges (sect. 87 and the Housing (Maximum Charge for Enforcement Action) Order 1996).

In demanding payment the LA must use prescribed Form No. 5.

Definitions

Dwelling-house and flat – as for repair notices; see page 498
    HIMO see page 529.

REPAIR OF HOUSES

References

Housing Act 1985 Part 6.
The Housing (Fitness Enforcement Procedures) Order 1996.
The Housing (Maximum Charge for Enforcement Action) Order 1996.
The Housing (Deferred Action and Charge for Enforcement) (Forms) Regulations 1996.
Housing (Prescribed Forms) (No. 2) Regulations 1990 as amended 1997.
DoE Circular 17/96 Annex B Fitness Enforcement – Code of Guidance for Dealing with Unfit Premises.

Scope

This procedure deals with three situations:

FC120 Repair of houses

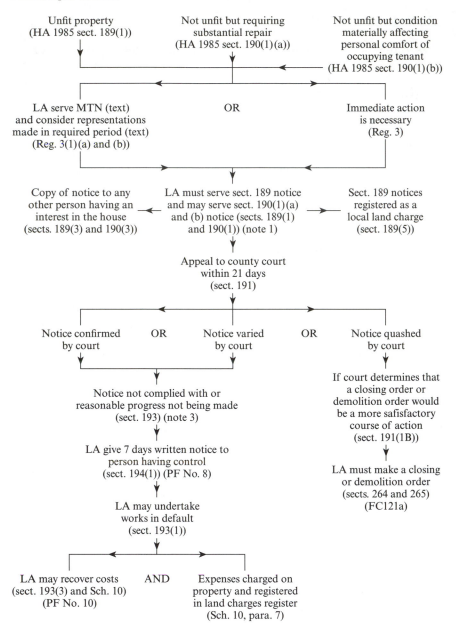

Unfit property
(HA 1985 sect. 189(1))

Not unfit but requiring
substantial repair
(HA 1985 sect. 190(1)(a))

Not unfit but condition
materially affecting
personal comfort of
occupying tenant
(HA 1985 sect. 190(1)(b))

LA serve MTN (text)
and consider representations
made in required period (text)
(Reg. 3(1)(a) and (b))

OR

Immediate action
is necessary
(Reg. 3)

Copy of notice to any
other person having an
interest in the house
(sects. 189(3) and 190(3))

LA must serve sect. 189 notice
and may serve sect. 190(1)(a)
and (b) notice (sects. 189(1)
and 190(1)) (note 1)

Sect. 189 notices
registered as a
local land charge
(sect. 189(5))

Appeal to county court
within 21 days
(sect. 191)

Notice confirmed
by court

OR

Notice varied
by court

OR

Notice quashed
by court

Notice not complied with or
reasonable progress not being made
(sect. 193) (note 3)

If court determines that
a closing order or
demolition order would
be a more safisfactory
course of action
(sect. 191(1B))

LA give 7 days written notice to
person having control
(sect. 194(1)) (PF No. 8)

LA must make a closing
or demolition order
(sects. 264 and 265)
(FC121a)

LA may undertake
works in default
(sect. 193(1))

LA may recover costs
(sect. 193(3) and Sch. 10)
(PF No. 10)

AND

Expenses charged on
property and registered
in land charges register
(Sch. 10, para. 7)

Notes
1. The prescribed forms are Nos. 1–5 inclusive of the 1990 Regulations as amended 1997.
2. The LA may carry out the works at the request of the person served and recover costs (sect. 191A).
3. Non-compliance with the notice constitutes an offence for which the maximum penalty is level 4 (sect. 198A). In addition or as an alternative to prosecutions the LA may undertake works in default.
4. Regulation numbers refer to the Housing (Fitness Enforcement Procedures) Order 1996.
5. The LA may recover reasonable charges for operating this procedure – see text.

(a) dwelling-houses, flats or HIMOs which are unfit for human habitation (sect. 189);
(b) dwelling-houses, flats or HIMOs which are fit but which require substantial repairs in order to bring them up to a reasonable standard having regard to their age, character and locality (sect. 190(1)(a) and (1A)(a)); and
(c) dwelling-houses, flats or HIMOs which are fit but in such a state of disrepair as to interfere materially with the personal comfort of the occupying tenants (sect. 190(1)(b) and (1A)(b)).

In relation to (b) and (c), notices may only be served where there is an occupying tenant or if the premises are in a renewal area (sect. 190(1B)).

Notices under sect. 189 may be useful in ensuring that HIMOs are provided with adequate lighting, ventilation and heating since these cannot be included in notices under sect. 352 (FC125).

Decision to serve repairs notice

In addition to the general guidance (pages 499–500) in deciding whether to serve a repairs notice under either sect. 189 or 190 the Code of Guidance requires the LA to consider:

(a) in cases where the work is likely to qualify for renovation grant assistance, whether the LA should exercise its discretion to provide a grant;
(b) the circumstances and wishes of the owner and occupiers; and
(c) the suitability of the premises for inclusion in a group repair scheme and the extent to which any such proposals have been developed (CoG para. 27).

Minded to take action notice (MTN)*

Except in the case where the LA considers immediate action is necessary (e.g. imminent risk to health and safety) the LA must first serve a MTN which gives the reasons why it is necessary to consider the service of a repairs notice under sect. 189 or sect. 190 and seeks any representations within not less than 14 days.

Oral representations may also be made if requested within 7 days of the service of the MTN. The LA must properly consider any representations made. If such procedure is not followed this is a ground of appeal against any subsequent repairs notice (The Housing (Fitness Enforcement Procedures) Order 1996 Regs 3 and 4).

Person responsible

Notices in respect of single occupancy houses are to be served on the person having control of the house whereas for HIMOs they may be served on the

---

* The DLTR intend to repeal this provision from a date yet to be determined.

person managing the house as an alternative to service on the person having control (sects. 189(1) and (2) and 190(1)).

Repairs notices

The notices must in each case specify the works required to remedy the particular conditions to which the notice relates and must specify a reasonable time, not less than 28 days, by which the works must be commenced and a reasonable period within which they must be completed (sects. 189(2) and 190(2)).

The notice becomes operative, if no appeal is brought, 21 days from the date of service (sects. 189(4) and 190(4)).

Notices under sect. 189 dealing with unfitness may require works of improvement as well as repair (sect. 189(2)(a)), while notices under sect. 190 cannot deal with internal decorative repair (sect. 190(2)).

Appeals

Any person aggrieved by a repairs notice may appeal against the notice to a County Court within 21 days. Grounds of appeal are not exhaustively specified but the following are specific grounds:

(a) that the works required are the responsibility of someone else who is an owner;
(b) that in the case of notices served under sect. 189 dealing with an unfit house, a demolition or closing order would have been a more satisfactory course of action (sect. 191(1)(1A)(1B)).

Where an appeal is made the notice does not become operative until after the County Court decision has been given and the period in which an appeal to the Court of Appeal is allowed has elapsed, or until the decision of the Court of Appeal is given (sect. 191(4)).

Enforcement

If the notice is not complied with, the LA may undertake the works in default after giving at least 7 days' notice and recover its expenses, including a power to sequester rents.

Non-compliance with the notice is deemed as being:

(a) where there is no appeal, after the date and period specified in the notice for commencement and completion;
(b) in the event of an appeal not withdrawn, after such dates and periods specified by the court for compliance;
(c) in relation to a withdrawn appeal, 21 days after the notice becomes operative and within the period specified in the notice.

The LA may also carry out works and recover expenses where it considers that, after works have commenced, reasonable progress is not being made towards their completion (sect. 193).

Where the person served with a notice under either sects. 189 or 190 asks the LA to undertake the works on his behalf, the LA may do this and recover costs (sect. 191A).

In addition to, or as an alternative to, carrying out the required works in default, the LA may prosecute for non-compliance with the notice for which the maximum penalty is a fine not exceeding level 4 on the standard scale (sect. 198A).

Obstruction

In addition to the general provisions relating to powers of entry (page 497) once a LA has given notice of its intention to carry out works in default following expiration of one of these notices, there is a penalty of max. level 3 for anyone convicted of obstructing the LA by attempting to carry out works themselves 7 days or more from the service of the notice of intention (sects. 194(2) and 198).

Charges

The LA may make a reasonable charge for its costs in taking action through repairs notices under either sect. 189 or 190. The costs may relate to:

(a) determining whether to serve the notice;
(b) identifying the works; and
(c) serving the notice.

The maximum charge is set at £300. A court in allowing an appeal may also reduce, quash or require payment of the charges (sect. 87 HGCRA 1996 and the Housing (Maximum Charge for Enforcement Action) Order 1996).

In demanding payment the LA must use prescribed form No. 5 (1996 Forms Regs).

Definitions

The following definitions are appropriate to this procedure:

House, flat or dwelling-house

A dwelling-house is a house if, and only if (or so much of it as does not consist of land included by virtue of sect. 184) is a structure reasonably so called; so that:

(a) where a building is divided horizontally, the flats or other units into which it is divided are not houses;

(b) where a building is divided vertically, the units into which it is divided may be houses;

(c) where a building is not structurally detached, it is not a house if a material part of it lies above or below the remainder of the structure;

(d) a dwelling-house which is not a house is a flat (sect. 183 HA 1985).

House in multiple occupation

See page 529.

Occupying tenant, in relation to a dwelling-house, means a person (other than an owner-occupier) who:

(a) occupies or is entitled to occupy the dwelling-house as a lessee; or

(b) is a statutory tenant of the dwelling-house; or

(c) occupies the dwelling-house as a resident under a restricted contract; or

(d) is a protected occupier, within the meaning of the Rent (Agriculture) Act 1976; or

(e) is a licensee under an assured agricultural occupancy.

Owner, in relation to premises:

(a) means a person (other than a mortgagee not in possession) who is for the time being entitled to dispose of the fee simple in the premises, whether in possession or reversion; and

(b) includes also a person holding or entitled to the rents and profits of the premises under a lease of which the expired term exceeds 3 years.

Owner-occupier, in relation to a dwelling-house, means the person who, as owner or lessee under a long tenancy, within the meaning of Part 1 Leasehold Reform Act 1967, occupies or is entitled to occupy the dwelling-house.

Person having control subject to sects. 189(1B), 190(1C) and 191:

(a) in relation to a dwelling-house or HIMO means the person who receives the rackrent of the premises (that is to say, a rent which is not less than two-thirds of the full net annual value of the premises), whether on his own account or as agent or trustee for another person, or who would so receive it if the premises were let at such a rackrent; and

(b) in relation to a part of a building to which relates a repair notice served under subsection (1A) of sect. 189 or sect. 190 means a person who is an owner in relation to that part of the building (or the building as a whole) and who, in the opinion of the authority by whom the notice is served, ought to execute the works specified in the notice.

Premises includes a dwelling-house or part of a building and, in relation to any premises any reference to a person having control shall be construed accordingly (sect. 207).

CLOSING ORDERS, DEMOLITION ORDERS AND DETERMINATIONS TO PURCHASE

References

Housing Act 1985 sects. 264–282.
The Housing (Fitness Enforcement Procedures) Order 1996.
The Housing (Maximum Charge For Enforcement Action) Order 1996.
The Housing (Deferred Action and Charge for Enforcement Action) (Forms) Regulations 1996.
Housing (Prescribed Forms) Regulations 1990 as amended 1997.
DoE Circular 17/96 Annex B Fitness Enforcement – Code of Guidance for Dealing with Unfit Premises.

Scope

This procedure deals with:

(a) the making of COs for an unfit dwelling-house or HIMO or for buildings consisting of or containing flats (sect. 264(1) and (2)); and

(b) the making of DOs for unfit dwelling-houses or HIMOs and for buildings containing one or more unfit flats (sect. 265(1) and (2)); and

(c) determination to purchase by the LA for use for temporary accommodation (sect. 300).

Before deciding upon the making of COs or DOs, the LA must be satisfied that this is the most satisfactory course of action after taking account of the guidance from the SoS (pages 499–500).

Closure of part of a dwelling-house or HIMO is not permitted except under the separate procedure relating to inadequate means of escape in the case of fire (FC128).

Minded to take action notices (MTNs)*

Except where a LA considers that immediate action is necessary (e.g. imminent risk to health and safety) before making a closing or demolition order the LA must first serve a MTN. This must give the reasons why the LA is considering such action and seeks any written representations within not less than 14 days. Oral representations may also be made if requested within 7 days of the service of the MTN. The LA must properly consider any representations made.

If such a procedure is not followed this is a ground of appeal against any subsequent CO or DO (The Housing (Fitness Enforcement Procedures) Order 1996 Regs. 3 and 4).

* The DLTR intend to repeal this provision from a date yet to be determined.

FC121a Unfit premises: DOs, COs

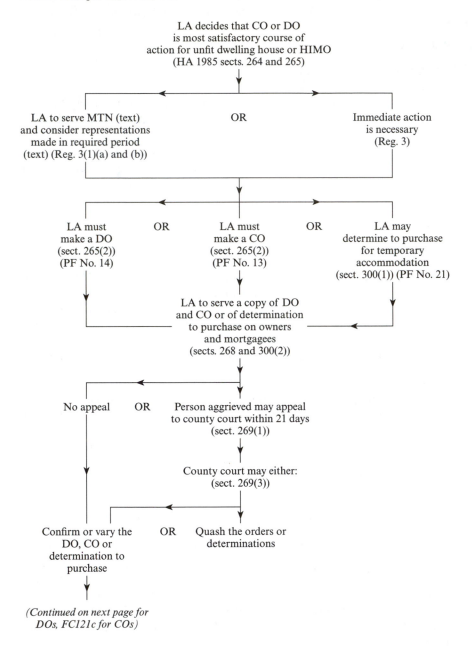

LA decides that CO or DO
is most satisfactory course of
action for unfit dwelling house or HIMO
(HA 1985 sects. 264 and 265)

LA to serve MTN (text)
and consider representations
made in required period
(text) (Reg. 3(1)(a) and (b))

OR

Immediate action
is necessary
(Reg. 3)

LA must
make a DO
(sect. 265(2))
(PF No. 14)

OR

LA must
make a CO
(sect. 265(2))
(PF No. 13)

OR

LA may
determine to purchase
for temporary
accommodation
(sect. 300(1)) (PF No. 21)

LA to serve a copy of DO
and CO or of determination
to purchase on owners
and mortgagees
(sects. 268 and 300(2))

No appeal

OR

Person aggrieved may appeal
to county court within 21 days
(sect. 269(1))

County court may either:
(sect. 269(3))

Confirm or vary the
DO, CO or
determination to
purchase

OR

Quash the orders or
determinations

*(Continued on next page for
DOs, FC121c for COs)*

Notes

1. Regulation numbers refer to the Housing (Fitness Enforcement Procedures) Order 1996.
2. PF numbers refer to the Housing (Prescribed Forms) Regulations 1990 as amended 1997.

FC121b Unfit premises: demolition orders

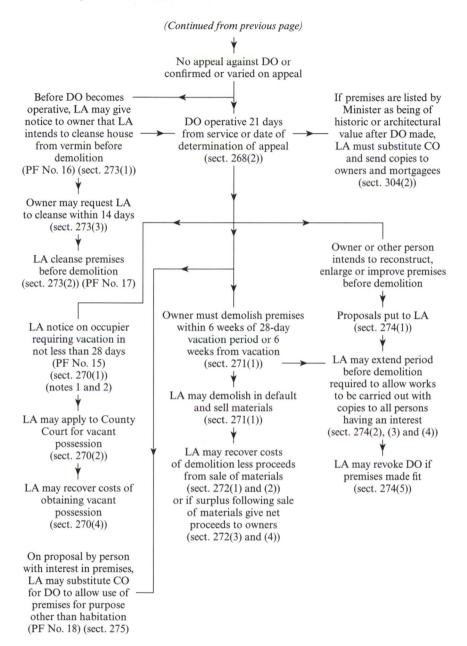

(Continued from previous page)

No appeal against DO or
confirmed or varied on appeal

DO operative 21 days
from service or date of
determination of appeal
(sect. 268(2))

Before DO becomes
operative, LA may give
notice to owner that LA
intends to cleanse house
from vermin before
demolition
(PF No. 16) (sect. 273(1))

Owner may request LA
to cleanse within 14 days
(sect. 273(3))

LA cleanse premises
before demolition
(sect. 273(2)) (PF No. 17)

LA notice on occupier
requiring vacation in
not less than 28 days
(PF No. 15)
(sect. 270(1))
(notes 1 and 2)

LA may apply to County
Court for vacant
possession
(sect. 270(2))

LA may recover costs of
obtaining vacant
possession
(sect. 270(4))

On proposal by person
with interest in premises,
LA may substitute CO
for DO to allow use of
premises for purpose
other than habitation
(PF No. 18) (sect. 275)

If premises are listed by
Minister as being of
historic or architectural
value after DO made,
LA must substitute CO
and send copies to
owners and mortgagees
(sect. 304(2))

Owner or other person
intends to reconstruct,
enlarge or improve premises
before demolition

Proposals put to LA
(sect. 274(1))

LA may extend period
before demolition
required to allow works
to be carried out with
copies to all persons
having an interest
(sect. 274(2), (3) and (4))

LA may revoke DO if
premises made fit
(sect. 274(5))

Owner must demolish premises
within 6 weeks of 28-day
vacation period or 6
weeks from vacation
(sect. 271(1))

LA may demolish in default
and sell materials
(sect. 271(1))

LA may recover costs
of demolition less proceeds
from sale of materials
(sect. 272(1) and (2))
or if surplus following sale
of materials give net
proceeds to owners
(sect. 272(3) and (4))

Notes
1. The maximum penalty for occupation after DO is effective is level 5 and £5 daily (sect. 270(5)).
2. Where suitable alternative accommodation on reasonable terms is not otherwise available the LA must rehouse the occupant (LCA 1973 sect. 39).
3. PF numbers refer to the Housing (Prescribed Forms) Regulations 1990 as amended 1997.

FC121c Unit premises: closing orders

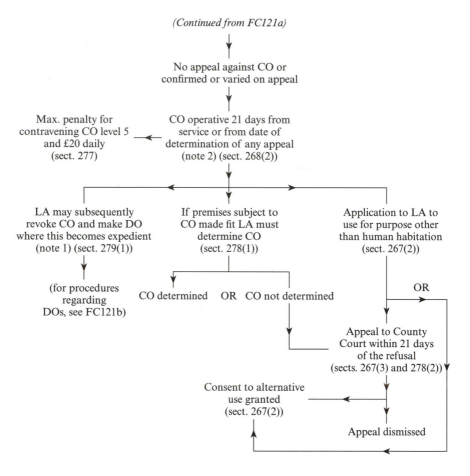

(Continued from FC121a)

No appeal against CO or
confirmed or varied on appeal

Max. penalty for
contravening CO level 5
and £20 daily
(sect. 277)

CO operative 21 days from
service or from date of
determination of any appeal
(note 2) (sect. 268(2))

LA may subsequently
revoke CO and make DO
where this becomes expedient
(note 1) (sect. 279(1))

If premises subject to
CO made fit LA must
determine CO
(sect. 278(1))

Application to LA to
use for purpose other
than human habitation
(sect. 267(2))

(for procedures
regarding
DOs, see FC121b)

CO determined   OR   CO not determined

OR

Appeal to County
Court within 21 days
of the refusal
(sects. 267(3) and 278(2))

Consent to alternative
use granted
(sect. 267(2))

Appeal dismissed

Notes
1. This cannot be done if the premises is listed or is a building subject to a closing order because it is to be listed or where the dwelling-house concerned is a flat or flat in multiple occupation (sect. 279(2) and (2A)).
2. Where suitable alternative accommodation on reasonable terms is not otherwise available the LA must rehouse the occupant (LCA 1973 sect. 39).

FC121d Unfit premises: determination to purchase

(Continued from FC121a)

↓

No appeal against
determination to purchase or
confirmed or varied on appeal

↓

Determination to purchase by
LA operative 21 days from
service or from date of
determination of appeal
(sect. 268(2))

↓

LA may purchase by
agreement or by CPO
confirmed by Minister
(sect. 300(3))

↓

House used by LA for
providing accommodation
adequate for the time being
(sect. 300(1))

↓

LA may carry out works to
keep house adequate
(sect. 302(b))

↓

LA may subsequently demolish
after temporary use for
accommodation

Demolition orders

(a) **Decision to make DO.** In deciding whether to make a DO the LA should consider:

  (i) whether the building is listed as being of architectural value or historic interest and if so whether a CO should be considered;

  (ii) the availability of accommodation for rehousing the occupants;

  (iii) the local environment, the suitability of the area for residential use and the impact of a cleared site (CoG. para. 37).*

(b) **Application.** Where the LA is satisfied having considered the code of guidance that a DO is the most satisfactory course of action it must make a DO unless:

  (i) it determines to purchase (FC121d); or

  (ii) the house is listed as being of architectural value or historic interest or is protected by notice pending listing (sect. 304).

(c) **Content.** DOs must require:

  (i) vacation within a period specified in the order but not less than 28 days from the date of operation of the order; and

  (ii) demolition of the premises within 6 weeks of the expiration of the period in (i) or the vacation of the premises, whichever is longer, or within such longer period as the LA may agree (sect. 267).

(d) **Service.** The DO is to be served on any person who is an owner or mortgagee (sect. 268).

(e) **Appeals.** Any person aggrieved by a DO, other than an occupant under a lease or agreement with an unexpired term of 3 years or less, may appeal to the County Court within 21 days of the service of the order. The court may vary, quash or confirm the order (sect. 269).

Closing orders

(a) **Decision to make CO.** In considering whether a CO is the most satisfactory course of action the LA must consider the following issues in addition to the general guidance on pages 499–500:

  (i) if the premises are listed or protected, the service of a DAN (FC119a);

  (ii) the position of the building in relation to neighbouring buildings;

  (iii) the potential alternative uses for the building;

  (iv) the existence of any conservation or renewal area and the proposals generally for the area;

  (v) the effect of closure on the cohesion and well-being of the community and the appearance of the locality; and

  (vi) the availability of accommodation for rehousing (CoG. para. 35).

(b) **Application.** A CO is an order prohibiting the use of the premises to which it relates for any purpose not approved by the local housing authority (sect. 267(2)).

---

* This is additional to the general guidance on pages 499–500.

(c) **Content.** A CO prohibits the use of the premises, or part, concerned for human habitation or otherwise without the consent of the LA. The approval by a LA of a use must not be unreasonably withheld (sect. 267(2) and (3)).

(d) **Service.** The CO is to be served on any owner and mortgagee (sect. 268).

(e) **Determination.** Once the LA is satisfied that the premises, or any part to which the CO relates, has been made fit for human habitation, it must determine the CO (sect. 278).

(f) **Appeals.** Any person aggrieved, other than an occupant with an unexpired term or agreement of 3 years or less, may appeal to the County Court within 21 days of the decision against either refusal by the LA to approve a use for the premises or to determine the CO either wholly or in part (sect. 269).

Determination to purchase

As an alternative to the making of a DO or CO, the LA may determine to purchase the premises where it appears that it is, or can be, rendered capable of providing accommodation which will be adequate for the time being (sect. 300(1)).

The provisions for the service of the notice of intention to purchase and appeals against it are the same as for DOs (page 516).

Compensation

The owner of premises subject to DOs and COs is entitled to compensation based on the diminution of the compulsory purchase value of the owners interest as a result of making the order. The tenants may be eligible for home loss and disturbance payments (sect. 584(A) and (B), LCA 1973 Part 3).

Charges

The LA may make a reasonable charge for its costs in making a CO or DO. The recoverable costs relate to determining whether to make the order and the service of related notices. The maximum charge is £300.

A court in allowing an appeal may also reduce, quash or require the payment of any LA charge (sect. 87 HGCRA 1996 and the Housing (Maximum Charge for Enforcement Action) Order 1996).

In demanding payment the LA must use the Form 5 as prescribed in the Housing (Deferred Action and Charge for Enforcement Action) (Forms) Regs. 1996.

Definitions

Owner in relation to premises:

(a) means a person (other than a mortgagee not in possession) who is for the time being entitled to dispose of the fee simple in premises whether in possession or in reversion; and

(b) includes the premises under a lease of which the unexpired term exceeds 3 years.

Premises, in relation to a DO or CO, means the dwelling-house, HIMO, building or part of a building in respect of which the CO or, as the case may be, DO is made (sect. 322).

CLEARANCE AREAS

References

Housing Act 1985 sect. 289–299.
Housing Grants, Construction and Regeneration Act 1996 part 4.
Land Compensation Act 1973 sect. 39 (as amended).
Housing (Prescribed Forms) Regulations 1990 as amended 1997.
DoE Circular 17/96 paras. 3.22–3.28 and Annex B Fitness Enforcement – Code of Guidance for Dealing with Unfit Premises.

Scope

Before declaring a CA the LA must be satisfied that this is the most satisfactory method of dealing with conditions in the area, having regard to the statutory Code of Guidance (sect. 604A).

In addition to the general requirements of the Code (pages 499–500), in respect of proposed CAs, LAs must consider:

(a) the degree of concentration of unfit premises within the area;
(b) the density of the buildings and the street pattern;
(c) the overall availability of housing in the wider neighbourhood;
(d) the proportion of fit premises and non-residential premises in sound condition which would need to be cleared;
(e) the need for the LA to acquire surrounding land;
(f) the existence of listed or protected buildings;
(g) the results of the statutory consultations (see below);
(h) the arrangements for displaced occupants and if occupants are satisfied with them;
(i) the impact of clearance on commercial premises;
(j) the suitability of the proposed after-use of the site, the needs of the wider neighbourhood and the socio-environmental benefits, the degree of support from local residents and the extent the after-use would attract private investment (CoG Annex B para. 3a).

FC122 Clearance areas

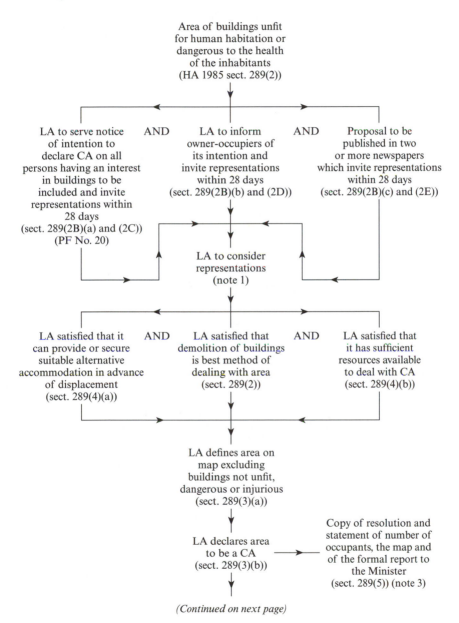

Area of buildings unfit
for human habitation or
dangerous to the health
of the inhabitants
(HA 1985 sect. 289(2))

LA to serve notice **AND** LA to inform **AND** Proposal to be
of intention to owner-occupiers of published in two
declare CA on all its intention and or more newspapers
persons having an interest invite representations which invite representations
in buildings to be within 28 days within 28 days
included and invite (sect. 289(2B)(b) and (2D)) (sect. 289(2B)(c) and (2E))
representations within
28 days
(sect. 289(2B)(a) and (2C))
(PF No. 20)

LA to consider
representations
(note 1)

LA satisfied that it **AND** LA satisfied that **AND** LA satisfied that
can provide or secure demolition of buildings it has sufficient
suitable alternative is best method of resources available
accommodation in advance dealing with area to deal with CA
of displacement (sect. 289(2)) (sect. 289(4)(b))
(sect. 289(4)(a))

LA defines area on
map excluding
buildings not unfit,
dangerous or injurious
(sect. 289(3)(a))

LA declares area Copy of resolution and
to be a CA statement of number of
(sect. 289(3)(b)) occupants, the map and
of the formal report to
the Minister
(sect. 289(5)) (note 3)

(Continued on next page)

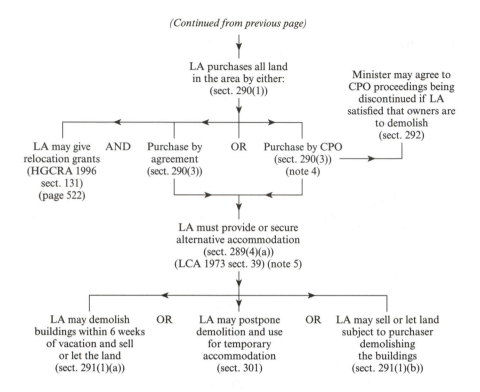

(Continued from previous page)

LA purchases all land
in the area by either:
(sect. 290(1))

Minister may agree to
CPO proceedings being
discontinued if LA
satisfied that owners are
to demolish
(sect. 292)

LA may give AND Purchase by OR Purchase by CPO
relocation grants agreement (sect. 290(3))
(HGCRA 1996 (sect. 290(3)) (note 4)
sect. 131)
(page 522)

LA must provide or secure
alternative accommodation
(sect. 289(4)(a))
(LCA 1973 sect. 39) (note 5)

LA may demolish OR LA may postpone OR LA may sell or let land
buildings within 6 weeks demolition and use subject to purchaser
of vacation and sell for temporary demolishing
or let the land accommodation the buildings
(sect. 291(1)(a)) (sect. 301) (sect. 291(1)(b))

Notes

1. After consideration of the representations the LA may decide:
   (a) to declare the area to be a CA; or
   (b) declare the CA but exclude certain residential buildings from the proposal; or
   (c) not declare a CA in which case other suitable action must be taken, i.e. service of repair notices or the making of COs and DOs (sect. 289(2F)).
2. Guidance on procedural requirements is given in DoE Circular 17/96.
3. The Minister has no powers of intervention at this stage, these powers arise during any subsequent CPO process.
4. Guidance on the procedural requirements of the CPOs is given in DoE Circular 17/96 Annex E.
5. Tenants are also entitled to home loss payments under sect. 29 LCA 1973 and/or disturbance payment under sect. 37.

Definition of CA

A CA is an area which is to be cleared of all buildings in accordance with this procedure (sect. 289(1)).

Consultation

Before proceeding to make a CA the LA must consult those people who will be directly affected by the service of a notice on them, i.e. on every person who has an interest, including freeholders, leaseholders and mortgagees, on those with an interest in any flats and their freeholders and subleaseholders and on all residential occupiers (sect. 289(2B)(a) and (b), (2C) and (2D)).

The LA must also place notices in at least two local newspapers circulating in the area (sect. 289(2B)(c) and (2E)).

In all cases the LA must invite representations within not less than 28 days (sect. 289(2B)(a), (b) and (c), (2C), (2D) and (2E)).

Other criteria

Before declaring an area to be a CA the LA must be satisfied that:

(a) the houses are unfit for human habitation (page 497), or by reason of their bad arrangement, or the narrowness or bad arrangement of the streets, are dangerous or injurious to the health of the inhabitants;
(b) any buildings other than houses are for a like reason dangerous or injurious to the health of the inhabitants of the area;
(c) that the most satisfactory way of dealing with the conditions, after having regard to the Code of Guidance including an economic assessment, is the demolition of all the buildings;
(d) the LA can provide or secure suitable accommodation for persons who will be displaced; and
(e) the resources of the LA are sufficient to implement the CA (HA 1985 sect. 289).

Clearance area map

This must be prepared before the resolution declaring a CA is passed and must exclude buildings which are not unfit or dangerous or injurious to health (sect. 289(3)(a)).

Purchase

The CA is implemented by the LA purchasing all land in the area by either:

(a) agreement; or
(b) CPO (sect. 290).

The LA may also acquire land which surrounds the CA and is needed to secure an area of convenient shape and dimension and land necessary for satisfactory development of the area (sect. 290(2)).

CPOs made under sect. 290 are regulated by the provisions of the Acquisition of Land Act 1981, the Compulsory Purchase of Land Regulations 1994 and the Compulsory Purchase of Non-Ministerial Acquiring Authorities (Inquiries Procedure) Rules 1990. DoE Circular 17/96 Annex E1 contains guidance. Advice on CPOs generally is given in 'The Compulsory Purchase Procedure Manual' published by TSO in November 2001.

A LA submitting a CPO to the Secretary of State for approval will be expected to deal with the following issues in its case:

(a) the justification for the CA having regard to the Code of Guidance Annex B;
(b) the principal grounds for unfitness;
(c) justification for any added lands;
(d) proposals for rehousing and for relocating commercial and industrial premises; and
(e) the proposed use for the cleared site (Circular 17/96 Annex E1 para. 3).

Compensation

Full market value is payable for all dwellings purchased via a CPO.

Home loss and disturbance payments are also required for those displaced (LCA 1973 sects. 29 and 37).

Relocation grants

In addition to market value compensation the HGCRA 1996 makes provision for LAs to pay relocation grants to assist low-income house owners in clearance areas where the LA considers this to be the most satisfactory course of action. The framework for these grants is provided for by sects. 131–40 of the 1996 Act and in the Relocation Grants Regulations 1997, the Relocation Grants (Forms of Application) Regulations 1997 (as amended) and in DoE Circular 17/97 which also contains statutory guidance.

This form of grant aid will be repealed from a date to be determined as a result of the provisions of the Regulatory Reform (Housing Assistance) (England and Wales) Order 2002 and will be replaced by a new general power for LAs to provide assistance across a range of housing functions – see pages 561 and 563.

Demolition

Having secured the land by agreement or CPO, the LA must arrange for the vacation of the buildings and demolition within 6 weeks of vacation or such longer period as the LA may consider reasonable. The land may subsequently be sold, let or exchanged.

Alternatively, the LA may sell or exchange the land with a condition that the purchaser secures the demolition of the buildings (sect. 291).

Temporary use of residential buildings

Although purchased for demolition, if the LA considers that any of the residential buildings can provide adequate temporary accommodation with or without works being carried out, it may postpone the demolition for such period as it may determine (sect. 301(1)).

The demolition of other buildings in the area may also be postponed where:

(a) the demolition of residential buildings has been postponed for temporary accommodation;
(b) they are required to support a residential building retained for that purpose; and
(c) there is some other special reason connected with the provision of temporary accommodation in the area (sect. 301(2)).

Definitions

Residential building. Buildings which are dwelling-houses or houses in multiple occupation or contain one or more flats (sect. 289(2)(a)).

Dwelling-house includes any yard, garden, outhouse and appurtenances belonging to it or usually enjoyed with it (sect. 322(2)).

GROUP REPAIR SCHEMES (GRS)

References

The Housing Grants, Construction and Regeneration Act 1996 sects. 60–75.
The Group Repair (Qualifying Buildings) Regulations 1996.
The Group Repair Schemes (England) General Approvals 1996 (at Annex D2 of circular 17/96).
DoE Circular 17/96 paras. 3.18–3.21 and Annexes D1, D2 and D3.
NB. Resulting from the Regulatory Reform (Housing Assistance) (England and Wales) Order 2002 these provisions are to be repealed from a date to be determined except in respect of schemes of which the date of approval is before that date. They are replaced by a new general power given to LAs to give assistance across a wide range of housing functions – see pages 561 and 563.

Scope

The object of a GRS is to secure the external fabric (definitions) of a group of properties so that they are in reasonable repair and are structurally stable. They may be pursued both within and outside of renewal areas (sect. 60 (1)).

FC123 Group repair schemes (GRS)

Works will put buildings
into reasonable repair
and/or make them
structurally stable
(sect. 62)

LA consider GRS to
be most satisfactory
course of action
(GA para. 5(e))

Buildings to be
included are qualifying
buildings – text
(sect. 61)

LA may prepare GRS ⟶ Consultation with
(sect. 60(1)) owners and residents
(paras. 16/17)

Scheme complies with
GA, LA may proceed
with written consent
of participants (text)
(sect. 63(1))

OR

Scheme does not
match GA criteria

Specific approval
of SoS required
(sect. 63)

Scheme approved
by SoS
(sect. 63)

OR

SoS approval not
given – scheme
does not proceed
(sect. 63)

LA to obtain scheme consent
from all participants
(sect. 65)

Scheme may be varied
before completion if it
accords with GA or SoS
gives specific approval and
with consent of participants
(sect. 68)

Scheme works
implemented by LA
(sect. 62)

LA to issue completion
certicate to each participant
(sect. 66)

LA secure contributions   AND
from participants
(sect. 67)

Assisted participants   AND
may discharge conditions
by paying balance
of cost at any time
(sect. 73)

Assisted participants
in breach of
conditions may be
requested to repay
balance of cost
(sects. 69–71)

Notes
1. GA refers to the Group Repair Schemes (England) General Approvals 1996.
2. This procedure is to be repealed from a date to be determined.

Qualifying buildings

To be included in a GRS a building (whole or part of a terrace of houses or other units) must have the whole or part:

(a) of its exterior not in reasonable repair; or
(b) of the structure unstable.

The Scheme must include at least one primary building, e.g. a terrace, comprising at least two dwellings. Additional buildings can be included if they comprise at least one dwelling and carrying out works at the same time as to the primary building is the most effective way of securing the repair and stability of them both (The Group Repair (Qualifying Building) Regulations 1996).

In determining what is 'reasonable repair' the LA must have regard to the age and character of the building and of its locality and must ignore internal decorative condition (sect. 96). An exterior of a building is not regarded as in reasonable repair unless it is substantially free from rising or penetrating damp (sect. 62(6)).

Selection of Schemes

LAs need to be satisfied following appraisal (including an economic appraisal detailed in Annex D3 or circ. 17/96) that a GRS is the most satisfactory course of action.

Non-economic factors to be considered are:

1. the existence of other problems in the area;
2. the overall condition of the properties, their layout and density;
3. the need for the particular type of housing;
4. the likelihood of improvement without a GRS;
5. alternative uses for the site;
6. the views of residents;
7. the social and demographic characteristics of the residents;
8. the effect of alternative courses of action on the community; and
9. visual impact of the GRS on the wider street environment (circ. 17/96 para. 319, Annex D2 paras. 4 and 5 and Annex D3).

Approval of Schemes

Schemes require the approval of the SoS but there is a general consent to proceed provided the Scheme complies with specified criteria (see below). Otherwise the Scheme must be submitted to the SoS with the following information:

(a) a plan indicating tenure, vacant properties, non-residential uses and previous public sector input, e.g. grants;
(b) the works proposed and justification for proceeding outside of the general consent;

(c) detailed costings for each property;

(d) total Scheme costs indicating public subsidy and private sector contributions; and

(e) an appraisal of the alternative courses of action having regard to the
economic appraisal in Annex D3 (sect. 63, the Group Repair Schemes
(England) General Approvals 1996 and circ. 17/96 Annex D1 para. 6).

Approval criteria

The conditions under which the general consent operates are:

(a) where all the qualifying buildings are in a renewal area:
   (i) there must be at least four houses;
  (ii) at least 60% must be affected by lack of reasonable external
        repair or structural instability;
 (iii) there must be no flats;
  (iv) the inclusion of buildings which are in reasonable repair or are
        not unstable must be necessary to give satisfactory visual effect
        to scheme work on houses that are so affected;
   (v) that the GRS is the most satisfactory course of action; and
  (vi) that the estimate of fees to be incurred does not exceed 15% of
        scheme costs.

(b) For buildings outside of a renewal area and in addition to the conditions above at least 25% of households must be in receipt of one or
more prescribed social benefit payments (the Group Repair Schemes
(England) General Approvals 1996).

Preparation of Scheme

The GRS is prepared by the LA in close liaison with the owners and residents. No scheme may proceed unless all of the eligible participants (below)
have signified their agreement via a scheme consent (sects. 63(1) and 65(1)).

Eligible participants

Persons are eligible to participate in a GRS if they:

(a) are the owner of a scheme property;

(b) control access to any building or have the consent of an occupier for
access;

(c) provide a certificate of owner-occupation that they (or members of
their family) intend to live there for 5 years after completion of work;
or

(d) give a certificate of intended letting or future occupation (for HIMOs)
for 5 years.

Owners not providing a certificate of owner-occupation may participate
as non-assisted participants, as may certain designated public bodies. By

giving this scheme consent eligible participants agree to the Scheme's being implemented and to contributing to its costs (see below).

Work may be done to a property without the consent of the owner if the LA is unable to trace that person or where work is necessary so that work to the rest of the building may be completed (sect. 64 and circ. 17/96 Annex D1 paras. 18–21).

Implementation

The LA arranges for the scheme works to be carried out and is respons-ible for letting and managing the contract, maintaining close liaison with participants.

Any variation to schemes in either participants, property or works requires the consent of new and existing participants and, unless in accord-ance with the criteria in the general consent (above), requires the approval of the SoS (sect. 68).

Certificates of completion

Once the LA is satisfied that the Scheme works have been completed satis-factorily it is required to issue a certificate of completion to each assisted (below) participant. This indicates the date from which the 5-year operation of conditions will operate and during which 'clawback' (below) is possible (sect. 66).

Contributions by participants

The costs of the Scheme are apportioned by the LA to each property on the basis of work carried out within its curtilage.

Unassisted participants are required to meet the full costs while assisted participants pay contributions according to the following circumstances:

1. owners of non-dwellings – 25% in a renewal area and 50% elsewhere;
2. owners of dwellings and HIMOS – not more than 25% in renewal areas and 50% elsewhere calculated by the use of the means test applied to renovation grants set out in Housing Renewal Grants Regu-lations 1996 (sect. 64 and circ. 17/96 Annex D1 paras. 24–28).

Payment of balance of costs ('clawback')

If an assisted participant:

(a) disposes of the property within 5 years;
(b) allows the dwelling to be unoccupied, not let or not kept available for letting or residential occupation in accordance with certificates of owner-occupation or letting; or
(c) fails to provide the LA with information about the fulfilment of certificate condition within 21 days;

the owner, on demand from the LA, must pay the difference between the cost of the works to that property and the amount of his contribution. The LA does have discretion to waive payments in some circumstances and owners may discharge the clawback provisions by paying the requisite sum at any time (sects. 69–73 and circ. 17/96 Annex D1 paras. 29–32).

Exchequer contributions

The Government pays to the LA a contribution of 60% of the costs of the Scheme works less the contributions from participants (the Housing Renewals Grants etc. (Exchequer Contributions) (England) Determination 1996).

Definitions

Exterior of a building means:

(a) any part of the building which is exposed to the elements of wind and rain or otherwise faces into the open air (including in particular roofs, chimneys, walls, doors, windows, rainwater goods and external pipework); and
(b) the curtilage of the building, including any wall within the curtilage which is constructed as a retaining wall or otherwise protects the structure of the building (sect. 62(4)).

Owner of a dwelling means the person who:

(a) is for the time being entitled to receive from a lessee of the dwelling ... a rent at an annual rate of not less than two-thirds of the net annual value of the dwelling; and
(b) is not himself liable as lessee of the dwelling, or of property which includes the dwelling, to pay such rent to a superior landlord (sect. 99(1)).

HOUSES IN MULTIPLE OCCUPATION

DEFINITIONS

In this section:

Dispossessed proprietor, in relation to a house subject to a control order, means the person by whom the rent or other periodical payments to which the local housing authority become entitled on the coming into force of the order would have been receivable but for the making of the order, and the successors in title of that person (sect. 399).

House includes any yard, garden, outhouses and appurtenances belonging to the house or usually enjoyed with it (sect. 399).

House in multiple occupation means a house which is occupied by persons who do not form a single household. For the purposes of this section, 'house' in the expression 'house in multiple occupation', includes any part of a building which:

(a) apart from this subsection would be regarded as a house; and
(b) was originally constructed or subsequently adapted for occupation by a single household;

and any reference in this Part to a flat in multiple occupation is a reference to a part of a building which, whether by virtue of this subsection or without regard to it, constitutes a house in multiple occupation (sect. 345).

For guidance on the interpretation of this definition see DoE Circular 5/90 and 12/93.

Lessee includes a statutory tenant of the premises, and references to a lessee or to a person to whom premises are let shall be construed accordingly (sect. 398(2)).

Owner (a) means a person (other than a mortgagee not in possession) who is for the time being entitled to dispose of the fee simple of the premises whether in possession or in reversion; and

(b) includes also a person holding or entitled to the rents and profits of the premises under a lease having an unexpired term exceeding 3 years (sect. 398(3)).

Person having control means the person who receives the rackrent of the premises, whether on his own account or as agent or trustee of another person, or who would so receive it if the premises were let at a rackrent (and for this purpose a 'rackrent' means a rent which is not less than two-thirds of the full net annual value of the premises) (sect. 398(5)).

Person having an interest or estate includes a statutory tenant of the premises (sect. 398(4)).

Person managing (a) means the person who, being an owner or lessee of the premises:

(i) receives, directly or through an agent or trustee, rents or other payments from persons who are tenants of parts of the premises, or who are lodgers; or
(ii) would so receive those rents or other payments but for having entered into an arrangement (whether in pursuance of a court order or otherwise) and another person who is not an owner or lessee of the premises by virtue of which that other person receives the rents or other payments and

(b) includes, where those rents or other payments are received through another person as agent or trustee, that other person (sect. 398(6) as amended by sect. 79(2) HA 1996).

Statutory tenant means a statutory tenancy or statutory tenant within the meaning of the Rent Act 1977 or the Rent (Agriculture) Act 1976 (sect. 622).

REGISTRATION SCHEMES

References

Housing Act 1985 sects. 346–351 as amended by the Housing Act 1996.
Houses in Multiple Occupation (Charges for Registration Schemes) Order 1997 as amended 1998.
Houses in Multiple Occupation (Charges for Registration Schemes) Order 1998.

DoE Circulars 5/90, 6/91, 12/93 and 3/97 which includes Model Registration Schemes.

Scope

A LA may make a scheme for the registration of some or all HIMOs in the whole or part of its area (sect. 346).

Such schemes may either involve a simple notification scheme, where the LA has no discretion about registration, or be a control scheme in which the LA is provided with powers of refusal and revocation (sect. 346A and 347).

Model schemes of both types have been issued by the SoS and are included at Annex C1 and C2 of circular 3/97 (sect. 346B).

Application of schemes

Both model schemes describe the same exclusions from any registration requirement, the principal ones of which are:

(a) a house occupied by persons who form only two households;
(b) a house occupied by no more than four persons who form more than two households;
(c) a house occupied by no more than three people in addition to the responsible person (person having control or managing) and his household;
(d) where the accommodation consists entirely of self-contained flats;
(e) where the house is owned by LAs, health service bodies, social landlords and universities and other higher education institutions.

Schemes which conform to the model do not require confirmation by the SoS but schemes which do not must be submitted for confirmation (sects. 346B(3) and (4)).

Publicity

Before making a scheme which does not require confirmation by the SoS the LA must publish a notice of its intent at least one month before so doing in at least one local newspaper. The LA must similarly publicize its intent to submit a scheme to the SoS for approval. Notices must also be published as soon as schemes are made or confirmed.

FC124 Registration schemes

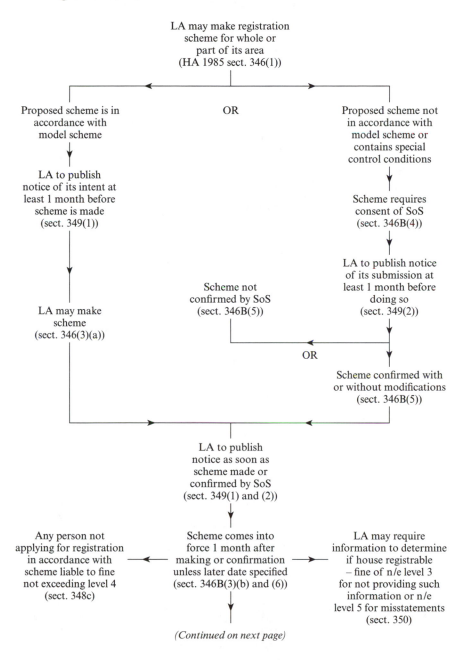

LA may make registration
scheme for whole or
part of its area
(HA 1985 sect. 346(1))

Proposed scheme is in
accordance with
model scheme

OR

Proposed scheme not
in accordance with
model scheme or
contains special
control conditions

LA to publish
notice of its intent at
least 1 month before
scheme is made
(sect. 349(1))

Scheme requires
consent of SoS
(sect. 346B(4))

LA to publish notice
of its submission at
least 1 month before
doing so
(sect. 349(2))

LA may make
scheme
(sect. 346(3)(a))

Scheme not
confirmed by SoS
(sect. 346B(5))

OR

Scheme confirmed with
or without modifications
(sect. 346B(5))

LA to publish
notice as soon as
scheme made or
confirmed by SoS
(sect. 349(1) and (2))

Any person not
applying for registration
in accordance with
scheme liable to fine
not exceeding level 4
(sect. 348c)

Scheme comes into
force 1 month after
making or confirmation
unless later date specified
(sect. 346B(3)(b) and (6))

LA may require
information to determine
if house registrable
– fine of n/e level 3
for not providing such
information or n/e
level 5 for misstatements
(sect. 350)

(Continued on next page)

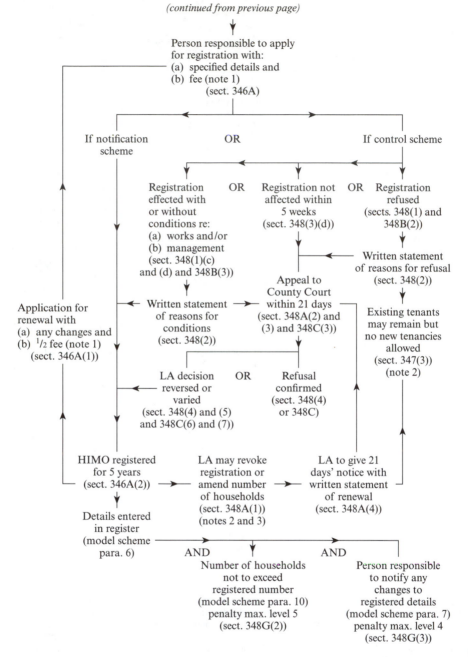

(continued from previous page)

Person responsible to apply
for registration with:
(a) specified details and
(b) fee (note 1)
        (sect. 346A)

If notification   OR   If control scheme
scheme

Registration   OR   Registration not   OR   Registration
effected with   affected within   refused
or without   5 weeks   (sects. 348(1) and
conditions re: (sect. 348(3)(d))   348B(2))
(a) works and/or
(b) management   Written statement
(sect. 348(1)(c)   of reasons for refusal
and (d) and 348B(3))   Appeal to   (sect. 348(2))
                           County Court
Application for   Written statement → within 21 days   Existing tenants
renewal with   of reasons for   (sect. 348A(2) and   may remain but
(a) any changes and   conditions   (3) and 348C(3))   no new tenancies
(b) ½ fee (note 1)   (sect. 348(2))   allowed
(sect. 346A(1))   (sect. 347(3))
                                                                   (note 2)
                   LA decision   OR   Refusal
                   reversed or   confirmed
                   varied   (sect. 348(4)
                   (sect. 348(4) and (5)   or 348C)
                   and 348C(6) and (7))

HIMO registered   LA may revoke   LA to give 21
for 5 years   registration or   days' notice with
(sect. 346A(2)) →   amend number   →   written statement
                           of households   of renewal
Details entered   (sect. 348A(1))   (sect. 348A(4))
in register   (notes 2 and 3)
(model scheme
para. 6)   AND   AND
                   Number of households   Person responsible
                   not to exceed   to notify any
                   registered number   changes to
                   (model scheme para. 10)   registered details
                   penalty max. level 5   (model scheme para. 7)
                   (sect. 348G(2))   penalty max. level 4
                                                     (sect. 348G(3))

Notes
1. Maximum fees are set by the Houses in Multiple Occupation (Fees for Registration Schemes) Order 1997 as amended 1998 – see text.
2. Where special control provisions exist, the LA has power to make occupancy directions (text).
3. The LA must publish a notice of the revocation (sect. 349(5)).

After publication of such a notice and for so long as any scheme remains in force the LA must keep a copy available for public inspection at no cost and also supply copies of the scheme or entries in the register at a reasonable fee (sect. 349).

Operation of scheme

Schemes become operative one month after being made or being confirmed or on such later date as may be specified by either the LA or the SoS (sect. 346B(3)(b) and (6)).

Control schemes

Such schemes provide the LA with powers to:

(a) refuse applications if the premises are unsuitable;
(b) refuse applications where the responsible person is not a fit and proper person;
(c) require works before registration;
(d) impose management conditions;
(e) alter during registration the number of households allowed;
(f) revoke a registration where conditions are breached or the responsible person is not fit and proper (sect. 347).

There is provision for a scheme to include special control provisions designed to prevent HIMOs from adversely affecting the amenity or character of the area through their existence or the behaviour of its residents. The inclusion of such conditions does however require the scheme to be confirmed by the SoS and this will only be forthcoming if both:

(a) a problem already exists with HIMOs seriously adversely affecting the area; and
(b) a problem is likely to arise in the future (sect. 348C).

Application

The responsible person must apply to the LA in writing (no prescribed form):

(a) indicating details of the property as specified in the scheme (para. 6 of Model Scheme) or details of change if for a renewal, and
(b) enclosing the required fee (sect. 346A(4)).

Fees

LAs may, at their discretion, make charges for dealing with applications for registration for schemes containing control provisions of an amount up to a maximum of £60 for each habitable room (bedrooms and living rooms). Charges must be made for registrations at a level up to £80 per house for non-control schemes and £60 per habitable room for control schemes. However, if a charge has been made on application for registration or renewal of registration, no further fee is payable on registration. On renewal of registration the fee payable is up to half of these figures (Houses in Multiple Occupation (Fees for Registration Schemes) Order 1997 as amended 1998 and the Houses in Multiple Occupation (Fees for Registration Schemes) Order 1998).

Registration

The period of registration and its renewal in all cases is 5 years (sect. 346A(2)). For notification schemes, the LA must effect the registration within 5 weeks and has no other discretion.

The LA has these options available to decide control scheme applications:

(a) refusal where:
    (i) the house is unsuitable and incapable of being made suitable;
    (ii) the responsible person is not a fit and proper person;
(b) registration conditional upon:
    (i) specific works being carried out in a specified period; and/or
    (ii) specified management arrangements;
(c) unconditional registration.

In the case of a refusal or the application of conditions the LA must give a written statement indicating its reasons for the decision. Applications which have not been dealt with within 5 weeks (or longer if mutually agreed) are deemed to be refused. Applicants may appeal to the county court within 21 days (sect. 348).

Where an application is refused (or revoked) persons in residence at the time may remain but no new lettings may be made until the house ceases to be a HIMO (sect. 347(3)).

Revocation/alteration of registration

In control schemes the LA may revoke registration where:

(a) the house is unsuitable and incapable of being made suitable;
(b) the responsible person is not a fit and proper person;
(c) works are required or are not executed within the specified time; and
(d) there has been a breach of any management conditions.

LAs may also alter the registered number of landlords/persons where works are required or the house becomes unsuitable for the present numbers. In the case of revocation and alteration to the register the LA must give a written statement of reasons and there is appeal to the county court within 21 days (sect. 348A)).

NOTICES TO RENDER HIMOs FIT FOR A
NUMBER OF OCCUPANTS

References

Housing Act 1985 sects. 352–3, 375–7 as amended by the Housing Act 1996.
The Housing (Enforcement Procedures for Houses in Multiple Occupation) Order 1997.
The Housing (Recovery of Expenses for section *352* Notices) Order 1997.
The Housing (Fire Safety in HIMOs) Order 1997.
The Housing (Prescribed Forms) Regulations 1990 as amended 1997.
DoE Circulars 5/90, 12/92 12/93 and 3/97.
Houses in Multiple Occupation: Guidance on Standards. DETR. 28 April 1999.

Scope

The procedure applies where a HIMO fails to meet one or more of the specified requirements **and**, having regard to a number of individuals or households, or both, living there, the premises are not reasonably suitable for occupation. The requirements are that:

(a) there are satisfactory facilities for the storage, preparation and cooking of food including an adequate number of sinks with a satisfactory supply of hot and cold water;
(b) it has an adequate number of suitably located water closets for the exclusive use of the occupants;
(c) it has, for the exclusive use of the occupants, an adequate number of suitably located fixed baths or showers and hand wash basins each of which is provided with a satisfactory supply of hot and cold water;
(d) subject to sect. 365, there are adequate means of escape from fire; and
(e) there are adequate other fire precautions (sect. 352(1) and (1A)).

The general repair of HIMOs may be dealt with under sect. 189 of the HA 1985 by the procedure in FC121a.

Preliminary notices*

Unless the LA considers that there is a need for immediate enforcement action, they must first serve a preliminary ('minded-to') notice which informs the person responsible:

(a) that the LA is considering serving a sect. 352 notice;
(b) the reasons why; and
(c) the work required.

The recipient may make written representations within 14 days or oral representations if they are requested within 7 days. The LA must have regard to them.

---

* The DLTR intend to repeal this provision from a date to be determined.

FC125 Notices to render HIMOs fit for a number of occupants

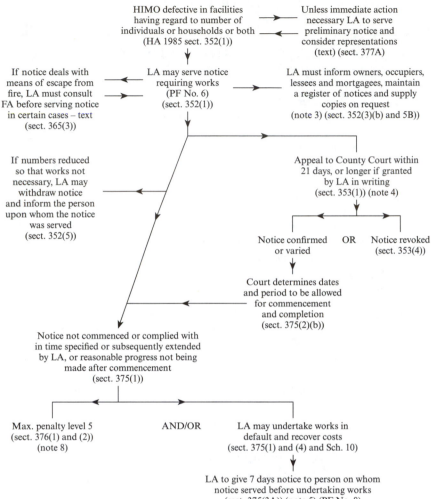

HIMO defective in facilities
having regard to number of
individuals or households or both
(HA 1985 sect. 352(1))

Unless immediate action
necessary LA to serve
preliminary notice and
consider representations
(text) (sect. 377A)

If notice deals with
means of escape from
fire, LA must consult
FA before serving notice
in certain cases – text
(sect. 365(3))

LA may serve notice
requiring works
(PF No. 6)
(sect. 352(1))

LA must inform owners, occupiers,
lessees and mortgagees, maintain
a register of notices and supply
copies on request
(note 3) (sect. 352(3)(b) and 5B))

If numbers reduced
so that works not
necessary, LA may
withdraw notice
and inform the person
upon whom the notice
was served
(sect. 352(5))

Appeal to County Court within
21 days, or longer if granted
by LA in writing
(sect. 353(1)) (note 4)

Notice confirmed   OR   Notice revoked
or varied   (sect. 353(4))

Court determines dates
and period to be allowed
for commencement
and completion
(sect. 375(2)(b))

Notice not commenced or complied with
in time specified or subsequently extended
by LA, or reasonable progress not being
made after commencement
(sect. 375(1))

Max. penalty level 5
(sect. 376(1) and (2))
(note 8)

AND/OR

LA may undertake works in
default and recover costs
(sect. 375(1) and (4) and Sch. 10)

LA to give 7 days notice to person on whom
notice served before undertaking works
(sect. 375(3A)) (note 5) (PF No. 8)

Notes

1. For directions to reduce overcrowding in HIMOs, see FC129.
2. For penalties for obstruction of owners by occupiers and of LA by owners or occupier in relation to carrying out of works, see sect. 377.
3. Notices are to be registered as local land charges (sect. 352(5A)).
4. If the appeal is lodged but subsequently withdrawn before being heard, the works must commence within 21 days after the date of the appeal being withdrawn and completed within the period specified in the notice (sect. 375(2)(c)).
5. If the works are carried out before the LA commences works, the LA may recover its costs up to that point including administrative costs (sect. 375(3B)).
6. The power of entry (page 497) to enforce these notices does not require prior notice of intention to enter.
7. For grants relating to the repair of HIMOs see FC132.
8. A penalty is incurred not only for an initial failure to comply, but also for a continuing failure after the initial conviction.

Failure to undertake this procedure is a ground of appeal against any sect. 352 notice served (sect. 377A and the Housing (Enforcement Procedures for HIMOs) Order 1997).

Notices – the form of notice is prescribed in the Housing (Prescribed Forms) Regulations 1990 as amended 1997

    (a) **Content.** The notice must specify both the requirements of sect. 352 which have not been met and the works required to make the house suitable either:
       (i) for occupation by the individuals or households accommodated there; or
      (ii) for a smaller number as specified in the notice.

Works to any premises outside of the house cannot be included in the notice (sect. 352(2)).

    (b) **Person responsible.** The notice is served on either:
       (i) the person having control of the house; or
      (ii) the person managing the house;
    and the LA must inform any other owners, tenants, lessees and mortgagees that the notice has been served (sect. 352(3)).
    (c) **Compliance.** The notice must specify:
       (i) a reasonable date, not less than 21 days, by which the works must be commenced; and
      (ii) a reasonable period within which the works must be completed (sect. 352(4)).
    (d) **General.** Notices are to be registered as local land charges and therefore new notices do not have to be served upon a change of ownership. The LA must retain a register of all such notices served. Any person is entitled to a copy of the notice on the payment of a reasonable fee (sect. 352(5A) and (5B)).

Restriction on service of sect. 352 notices

After a sect. 352 notice has been served and complied with, the LA cannot serve a further notice for similar requirements for 5 years unless there has been a change in circumstances (HA 1996 sect. 71).

Appeals

A person upon whom the sect. 352 notice is served may appeal to the County Court within 21 days from the service of the notice on any of the following grounds:

    (a) that the condition of the premises did not justify the authority, having regard to the requirements set out in subsection (1A) of that section, in requiring the execution of the works specified in the notice;

(b) in the case of a notice under subsection (2)(b) of that section (notice requiring works to render premises fit for smaller number of occupants), that the number of individuals or households, or both, specified in the notice is unreasonably low;

(c) that there has been some material informality, defect or error in, or in connection with, the notice;

(d) that the authority has refused unreasonably to approve the execution of alternative works, or that the works required by the notice to be executed are otherwise unreasonable in character or extent, or are unnecessary;

(e) that the date specified for the beginning of the works is not reasonable;

(f) that the time within which the works are to be executed is not reasonably sufficient for the purpose; or

(g) that some other person is wholly or partly responsible for the state of affairs calling for the execution of the works, or will as holder of an estate of interest in the premises derive a benefit from their execution, and ought to pay the whole or a part of the expenses of executing them;

(h) that the LA has not complied with the procedure relating to the need to serve a preliminary notice (sect. 353(2) and (3) and sect. 76 HA 1996).

Withdrawal

The notice may be withdrawn by the LA if it is satisfied that the number of individuals has been reduced to a level where works are no longer necessary and that those numbers will be maintained, whether or not a direction order is used. A further notice may be served later if required (sect. 352(5)).

Enforcement

The LA may itself undertake the required works and recover costs:

(a) when the works have not been commenced by the required date and no appeal has been made;

(b) where an appeal has been made but withdrawn, not less than 21 days after the notice has been withdrawn;

(c) where an appeal is heard, after such period as specified by the court has elapsed (sect. 375(1) and (2)).

Where works have been commenced but in the opinion of the LA reasonable progress is not being made, the LA may also complete the works (sect. 375(3)).

In all cases the LA must give at least 7 days' notice to the person upon whom the notice was served and any other owner before commencing works (sect. 372(3A)).

Where the LA is refused consent to undertake works in default, an application may be made to a magistrates' court for an order (sect. 377(2)(b)).

In addition to or as an alternative to undertaking works in default the LA may prosecute for non-compliance with the notice for which the penalty is up to level 4 (sect. 376).

Recovery of LA expenses

The LA is able to make a reasonable charge on the person served with a sect. 352 notice to recover the expenses associated with:

(a) determining whether a notice should be served;
(b) identifying the works required; and
(c) serving the notice (sect. 352A).

The maximum charge for any one notice is £300. PF No.10A is to be used to demand recovery of these costs (the Housing (Recovery of Expenses for Section 352 Notices) Order 1997 and the Housing (Prescribed Forms) (Amendment) Regulations 1997).

Direction orders

A direction order under sect. 354 (FC129) can be used in parallel with a notice under this procedure in order to establish a limit to the occupancy which, if it is a lower level than currently in occupation, can be achieved through normal tenant movement rather than necessitating eviction. The notice under sect. 352 may require works only with regard to the lower numbers specified in the direction order (sect. 352(2A)).

Fire precautions

The provision of an adequate means of escape from fire and of adequate other fire precautions, e.g. fire detection systems, fire warning systems and fire fighting equipment, can be a requirement of a notice under this procedure. However, where the HIMO does not have adequate means of escape an alternative procedure to secure the closure of part of the house is provided in sect. 368 (FC128) and, where this is used, the notice under sect. 352 may specify works required having regard to the closure of that part (sect. 352(2A), 365 and 368).

Before serving a notice under sect. 352 dealing with means of escape, the LA is required to consult with the FA (sect. 365(3)).

LA duty to ensure adequate fire precautions in HIMOs

LAs are under a duty to secure improvements to sub-standard fire safety precautions in houses of at least 3 storeys other than:

(a) certain exempted premises in article 2, e.g. children's houses;
(b) LA-owned property;

(c) those subject to a control order;
(d) where occupied by persons who form only 2 households;
(e) where occupied by no more than 4 people;
(f) where occupied by no more than 3 people in addition to the responsible person and members of his household;
(g) where the living accommodation is entirely of self-contained flats; and
(h) houses where the responsible person is a social landlord.

The LA may either use this procedure under sect. 352 or the later procedure FC128 dealing with closing orders where means of escape etc. are not adequate (sect. 365 and the Housing (Fire Safety in HIMOs) Order 1997).

Consultation with Fire Authority

Where the LA is proposing action either under sect. 352 or under sect. 368 (FC128) in relation to the following types of HIMO:

(a) those for which it has a duty ((1)(a)–(h) above);
(b) houses registered under the Childrens Act 1989;
(c) where the responsible person is a:
  (i) health service body;
  (ii) university or higher education institution;
  (iii) a further education establishment;
  (iv) approved under the Probation Services Act 1993;
  (v) where there is a valid fire certificate;

they must consult with the Fire Authority before taking the enforcement action (sect. 365 and the Housing (Fire Safety in HIMOs) Order 1997).

OVERCROWDING IN HIMOs

Reference

Housing Act 1985 sects. 358–364.
DoE Circular 12/93.
The Housing (Prescribed Forms) Regulations 1990 as amended 1997.

Scope

The procedure may be applied to any house which is in multiple occupation (for a definition, page 529). This notice procedure may be used either on its own or in combination with a direction order under sect. 354 (FC129).

Overcrowding

The LA must be satisfied that the house is, or is likely to be, occupied by an excessive number of persons having regard to the rooms available

FC126 Overcrowding in HIMOs

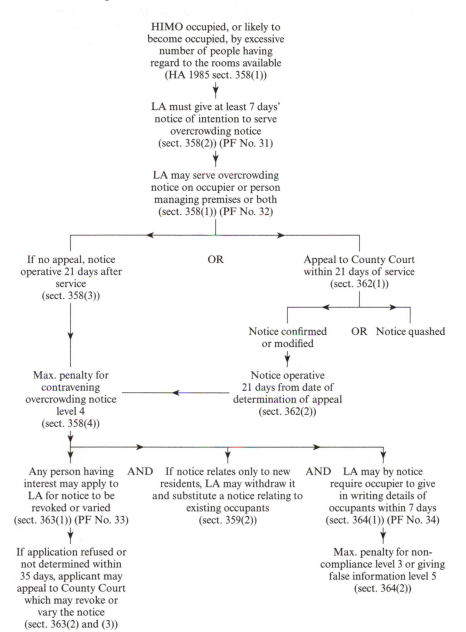

HIMO occupied, or likely to
become occupied, by excessive
number of people having
regard to the rooms available
(HA 1985 sect. 358(1))

LA must give at least 7 days'
notice of intention to serve
overcrowding notice
(sect. 358(2)) (PF No. 31)

LA may serve overcrowding
notice on occupier or person
managing premises or both
(sect. 358(1)) (PF No. 32)

If no appeal, notice
operative 21 days after
service
(sect. 358(3))

OR

Appeal to County Court
within 21 days of service
(sect. 362(1))

Notice confirmed
or modified

OR Notice quashed

Max. penalty for
contravening
overcrowding notice
level 4
(sect. 358(4))

Notice operative
21 days from date of
determination of appeal
(sect. 362(2))

Any person having
interest may apply to
LA for notice to be
revoked or varied
(sect. 363(1)) (PF No. 33)

AND

If notice relates only to new
residents, LA may withdraw it
and substitute a notice relating to
existing occupants
(sect. 359(2))

AND

LA may by notice
require occupier to give
in writing details of
occupants within 7 days
(sect. 364(1)) (PF No. 34)

If application refused or
not determined within
35 days, applicant may
appeal to County Court
which may revoke or
vary the notice
(sect. 363(2) and (3))

Max. penalty for non-
compliance level 3 or giving
false information level 5
(sect. 364(2))

Notes
1. For provision of facilities in HIMOs, see FC125.
2. Prescribed forms 31–34 inclusive of the Housing (Prescribed Forms) (No. 2) Regulations 1990 as amended 1997 relate to this procedure.
3. For direction orders to reduce overcrowding in HIMOs, see FC129.

(sect. 358). There is no statutory test to determine whether or not a HIMO is overcrowded as there is under Part 10 of the Act dealing with overcrowding in dwellings. For HIMOs under this procedure the test is subjective and depends on the LA's judgement of what is an excessive number.

Notice of intention

Before serving an overcrowding notice, the LA must give at least 7 days' written notice to the occupier of the premises and any person appearing to have the control and management of it. The LA must also, as far as reasonably practicable, notify every person living in the house. In both cases the LA must afford such persons an opportunity of making representations (sect. 358(2)).

Overcrowding notices

(a) **Person responsible.** Notices may be served on the occupier or on the person managing the premises (sect. 358(1)).

(b) **Content.** These notices:

(i) must state the maximum number of persons suitable to use each room as sleeping accommodation at any one time or state that specified rooms are unsuitable for sleeping accommodation; and

(ii) may prescribe special maxima for rooms where some or all of the occupants are below a specified age (sect. 359);

and in addition must include either:

(iii) a requirement to refrain from permitting occupation in excess of that specified in the overcrowding notice or permitting such number of persons to use the house for sleeping accommodation that it is not possible, without contravening the notice, to avoid persons of opposite sexes over 12 years old (other than persons living together as man and wife) occupying accommodation in the same room (sect. 360); or

(iv) a requirement to refrain from permitting new residents to take up occupation otherwise than in accordance with the notice or in circumstances where it is not possible without contravening the notice to avoid persons of opposite sexes over 12 years old (other than persons living together as man and wife) occupying accommodation in the same room (sect. 361).

Where a requirement in accordance with (iv) above is specified, the LA may at any time withdraw that notice and serve a new notice containing a requirement as in para. (iii) above (sect. 359(2)).

Requests for information

At any time following the operation of an overcrowding notice the LA may by written notice require the occupier to give it the following information in writing within 7 days:

(a) the number of individuals using any part of the house for sleeping accommodation on a date specified in the LA notice;
(b) the number of families or households to which those individuals belong;
(c) the names of those individuals and of the heads of those families or households; and
(d) the rooms used by those individuals and families or households (sect. 364).

NOTICES TO REMEDY NEGLECT OF MANAGEMENT

References

Housing Act 1985 sects. 369–378 (as amended 1996).
Housing (Management of Houses in Multiple Occupation) Regulations 1990.
The Housing (Enforcement Procedures for HIMOs) Order 1997.
The Housing (Prescribed Forms) Regulations 1990 as amended 1997.
DoE Circular 5/90, 12/93, and 3/97.

Scope

Where the LA considers that the condition of a HIMO fails to comply with the conditions specified in the Housing (Management of HIMOs) Regulations 1990, it may serve a notice requiring the works to be completed (sect. 372(1)).

While failure to comply with the regulations is itself a criminal offence, the DoE expect LAs will normally serve notice requiring works to be done before a prosecution for neglect is taken (Circular 5/90 para. 22 and 12/93 para. 4.4).

Standards of management

The regulations make detailed requirements relating to:

(a) Water supply and drainage
(b) Supply of gas and electricity
(c) Repair etc. of parts in common use
(d) Maintenance etc. of installations in common use
(e) Repair and cleanliness etc. of living accommodation
(f) Windows and ventilation
(g) Means of escape from fire

FC127 Notices to remedy neglect of management

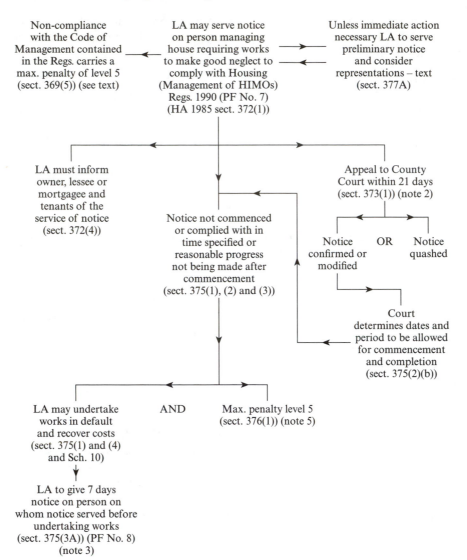

Notes

1. The power of entry (page 497) applies here, but 24 hours' notice of intention to enter need not be given.
2. If the appeal is lodged but subsequently withdrawn before being heard, the works must commence within 21 days after the date of the appeal being withdrawn and completed within the period specified in the notice (sect. 375(2)(c)).
3. If the works are carried out before the LA commences works, the LA may recover its costs up to that point including administration costs (sect. 375(3B)).
4. For grants relating to the repair of HIMOs, see FC132.
5. Penalty arises not only for an initial failure to comply but also for further failures after conviction.

(h) Outbuildings and yards
(i) Disposal of refuse and litter
(j) General safety of residents
(k) Information notices
(l) Information to be supplied to the LA
(m) Duties of residents.

Preliminary notices*

Unless the LA considers there is a need for immediate enforcement action, they must first serve a preliminary ('minded-to') notice which informs the person responsible:

(a) that the LA is considering serving a sect. 372 notice;
(b) the reasons why; and
(c) the work required.

The recipient may make written representations within 14 days or oral representations if they are requested within 7 days. The LA must have regard to them.

Failure to undertake this procedure is a ground of appeal against any sect. 372 notice served (sect. 377A and the Housing (Enforcement Procedures for HIMOs) Order 1997).

Person responsible

The notice is served on the person managing the house but, where it has not been practicable to ascertain that person's name and address, it may be served by addressing it to 'the manager of the house' and delivering it to some person on the premises (sect. 372(1) and (2)).

The LA must inform all owners, tenants, lessees or mortgagees that the notice has been served (sect. 372(4)).

Content of notice

The notice must specify the requirements of the regulations not complied with and the works required to make good the neglect (sect. 372(1)).

Compliance with the notice

The notice must specify:

(a) a reasonable date, not less than 21 days, upon which the works must be commenced; and
(b) a reasonable period within which to complete the works (sect. 372(3)).

* The DLTR intend to repeal this provision from a date yet to be determined.

Appeals

The person on whom the notice is served may appeal to the County Court within 21 days of service (or longer if agreed in writing by the LA) on any of the following grounds:

(a) the condition of the house did not justify the service of the notice;
(b) there has been some material informality, defect or error in, or in connection with, the notice;
(c) the LA has unreasonably refused to accept alternative works, or the works required are unreasonable in character or extent or are unnecessary;
(d) the date specified for beginning the works is not reasonable;
(e) the time allowed for the completion of the works is not reasonably sufficient;
(f) that some other person is responsible or should pay all, or part of the costs of compliance (sect. 373(1)).

Enforcement

The LA may itself undertake the required works and recover costs:

(a) when the works have not been commenced by the required date and no appeal has been made;
(b) where an appeal has been made but withdrawn, not less than 21 days after the notice has been withdrawn;
(c) where an appeal is heard, after such period as specified by the court has elapsed (sect. 375(1) and (2)).

Where works have commenced but in the opinion of the LA reasonable progress is not being made, the LA may also complete the works (sect. 375(3)).

In all cases, the LA must give at least 7 days' notice to the person upon whom the notice was served and any other owner before commencing works (sect. 372(3A)).

Where the LA is refused consent to undertake works in default, an application may be made to a magistrates' court for an order (sect. 377(2)(b)).

MEANS OF ESCAPE IN CASE OF FIRE: CLOSURE OF PART OF HOUSE

References

Housing Act 1985 sects. 365 and 368 as amended 1996.
The Housing (Fire Safety in Houses in Multiple Occupation) Order 1997.
DoE circular 3/97.
The Housing (Prescribed Forms) Regulations 1990 as amended 1997.

FC128 Means of escape in case of fire: closure of part of house

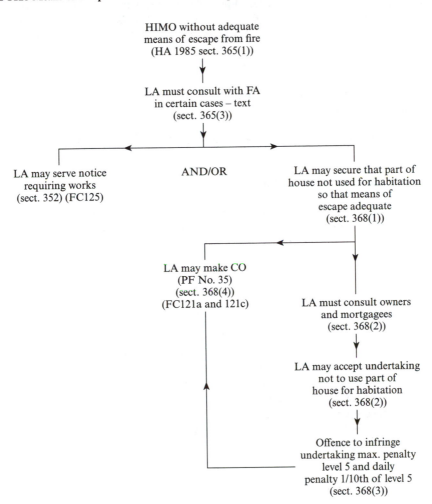

HIMO without adequate
means of escape from fire
(HA 1985 sect. 365(1))

LA must consult with FA
in certain cases – text
(sect. 365(3))

LA may serve notice
requiring works
(sect. 352) (FC125)

AND/OR

LA may secure that part of
house not used for habitation
so that means of
escape adequate
(sect. 368(1))

LA may make CO
(PF No. 35)
(sect. 368(4))
(FC121a and 121c)

LA must consult owners
and mortgagees
(sect. 368(2))

LA may accept undertaking
not to use part of
house for habitation
(sect. 368(2))

Offence to infringe
undertaking max. penalty
level 5 and daily
penalty 1/10th of level 5
(sect. 368(3))

Scope

The procedure relates to a HIMO which is not provided with such means of escape from fire as the LA considers to be necessary having regard to the number of individuals or households or both, and is not reasonably suitable for occupation by those individuals or households. It may be used along with or in addition to notices under sect. 352 (FC125) requiring works to be carried out (sect. 365(1)).

LA duty to ensure adequate fire precautions and consultation with the FA

The same provisions apply here as with sect. 352 notices – see pages 539–540.

Closing orders and undertakings

If the LA considers that the means of escape would be adequate if part of the house were not used for human habitation, it may secure that that part is not used by either a CO or accepting an undertaking (sect. 368(1)).

The LA may, in addition to securing that a part is not used, require works to provide an adequate means of escape for the remainder under sect. 352 (FC125).

(a) **Undertakings.** The LA may accept an undertaking that a part of the house will not be used for habitation after consulting with both the owner and any mortgagee and with the FA (sects. 368(2) and 365(3)). Contraventions of an undertaking bring liability to fines not exceeding level 5 and to a daily penalty (sect. 368(3)) and the LA may also make a CO (sect. 368(4)).

(b) **Closing orders.** If the LA does not accept an undertaking, or where one has been made, it is contravened, the LA may then make a CO. The procedure for the making of COs is set out in FC121a and 121b (sect. 368(4)).

Where the LA is satisfied after the making of a CO that the means of escape is adequate, owing to a change of circumstances, and will remain adequate, the CO must be determined under sect. 278(1) (sect. 368(5)(b)).

DIRECTION ORDERS

Reference

Housing Act 1985 sects 354–357 as amended 1996.
DoE Circulars 5/90 and 12/93.
The Housing (Prescribed Forms) Regulations 1990 as amended 1997.

FC129 Direction orders

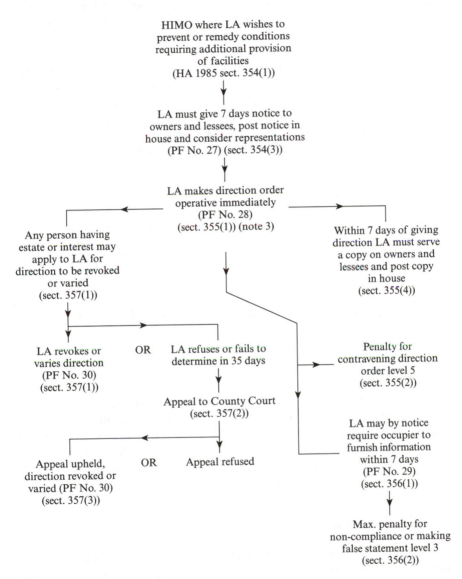

HIMO where LA wishes to
prevent or remedy conditions
requiring additional provision
of facilities
(HA 1985 sect. 354(1))

LA must give 7 days notice to
owners and lessees, post notice in
house and consider representations
(PF No. 27) (sect. 354(3))

LA makes direction order
operative immediately
(PF No. 28)
(sect. 355(1)) (note 3)

Any person having
estate or interest may
apply to LA for
direction to be revoked
or varied
(sect. 357(1))

Within 7 days of giving
direction LA must serve
a copy on owners and
lessees and post copy
in house
(sect. 355(4))

LA revokes or
varies direction
(PF No. 30)
(sect. 357(1))

OR

LA refuses or fails to
determine in 35 days

Penalty for
contravening direction
order level 5
(sect. 355(2))

Appeal to County Court
(sect. 357(2))

LA may by notice
require occupier to
furnish information
within 7 days
(PF No. 29)
(sect. 356(1))

Appeal upheld,
direction revoked or
varied (PF No. 30)
(sect. 357(3))

OR

Appeal refused

Max. penalty for
non-compliance or making
false statement level 3
(sect. 356(2))

Notes
1. For notices requiring provision of amenities, see FC125.
2. There is no right of appeal against the making of a direction order.
3. The order is to be registered as a local land charge (sect. 354(8)).

Scope

The purpose of this procedure is either to prevent the occurrence of, or remedy, a situation in a HIMO where the amenities are so defective in relation to the number of individuals or households (or both) that action would have to be taken to provide better facilities under sect. 352. The facilities concerned are listed (a)–(e) on page 535 and include means of escape from fire and fire precautions (sect. 352(1)).

The procedure may be used as an alternative to sect. 352 procedure or in combination with it and is also without prejudice to any action which may be taken under sect. 358 in relation to overcrowding in HIMOs (sect. 354(7)). The value of using both sects 352 and 354 together is that not only is a schedule of works specified for a particular number of occupants, but also the direction prohibits future reoccupation by a greater number of residents.

Notice of intention

Not less than 7 days before giving a direction the LA must:

(a) serve notice of intention on the owner and every other lessee; and
(b) post a copy of that notice in the house in a position accessible to the tenants (sect. 354(3)).

The LA must allow representations to be made to them by those served (sect. 354(3)).

Effect of directions

The direction has effect so as to make it the duty of the occupier of the house, or of any other person entitled or authorized to permit individuals to take up residence, not to permit any individual to take up residence in that house or part unless the number of individuals or households then occupying the house or part would not exceed the limit specified in the direction (sect. 355(1) as amended by HA 1996).

Content of directions

The LA must specify the highest number of individuals or households (or both) who should occupy the house, or any part of it, having regard to the facilities available ((a)–(e) page 535). In doing this the LA must have regard to the desirability of applying separate limits where different parts of the house are occupied by different persons (sect. 354(1) and (2)).

Not less than 7 days after making a direction order the LA must serve a copy on the owner and other lessees and post a copy in the house (sect. 354(4)).

Notices requiring information

At any time following a direction, the LA may serve a notice on any occupier requiring them to provide any or all of the following information in writing within 7 days:

(a) the number of individuals living in the house on a date specified in the notice;
(b) the number of families or households to which those individuals belong;
(c) the names of those individuals and of the heads of each of the families, or households;
(d) the rooms used by those individuals, families or households (sect. 356(1)).

It is an offence not to comply with these notices at a maximum penalty of level 3 (sect. 356(2)).

Appeals

There is no provision for appeal against the making of the direction order.

Revocation or variation

Any person having an estate or interest in the house may ask the LA to revoke the order or modify it to allow more people to live there. Refusal or failure to determine the request within 35 days leads to a right of appeal to the County Court (sect. 357).

CONTROL ORDERS

Reference

Housing Act 1985 sects. 379–393.
DoE Circulars 5/93 and 12/93.
The Housing (Prescribed Forms) Regulations 1990 as amended 1997.

Scope

A LA may make a control order in respect of any HIMO where it is satisfied that:

(a) a notice under sect. 352 or 372 requiring works, or a direction order (sect. 355) are operative in respect of the house; or
(b) the conditions of the house call for action to be taken under any of the provisions in (a); **and**
(c) living conditions in the house are such that a control order is necessary to protect safety, welfare or health of the occupants (sect. 379(1)).

FC130 Control orders

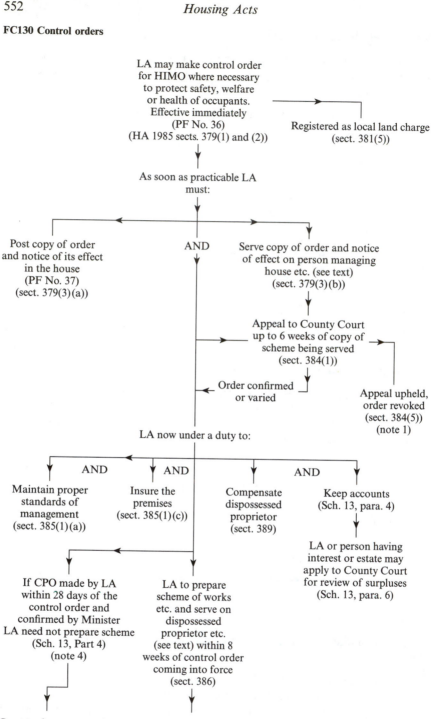

LA may make control order
for HIMO where necessary
to protect safety, welfare
or health of occupants.
Effective immediately
(PF No. 36)
(HA 1985 sects. 379(1) and (2))

Registered as local land charge
(sect. 381(5))

As soon as practicable LA
must:

Post copy of order
and notice of its effect
in the house
(PF No. 37)
(sect. 379(3)(a))

AND

Serve copy of order and notice
of effect on person managing
house etc. (see text)
(sect. 379(3)(b))

Appeal to County Court
up to 6 weeks of copy of
scheme being served
(sect. 384(1))

Order confirmed
or varied

Appeal upheld,
order revoked
(sect. 384(5))
(note 1)

LA now under a duty to:

AND   AND   AND

Maintain proper
standards of
management
(sect. 385(1)(a))

Insure the
premises
(sect. 385(1)(c))

Compensate
dispossessed
proprietor
(sect. 389)

Keep accounts
(Sch. 13, para. 4)

LA or person having
interest or estate may
apply to County Court
for review of surpluses
(Sch. 13, para. 6)

If CPO made by LA
within 28 days of the
control order and
confirmed by Minister
LA need not prepare scheme
(Sch. 13, Part 4)
(note 4)

LA to prepare
scheme of works
etc. and serve on
dispossessed
proprietor etc.
(see text) within 8
weeks of control order
coming into force
(sect. 386)

(Continued on next page)   *(Continued on next page)*

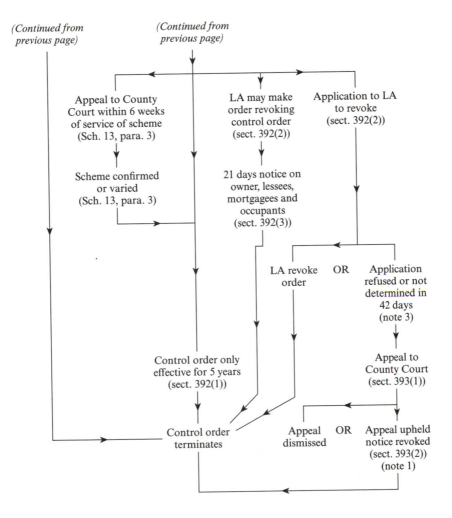

Notes

1. At the same time as ordering the revocation of an order a County Court, or a higher court on further appeal, may postpone the operation of the revocation to allow the LA to complete works etc. (Sch. 13, paras. 17(3) and 20(1)).
2. For consequences of the cessation of a control order, see Sch. 13.
3. The LA is obliged to inform the applicant of the decision and of their reasons for rejecting the grounds for revocation (sect. 392(4)).
4. For guidance on CPOs following a control order see DoE Circulars 12/93, 14/94 and 17/96. CPOs to follow control orders would be made under Part 2 of the 1985 Act (provision of housing accommodation).

The LA may exclude from the effect of any control order a part of the house occupied by a person with an estate or interest in the whole house (sect. 380).

The effect of this is that a control order allows the LA to take possession of a HIMO for up to 5 years and take whatever steps are necessary (other than selling the property) to bring it up to a satisfactory standard.

Publicity

As soon as practicable after making the control order the LA must:

(a) post a copy of the order and a notice of its effect in a conspicuous position in the house; and
(b) serve a copy of the order and notice of its effect on any person managing or having control of the house and on any owner, lessee or mortgagee (sect. 379(3)).

The notice of effect must contain:

(a) a statement of the general effect of the control order;
(b) the principal grounds upon which the LA made the order; and
(c) the rights of appeal (sect. 379(4)).

Effect of control order

The order becomes operative as soon as it has been made by the LA (sect. 379(2)). The principal effects of the order are:

(a) it gives the LA powers as if it had an estate or interest in the house including the management of the property, collection of rents, carrying out of works etc. (sect. 381);
(b) it preserves the position of tenants of the house including protection under the Rent Acts (sect. 382);
(c) it provides that the duties of occupants under the Housing (Management of HIMOs) Regulations 1990 shall still apply (sect. 382(5));
(d) it may exclude from control any part of the house occupied by a person having an interest or estate in the whole of the house (sect. 382); and
(e) it gives the LA right to the possession of any furniture which the resident has a right to use for payment (sect. 383).

The order operates for 5 years only (sect. 392(1)).

Duties of LA

Following the making of a control order the LA comes under a duty to:

(a) enter into the premises and take all such immediate steps as appear to be necessary to protect safety, welfare, or health of the occupants (sects. 379(2) and 385(1)(b));

(b) exercise the powers conferred upon them by the order so as to maintain proper standards of management and remedy any matters which it would have required to be remedied under any other provisions of the 1985 Act (sect. 385(1)(a));
(c) make reasonable provisions for the insurance of the premises (sect. 385(1)(c));
(d) pay compensation to the dispossessed proprietor (sect. 389);
(e) prepare a scheme of works etc. (below) (sect. 386); and
(f) keep accounts showing surpluses of revenue (Sch. 13, para. 4).

Scheme of works etc.

Following the making of the control order, the LA must produce a scheme of works and, not later than 8 weeks from the making of the order, serve a copy of the scheme on:

(a) the dispossessed proprietor; **or**
(b) an owner or lessee or mortgagee of the house; **and**
(c) on any other person on whom the control order was served (sect. 386(1)).

The scheme must include:

(a) particulars of works involving capital expenditure which the LA would have required to be carried out under any housing or public health regulations including Part II of the HA 1985 relating to multiple occupation;
(b) an estimate of the cost of those works;
(c) a specification of the highest number of individuals or households who should live in the house from time to time having regard to:
  (i) the facilities available set out in sect. 352(1) (page 535);
  (ii) the existing condition of the house and its future condition as the works progress;
(d) an estimate of balances which will accrue to the LA out of the net rent and other payments received from occupants after deducting:
  (i) compensation payable under sect. 389 and Sch. 13, Part 2; and
  (ii) all expenditure, other than capital works included in the scheme, together with appropriate establishment charges (Sch. 13, paras. 1 and 2).

The scheme may be modified from time to time so as to increase the surpluses but cannot reduce them. The LA may carry out other works which were not included in the scheme (sect. 386(3)), but the cost of these works is not recoverable from persons having an estate or interest in the house.

Recovery of capital expenditure

During the operation of the control order the LA must keep accounts showing:

(a) expenditure on capital works included in the scheme; and

(b) surpluses on the revenue account;

and any excess of expenditure on capital works over revenue income is a charge against the premises and all estates and interest in the premises (Sch. 13, paras. 4 and 16).

Appeals

Appeals against the control order may be made to the County Court, up to 6 weeks after the service of copies of the scheme, by any person having an estate or interest in the house or any other person who may be prejudiced by the making of the order. The grounds of appeal are:

(a) the condition of the house was not such as to call for a control order (even though orders, notices or directions under Part II may have been issued);

(b) it was not necessary to make a control order to protect the safety, welfare or health of the occupants;

(c) where part of the house was occupied by a dispossessed proprietor, that it was practicable and reasonable for the LA to have excluded that part from the order;

(d) the control order is invalid because a requirement of the Act has not been complied with or there is some informality, defect or error in the order. In this case it must also be shown that the interests of the appellant have been substantially prejudiced (sect. 384).

Appeals against the scheme may be made to a County Court within 6 weeks from the date of service of the scheme by any person having an estate or interest in the house on any of the following grounds:

(a) any of the works listed are unreasonable in character or extent or are unnecessary, having regard to the condition of the house or other circumstances;

(b) that any of the works do not involve expenditure which ought to be regarded as capital expenditure;

(c) that the number of individuals or households specified is unreasonable;

(d) the estimate of surpluses is unduly low whether by rents charged by the LA or otherwise (Sch. 13, para. 3).

Termination of control order

This may come about in any of five ways:

(a) revocation by County Court on appeal (Sch. 13, para. 17);

(b) revocation by higher court on appeal from County Court (Sch. 13, para. 20);

(c) expiration of the 5 years' period for which control orders are effective (sect. 392(1));

(d) revocation by the LA either on its own initiative or following application to the LA by any person (sect. 392(2)); and

(e) revocation by a County Court on appeal against a refusal by the LA to revoke an application (sect. 393(1)).

Power of entry

Having taken possession of the house through the control order, the LA has the right of any person having an estate or interest in the premises but these powers are extended to include a right of entry for any authorized officer of the LA at all reasonable times for:

(a) survey and examination; and
(b) carrying out works (sect. 387(1) and (2)).

Having received notice of intended entry, if the occupant obstructs any officers, agents, servants or workmen of the LA in carrying out these powers, the LA may apply for an order to the magistrates' court. Failure to comply is punishable by a maximum fine of level 5 and not exceeding one-tenth of that amount daily (sect. 387(4) and (5)).

HOUSE AND AREA IMPROVEMENT

RENEWAL AREAS

Reference

Local Government and Housing Act 1989 Part 7.
The Regulatory Reform (Housing Assistance) (England and Wales) Order 2002.
DoE Circular 17/96 paras. 3.5–3.17 and Annexes C1–C4.

Renewal areas

RAs may be declared by a LA when satisfied that:

(a) the living conditions in an area are unsatisfactory; and
(b) the conditions can best be dealt with by declaring the area to be a RA (sect. 89(1)).

RAs provide LAs with scope to:

(a) improve the housing and general environment of an area where social and environmental problems are combined with poor housing;
(b) develop effective partnerships with both residents and private sector interests;
(c) bring about wide-based regeneration; and
(d) secure maximum impact by increasing community and market confidence and help to reverse the process of decline.

FC131 Renewal areas

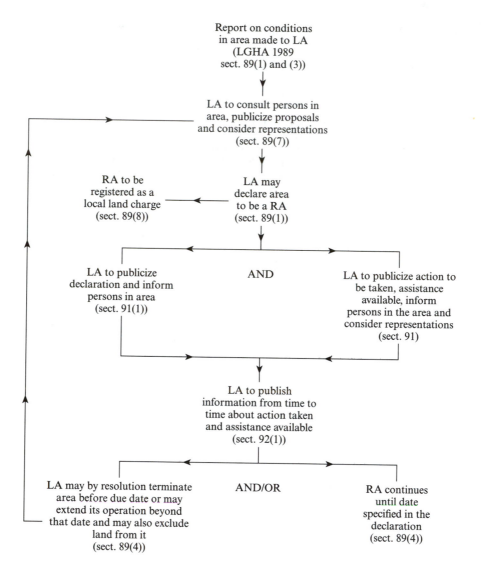

Report on conditions
in area made to LA
(LGHA 1989
sect. 89(1) and (3))

LA to consult persons in
area, publicize proposals
and consider representations
(sect. 89(7))

RA to be
registered as a
local land charge
(sect. 89(8))

LA may
declare area
to be a RA
(sect. 89(1))

LA to publicize
declaration and inform
persons in area
(sect. 91(1))

AND

LA to publicize action to
be taken, assistance
available, inform
persons in the area and
consider representations
(sect. 91)

LA to publish
information from time to
time about action taken
and assistance available
(sect. 92(1))

LA may by resolution terminate
area before due date or may
extend its operation beyond
that date and may also exclude
land from it
(sect. 89(4))

AND/OR

RA continues
until date
specified in the
declaration
(sect. 89(4))

In considering whether an area should be designated as a RA, the LA is required to take account of guidance issued by the SoS (sect. 89(5))

Such detailed guidance is contained in paras. 3.5–3.17 of Circular 17/96 and recommends the carrying out of a neighbourhood renewal assessment which is detailed in Annex C3. That assessment process includes:

(a) An initial decision that areas of private sector housing could benefit from being declared a RA.
(b) An approximate definition of the area.
(c) Setting up a core team of officers.
(d) Establishing aims and objectives.
(e) Generating options which should incorporate the views of members, residents, community groups, businesses, etc.
(f) Information gathering from an environmental and housing survey and surveys of residents, landlords, commercial users, etc.
(g) An economic assessment of the various options.
(h) Options appraisal.
(Annex Cl)

The report considered by the LA in relation to a proposal to declare an RA must include particulars of the following:

(a) living conditions in the area;
(b) the ways in which the conditions may be improved;
(c) the powers available to the authority if the area is to be declared a renewal area;
(d) detailed proposals for the exercise of those powers;
(e) the cost of the proposals;
(f) the financial resources available;
(g) the representations made in relation to the proposals.
(sect. 89(3))

Consultation

Before declaring a RA, the LA must bring the proposals to the attention of persons residing or owning property in the area and indicate how representations can be made (sect. 89(7)). Para. 2 of Annex Cl to DoE Circular 17/96 indicates that this should involve:

(a) publishing the proposal in at least two newspapers circulating in the locality (at least one being a local newspaper);
(b) posting notices in the area for at least 7 days;
(c) in the case of both (a) and (b) identifying the opportunity to make representations within 28 days;
(d) not more than 7 days after the publication of the newspaper notice, a statement of the reasons for the proposed declaration must be delivered to each address in the area;
(e) all representation must be considered and written replies sent where a point is not accepted.

Criteria for renewal areas

The approval of the SoS is not required but the code of guidance in circ. 17/96 contains the following indicators:

(a) it contains a minimum of 300 dwellings;
(b) 75% of the dwellings are in private ownership;
(c) 75% of the dwellings must require works for which renovation grants, common parts grants, HIMO grants or group repair assistance would be available. The properties need not be statutorily unfit;
(d) 30% of the households in the area appear to be eligible for specified social benefits; and
(e) the area is coherent and not fragmented.

(sect. 90(1) and circular 17/96 para. 3.7 and annex C2)

Procedure following declaration

Following declaration of the RA the LA must bring it to the attention of persons residing or owning property in the area and indicate how inquiries and representations may be made (sect. 91). The circular 17/96 at para. 3.13 suggests that this should include:

(1) Publicize the declaration by advertisement in one or more newspapers circulating in the locality (at least one to be local), posting notices in the area for at least 7 days and by delivering a statement about the declaration to each address in the area not more than 7 days after the newspaper notice.
(2) Use their best endeavours to ensure that the declaration is brought to the attention of owners and occupiers.
(3) Secure an advice and information service for those wishing to carry out works.

On-going publicity

Upon declaration and, the guidance suggests, at intervals of not more than 2 years, the LA must publish progress reports about the action which it is taking and give details of the assistance available for the carrying out of works (sect. 92(1)).

Powers available to the LA

The following powers are available to a LA within a RA:

(a) acquisition of land and property by agreement or through a CPO;
(b) authority to carry out work;
(c) power to extinguish rights of way;
(d) increased assistance to participants in group repair schemes (FC123).

The power under sect. 93(5)(b) for LAs to assist in the carrying out of work by others in renewal areas, e.g. by financial contribution etc., is to be replaced by a general power to assist in a wide range of housing functions including renewal areas – see page 563 (The Regulatory Reform (Housing Assistance) (England and Wales) Order 2002)

Under sect. 96(2) the SoS can make contributions towards expenses incurred by local authorities under Part VII of the Act in carrying out environmental improvements. In Annex C1 para. 8 the SoS has determined that the contribution shall not exceed 50% of the expenditure incurred based on a maximum eligible expense of £1000 per dwelling.

Matters eligible for contribution include:

(a) street works;
(b) landscaping;
(c) improvements to the exteriors and curtilage of dwellings;
(d) community facilities;
(e) miscellaneous environmental improvements.

The purpose in carrying out such work is for the authority to demonstrate its commitment to an area and its future such that confidence will be increased and residents will be encouraged to invest in their own properties (sects. 93 and 96).

Duration of renewal area

The RA will operate to the date specified at the time of declaration and may be extended or shortened by later resolution of the LA (sect. 89(4)).

Termination or alteration

The LA may terminate the operation of the RA before the due date (where for example the outline programme has been completed) and may exclude land at any time (sect. 95(1)).

If either is proposed the LA must bring it to the attention of persons residing or owning property in the area and consider representations (sect. 95).

Powers of entry within a renewal area

Authorized officers are required to give 7 days' notice to occupiers and owners (where known) before requiring entry for survey or examination. Obstruction of the AO is punishable by fines not exceeding level 3 (sect. 97).

GRANTS FOR WORKS OF IMPROVEMENT AND REPAIR

NB. The Regulatory Reform (Housing Assistance) (England and Wales) Order 2002 makes fundamental changes to the provisions dealt with in this

FC132 House renovation grants

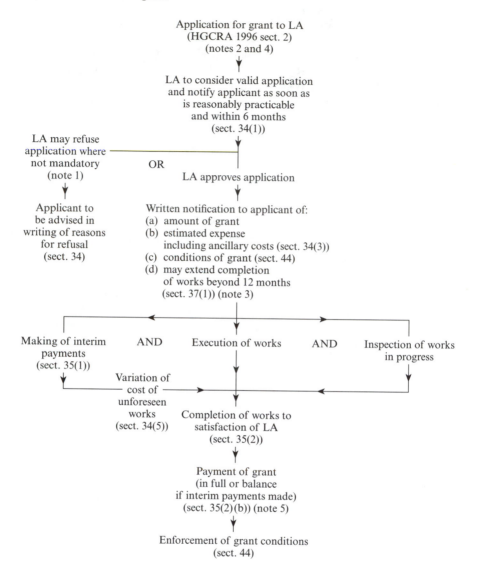

Application for grant to LA
(HGCRA 1996 sect. 2)
(notes 2 and 4)

LA to consider valid application
and notify applicant as soon as
is reasonably practicable
and within 6 months
(sect. 34(1))

LA may refuse
application where
not mandatory
(note 1)

OR

LA approves application

Applicant to
be advised in
writing of reasons
for refusal
(sect. 34)

Written notification to applicant of:
(a) amount of grant
(b) estimated expense
    including ancillary costs (sect. 34(3))
(c) conditions of grant (sect. 44)
(d) may extend completion
    of works beyond 12 months
    (sect. 37(1)) (note 3)

Making of interim
payments
(sect. 35(1))

AND   Execution of works   **AND**   Inspection of works
in progress

Variation of
cost of
unforeseen
works
(sect. 34(5))

Completion of works to
satisfaction of LA
(sect. 35(2))

Payment of grant
(in full or balance
if interim payments made)
(sect. 35(2)(b)) (note 5)

Enforcement of grant conditions
(sect. 44)

Notes
1. There is no appeal against a refusal of grant.
2. The form of application has been specified by the SoS.
3. Completion of the works is normally required within 12 months unless extended by the LA (sect. 37(1)).
4. LAs are encouraged by the DoE to establish a preliminary enquiry system to provide information before formal applications are made (Circular 17/96 Annex F paras. 2–4).
5. The LA is required to maintain a register of all grants paid (para. 9 Exchequer Contributions Determination 1996).
6. As from the date of implementation of the Regulatory Reform (Housing Assistance) Order 2002, this procedure will apply only to mandatory disabled facilities grants.

section and will be implemented over a period of 1 year from, probably, May 2002.

The Order confers new and wide ranging powers on LAs in England and Wales to improve living conditions in their areas by providing assistance in relation to living accommodation for:

(a) acquisition where the LA wish to purchase it as an alternative to adapting, improving or repairing it;
(b) adaptation or improvement;
(c) repair;
(d) demolition;
(e) construction of replacement accommodation.

LAs are required to consider a person's ability to meet a contribution or to repay the assistance. They must also provide a written statement of the conditions applicable to any assistance and may take security, including a charge on the property. These powers are not exercisable until the LA has adopted and published a policy relating to the way in which the powers will be used.

In the light of these powers, the existing provisions for the making of improvement grants, except for mandatory disabled facilities grants, will be repealed. LAs who have published policies referred to above will be able to use the new powers immediately afterwards even though the specific powers to give improvement grants may still be in operation. Approval for improvement grants given this may continue to be processed to their conclusion.

The provisions for mandatory disabled facilities grants will continue except that they will now also be available in relation to living accommodation in park homes and houseboats that were previously excluded.

References

The Housing Grants, Construction and Regeneration Act 1996 as amended.
The Housing Renewal Grants Regulations 1996 as amended 1997 and 2000.
The Housing Renewal Grant (Prescribed Forms and Particulars) Regulations 1996 as amended.
The Housing Renewal Grants (Services and Charges) Order 1996.
The Disabled Facilities Grants and Home Repair Assistance (Maximum Amounts) Order 1996 as amended.
The Housing Renewal Grants etc. (Exchequer Contributions) (England) Determination 1996.
The Home Repair Assistance Regulations 1996.
The Home Repair Assistance (Extension) Regulations 1998.
The Home Repair Assistance (Extension) (England) Regulations 1999.

The Housing Grants (Additional Purposes) (England) Order 2000.
(And the Welsh regulation equivalents)
The Regulatory Reform (Housing Assistance) (England and Wales) Order 2002.
DoE and DETR Circulars 17/96, 4/97, 17/97, 4/98, 6/99 and 3/2000.

Scope of grants

Financial assistance from LAs is available, other than to LAs and certain other specified public bodies, towards works of:

    (a) provision of dwellings or HIMOs by the conversion of houses or buildings;

    (b) the improvement or repair of dwellings, HIMOs and the common parts of buildings containing one or more flats;

    (c) the provision of facilities for disabled people in dwellings including in park homes and houseboats.

    (d) the improvement or repair of the common parts of a building containing one or more flats (sects. 1 and 3).

Types of grants

    (a) **Renovation grant** for the improvement and/or repair of houses (up to fitness standard or for other works additional to those required by enactment), and for the conversion of houses and other buildings into flats for letting.

    (b) **Common parts grant** for the improvement and/or repair of the common parts of buildings containing one or more flats.

    (c) **HIMO** grant for the improvement and/or repair of HIMOs (i.e. works to make the house suitable for the number of occupants, to make it fit or to go beyond the fitness standard) and for the conversion of buildings into HIMOs.

    (d) **Disabled facilities grant** for adapting, or providing facilities for, the home of a disabled person to make it more suitable for that person to live in. The grant is available for a wide range of essential adaptations (grant is mandatory) and includes work in dwellings which are part of park homes and houseboats.

There are separate provisions for Home Repair Assistance as these are dealt with on page 567.

Applications

An application shall be in writing and specify:

    (a) the premises;

    (b) particulars of the work;

   (c) unless the LA direct otherwise, at least two estimates from different contractors;

   (d) particulars of any preliminary or ancillary services and charges, e.g. architectural fees; and

   (e) other particulars, as described in the prescribed application form (sect. 2(2)).

Forms of applications and necessary particulars to be included are prescribed in the Housing Renovation etc. Grants (Prescribed Forms and Particulars) Regulations 1996 amended and the equivalent regulations for Wales.

Grant applications, other than for disabled facilities grants and HIMO grants in respect of a HIMO provided by conversion may not be considered in respect of properties built or provided by conversion less than 10 years before the date of the application (sect. 4).

Types of applicant

Before approving a grant, except in the case of common parts grants, the LA must be satisfied that the applicant has or proposes to acquire an owner's interest in all the 'land' on which it is proposed to carry out works. An 'owner's interest' means a freehold interest or a leasehold interest where there is at least 5 years to run. Applicants must be 18 years or over on the date of application.

A grant application from a tenant cannot be considered unless:

   (a) the terms of the tenancy require him to carry out the relevant work; or

   (b) the application is for a disabled facilities grant (sects. 3 and 7).

LAs, NHS Trusts and other specified public bodies may not apply for grants (sect. 3(2)).

Certificates and conditions as to future occupation

All applications for grants other than common parts grants and for certain types of ecclesiastical properties and charities, must be accompanied by a certificate of future occupation. There are three different types of certificates relating to renovation grants:

   (a) An 'owner occupation' certificate certifies that the applicant has or proposes to acquire an owner's interest in the dwelling and that he or a member of his family intends to live in the dwelling as his only or main residence for at least 5 years from the certified completion date (for disabled facilities grant the requirement is for an owner's certificate relating to occupancy by the disabled person for 5 years).

   (b) A 'tenant's certificate' certifies that the applicant is a tenant of the dwelling and that he or a member of his family intends to live in the dwelling as his only or main residence.

(c) A 'certificate of intended letting' certifies that the applicant has or intends to acquire an owner's interest and intends to let the dwelling as a residence for a period of at least 5 years after the certified completion of work.
(sects. 8 and 9)

The grant conditions remain in force for 5 years from the certified completion date and are binding not only on those who provide the certificate but also on any subsequent owners (sect. 44).

In the event of a breach of the conditions the LA may, at its discretion, demand repayment with interest of the whole or part of the grant (sect. 55(1)).

Grant conditions

LAs are able to impose grant conditions other than those relating to certificates of future occupation but only with the approval of the SoS. A general approval for this is detailed in Annex J4 of Circular 17/96 (sect. 52).

Purpose of grant

The purpose for which grants may be given are:

(a) Renovation grants, common parts grants and HIMO grants:
  (i) to make the dwelling fit;
  (ii) to put it into reasonable repair;
  (iii) to provide adequate thermal insulation;
  (iv) to provide adequate space heating;
  (v) to provide satisfactory internal arrangements;
  (vi) to provide a means of escape.
  (vii) to improve energy efficiency (sects 12, 17 and 27).
(b) Disabled facilities grant:
  (i) facilitating access by a disabled person;
  (ii) making the building safe for the disabled person and persons living with them;
  (iii) facilitating access to or providing a sleeping room, a room with a toilet, bath or shower and wash basin;
  (iv) for the preparation and cooking of food by the disabled person;
  (v) improving the heating system to meet the needs of that person;
  (vi) alterations to facilitate the use by the disabled person of power, heat and light;
  (vii) facilitating access and movement around the dwelling (sect. 23).

In addition grant may be paid for the provision of dwellings or HIMOs by the conversion of a house or other building (sect. 1(1)(b)).

Estimated expense/grant calculation

LAs are to consider four elements in calculating the grant:

1. which works are 'eligible works';
2. the amount of expenses which will be properly incurred in undertaking the works;
3. the costs attributable to preliminary or ancilliary services as defined by the Housing Renewal Grants (Services and Charges) Order 1996;
4. the amount of grant (sect. 34(2)).

The grant figure is subject to the means test set out in the Housing Renewal Grants Regulations 1996 as amended. These regulations do not apply to landlords where the LA is given discretion to enable the LA to decide the level of any grant (sect. 31(3)).

Maximum grants are prescribed by the SoS (sect. 33), e.g. the Disabled Facilities Grants and Home Repair Assistance (Maximum Amounts) Order 1996.

Grants approvals and payment

All grants, with the exception of those disabled facilities grants where the works are deemed to be essential, are discretionary.

A LA is required to notify grant applicants in writing not later than 6 months after the date of the valid application whether or not the application is approved or refused (sect. 34(1)). The grant can be paid after completion of the eligible work or by instalments as the works progress. Payment is conditional upon work being completed to the satisfaction of the LA and on submission of an invoice or receipt (sect. 35).

Home repair assistance

This Scheme is primarily to provide assistance in the form of either grant or materials for the repair, improvement or adaptation of a dwelling, houseboat or mobile home (sect. 76).

Applicants must:

(a) be aged 18 or over;
(b) live in, or intend to live in, the dwelling;
(c) have an owner's interest or be a tenant;
(d) have a power or duty to do the work;
(e) be in receipt (or the applicant's partner be in receipt) of an identified social security benefit or the applicant is elderly, disabled or infirm (sect. 77).

The Home Repair Assistance Regulations 1996 and the Extension Regulations of 1999 and 2000 outline the scheme and specify the form and content of applications while the Disabled Facilities Grants and Home Repairs Assistance (Maximum Amounts) Order 1996 determine the maximum amounts. The grants are discretionary in all cases.

Definitions

Disabled person means:

(a) a person who is registered in pursuance of arrangements made under sect. 29(1) of the National Assistance Act 1948 (disabled persons' welfare); or

(b) any other person for whose welfare arrangements have been made under that provision or, in the opinion of the social services authority, might be made under it (sect. 100).

Dwelling means a building or part of a building occupied or intended to be occupied as a separate dwelling, together with any yard, garden, outhouses and appurtenances belonging to it or usually enjoyed with it (sect. 101).

MISCELLANEOUS

OBSTRUCTIVE BUILDINGS

Reference

Housing Act 1985 sects. 283–288.

Obstructive building

This is defined as a 'building which, by virtue only of its contact with, or proximity to, other buildings, is dangerous or injurious to health' (sect. 283(1)).

The procedure does not, however, apply to buildings owned by statutory undertakers unless used as a dwelling, showroom or office, or to buildings owned by LAs (sect. 283(2)).

'Time and place' notices

The LA serves a notice on the owner or owners of the building which it considers to be an obstructive building notifying a time, being not less than 21 days after service of the notice, and place at which consideration will be given to demolition being ordered.

The owners are entitled to attend that meeting and to be heard (sect. 284(1) and (2)).

Obstructive building order

If, after giving the matter due consideration the LA is satisfied that the building is obstructive and that either the whole or part should be demolished, it may make an OBO requiring the vacation of the building, or part, and its subsequent demolition (sect. 284(3)).

FC133 Obstructive buildings

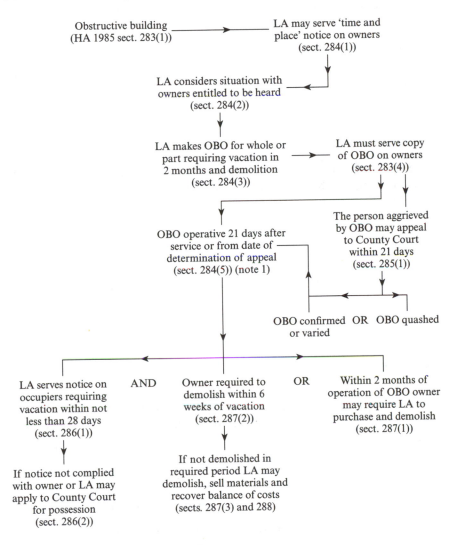

Obstructive building
(HA 1985 sect. 283(1)) → LA may serve 'time and place' notice on owners
(sect. 284(1))

LA considers situation with owners entitled to be heard
(sect. 284(2))

LA makes OBO for whole or part requiring vacation in 2 months and demolition
(sect. 284(3)) → LA must serve copy of OBO on owners
(sect. 283(4))

The person aggrieved by OBO may appeal to County Court within 21 days
(sect. 285(1))

OBO operative 21 days after service or from date of determination of appeal
(sect. 284(5)) (note 1)

OBO confirmed OR OBO quashed or varied

LA serves notice on occupiers requiring vacation within not less than 28 days
(sect. 286(1))

AND

Owner required to demolish within 6 weeks of vacation
(sect. 287(2))

OR

Within 2 months of operation of OBO owner may require LA to purchase and demolish
(sect. 287(1))

If notice not complied with owner or LA may apply to County Court for possession
(sect. 286(2))

If not demolished in required period LA may demolish, sell materials and recover balance of costs
(sects. 287(3) and 288)

Note

1. Maximum penalty for entering or allowing entry into building after OBO operative is level 2 and £5 daily (sect. 286(4)).

Subject to appeal, the OBO becomes operative 21 days after service or from the determination of the appeal. Vacation may be required within 2 months from the date of operation and demolition not less than 6 weeks from vacation (sect. 284(3) and (5)). Copies of the OBO must be served on every owner of the building.

Appeals

Any person aggrieved by the OBO, other than an occupant under a lease or agreement with 3 years or less to run, may appeal to the County Court within 21 days of the date of service (sect. 285(1) and (2)).

Purchase by LA

Once the OBO becomes operative but within 2 months of that date, any owner may request the LA to purchase and demolish the house. The LA cannot refuse such a request (sect. 287(1)).

Enforcement

If the demolition is not carried out, the LA may carry out the demolition, sell the materials resulting and recover costs (sects 287(3) and 288).

OVERCROWDING

References

Housing Act 1985 Part 10.
The Housing (Prescribed Forms) Regulations 1990 as amended.
    NB. The specific problems of overcrowding in HIMOs are dealt with in FC126.

Statutory overcrowding

A dwelling-house is deemed to be overcrowded when either:

(a) the number of persons sleeping there is such that any 2 persons 10 years old or more of opposite sexes, not being man and wife, must sleep in the same room (i.e. a room normally used in the locality either as a bedroom or as a living room and is available as sleeping accommodation); or

(b) the permitted number as calculated below is exceeded.

No account is taken of any child under 1 year and a child of an age between 1 year and under 10 years is counted as $1/2$ unit (sects. 325 and 326).

FC134 Overcrowding

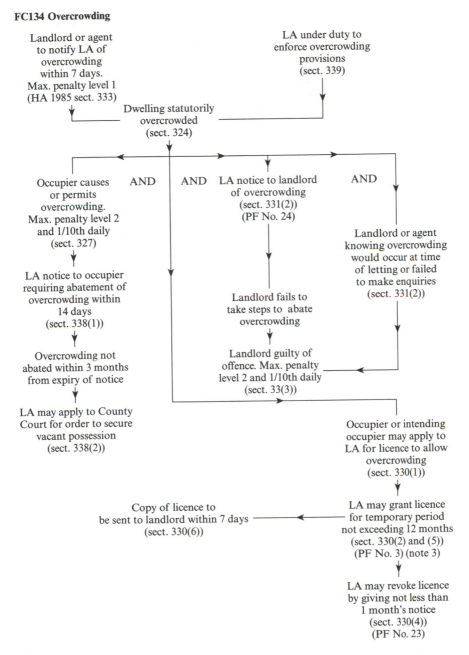

Landlord or agent to notify LA of overcrowding within 7 days. Max. penalty level 1 (HA 1985 sect. 333)

LA under duty to enforce overcrowding provisions (sect. 339)

Dwelling statutorily overcrowded (sect. 324)

Occupier causes or permits overcrowding. Max. penalty level 2 and 1/10th daily (sect. 327)

AND   AND

LA notice to landlord of overcrowding (sect. 331(2)) (PF No. 24)

AND

LA notice to occupier requiring abatement of overcrowding within 14 days (sect. 338(1))

Landlord or agent knowing overcrowding would occur at time of letting or failed to make enquiries (sect. 331(2))

Landlord fails to take steps to abate overcrowding

Overcrowding not abated within 3 months from expiry of notice

Landlord guilty of offence. Max. penalty level 2 and 1/10th daily (sect. 33(3))

LA may apply to County Court for order to secure vacant possession (sect. 338(2))

Occupier or intending occupier may apply to LA for licence to allow overcrowding (sect. 330(1))

Copy of licence to be sent to landlord within 7 days (sect. 330(6))

LA may grant licence for temporary period not exceeding 12 months (sect. 330(2) and (5)) (PF No. 3) (note 3)

LA may revoke licence by giving not less than 1 month's notice (sect. 330(4)) (PF No. 23)

Notes

1. For overcrowding in HIMOs, see FC126.
2. There are no appeal provisions relating to overcrowding notices.
3. The prescribed forms in this procedure are specified in the Housing (Prescribed Forms) (No. 2) Regulations 1990 except for the licence for temporary occupation which is contained with the Housing (Prescribed Forms) Regulations 1990.

Permitted number

This is the smaller of a number of persons calculated by the following two methods:

(a) where a house consists of:

1 room (i.e. room available as sleeping accommodation and of a type normally used in the locality either as a living room or as a bedroom)
  – 2 persons
2 rooms – 3 persons
3 rooms – 5 persons
4 rooms – 7^1/$_2$ persons
5 rooms or more 10 with additional 2 persons in respect of each room in excess of 5 (no regard is had to rooms of less than 50 sq. ft.);
or

(b) where the aggregate for all rooms is determined by reference to the following:

110 sq. ft. or more – 2 persons
90–100 sq. ft. – 1^1/$_2$ persons
70–90 sq. ft. – 1 person
50–70 sq. ft. – ^{1}/$_2$ person
Less than 50 sq. ft. – nil (sect. 326).

Offences by occupiers

The occupier of a dwelling-house who causes or permits it to be overcrowded is guilty of an offence unless:

(a) the overcrowding is caused only by virtue of a child attaining the age of 1 or 10 years and the occupier has applied to the LA for suitable alternative accommodation (defined in sect. 342) provided:
  (i) alternative accommodation is refused; or
  (ii) it subsequently becomes reasonably practicable for a person living in the house, but not a member of the family, to move; or
(b) the overcrowding is caused only because a member of the family is sleeping in the house temporarily; or
(c) the LA has issued a licence authorizing temporary overcrowding (below) (sect. 327).

Offences by landlords

The landlord is guilty of an offence if he has either:

(a) received a notice of overcrowding from the LA but fails to take steps reasonably open to him to abate the overcrowding; or
(b) when letting the house he had reasonable cause to believe it would become overcrowded or failed to make enquiries of the proposed occupier; or

(c) failed to notify the LA of overcrowding within 7 days of becoming aware of it (sects. 331 and 333).

Licence for temporary overcrowding

An occupier, or intending occupier, may apply to the LA for a licence authorizing the sleeping occupation of a house by a number of persons in excess of the permitted number.

Such a licence may be granted by the LA where it thinks it to be expedient, having had regard to any exceptional circumstances including any seasonal increase in the general population of this district.

Licences may not operate for in excess of 12 months and may be revoked by the giving of at least 1 month's notice. There is no appeal against a refusal by a LA to grant a licence or against revocation (sect. 330).

Definitions

In this procedure:

Agent in relation to the landlord of a dwelling:

(a) means a person who collects rent in respect of the dwelling on behalf of the landlord, or is authorized by him to do so; and

(b) in the case of a dwelling occupied under a contract of employment under which the provision of the dwelling for his occupation forms part of the occupier's remuneration, includes a person who pays remuneration on behalf of the employer, or is authorized by him to do so.

Dwelling means premises used or suitable for use as a separate dwelling.

Landlord in relation to a dwelling:

(a) means the immediate landlord of an occupier of the dwelling; and

(b) in the case of a dwelling occupied under a contract of employment under which the provision of the dwelling for his occupation forms part of the occupier's remuneration, includes the occupier's employer.

Owner in relation to premises:

(a) means a person (other than a mortgagee not in possession) who is for the time being entitled to dispose of the fee simple, whether in possession or in reversion; and

(b) includes also a person holding or entitled to the rents and profits of the premises under a lease of which the unexpired term exceeds 3 years (sect. 343).